Genetic Regulation of Plant Growth and Protection

Genetic Regulation of Plant Growth and Protection

Guest Editors

Yuxuan Hou
Zhiyong Li

MDPI

Basel • Beijing • Wuhan • Barcelona • Belgrade • Novi Sad • Cluj • Manchester

Guest Editors

Yuxuan Hou
State Key Laboratory of Rice
Biology and Breeding
China National Rice
Research Institute
Hangzhou
China

Zhiyong Li
State Key Laboratory of Rice
Biology and Breeding
China National Rice
Research Institute
Hangzhou
China

Editorial Office
MDPI AG
Grosspeteranlage 5
4052 Basel, Switzerland

This is a reprint of the Special Issue, published open access by the journal *International Journal of Molecular Sciences* (ISSN 1422-0067), freely accessible at: https://www.mdpi.com/journal/ijms/special_issues/growth_protection.

For citation purposes, cite each article independently as indicated on the article page online and as indicated below:

Lastname, A.A.; Lastname, B.B. Article Title. *Journal Name* **Year**, *Volume Number*, Page Range.

ISBN 978-3-7258-6478-2 (Hbk)
ISBN 978-3-7258-6479-9 (PDF)
https://doi.org/10.3390/books978-3-7258-6479-9

Contents

About the Editors . . . . **vii**

Preface . . . . **ix**

Jia Li, Xiaohong Tong, Zhiyong Li and Yuxuan Hou
Special Issue "Genetic Regulation of Plant Growth and Protection"
Reprinted from: *Int. J. Mol. Sci.* **2025**, *26*, 11535, https://doi.org/10.3390/ijms262311535 . . . . . **1**

Huimin Fang, Lili Song, Kangwei Liu, Yishu Gu, Yao Guo, Chao Zhang and Long Zhang
OsRNE Encodes an RNase E/G-Type Endoribonuclease Required for Chloroplast Development and Seedling Growth in Rice
Reprinted from: *Int. J. Mol. Sci.* **2025**, *26*, 2375, https://doi.org/10.3390/ijms26052375 . . . . . . **7**

Lu Liu, Shuangwei Song, Ning Liu, Zhiqin Wang, Yonglong Zhao, Naiqin Zhong, et al.
The Silencing of the *StPAM16-1* Gene Enhanced the Resistance of Potato Plants to the Phytotoxin Thaxtomin A
Reprinted from: *Int. J. Mol. Sci.* **2025**, *26*, 1361, https://doi.org/10.3390/ijms26031361 . . . . . . **25**

Panpan Li, Huihui Shang, Xia Xu, Junyi Gong, Jian-Li Wu and Xiaobo Zhang
A Novel Single Base Mutation in *OsSPL42* Leads to the Formation of Leaf Lesions in Rice
Reprinted from: *Int. J. Mol. Sci.* **2024**, *25*, 11871, https://doi.org/10.3390/ijms252211871 . . . . . **41**

Yinxia Chen, Zhize Wang, Weidan Nie, Tingjie Zhao, Yule Dang, Chenghao Feng, et al.
Study on the Function of *SlWRKY80* in Tomato Defense against *Meloidogyne incognita*
Reprinted from: *Int. J. Mol. Sci.* **2024**, *25*, 8892, https://doi.org/10.3390/ijms25168892 . . . . . . **63**

Juanjuan Chen, Xiaojiao Han, Linxiu Liu, Bingbing Yang, Renying Zhuo and Xiaohua Yao
Genome-Wide Detection of SPX Family and Profiling of *CoSPX-MFS3* in Regulating Low-Phosphate Stress in Tea-Oil *Camellia*
Reprinted from: *Int. J. Mol. Sci.* **2023**, *24*, 11552, https://doi.org/10.3390/ijms241411552 . . . . . **77**

Yazhou Shu, Wensheng Zhang, Liqun Tang, Zhiyong Li, Xinyong Liu, Xixi Liu, et al.
ABF1 Positively Regulates Rice Chilling Tolerance via Inducing Trehalose Biosynthesis
Reprinted from: *Int. J. Mol. Sci.* **2023**, *24*, 11082, https://doi.org/10.3390/ijms241311082 . . . . . **94**

Yuxuan Hou, Yan Liang, Changdeng Yang, Zhijuan Ji, Yuxiang Zeng, Guanghao Li and Zhiguo E
Complete Genomic Sequence of *Xanthomonas oryzae* pv. *oryzae* Strain, LA20, for Studying Resurgence of Rice Bacterial Blight in the Yangtze River Region, China
Reprinted from: *Int. J. Mol. Sci.* **2023**, *24*, 8132, https://doi.org/10.3390/ijms24098132 . . . . . . **114**

Guodong Li, Xinzhuan Yao, Zhouzhuoer Chen, Xingyu Tian and Litang Lu
The Overexpression of *Oryza sativa* L. *CYP85A1* Promotes Growth and Biomass Production in Transgenic Trees
Reprinted from: *Int. J. Mol. Sci.* **2023**, *24*, 6480, https://doi.org/10.3390/ijms24076480 . . . . . . **123**

Yan-Ning Xie, Ting Yang, Bin-Tao Zhang, Qian-Qian Qi, An-Ming Ding, Lian-Guang Shang, et al.
Systematic Analysis of BELL Family Genes in *Zizania latifolia* and Functional Identification of *ZlqSH1a/b* in Rice Seed Shattering
Reprinted from: *Int. J. Mol. Sci.* **2022**, *23*, 15939, https://doi.org/10.3390/ijms232415939 . . . . . **142**

Guimin Li, Yanxing Ma, Xiaoping Wang, Nini Cheng, Deyu Meng, Siyu Chen, et al.
CRISPR/Cas9 Gene Editing of *NtAITRs*, a Family of Transcription Repressor Genes, Leads to Enhanced Drought Tolerance in Tobacco
Reprinted from: *Int. J. Mol. Sci.* **2022**, *23*, 15268, https://doi.org/10.3390/ijms232315268 . . . . . **161**

Kun Zhou, Zhengliang Luo, Weidong Huang, Zemin Liu, Xuexue Miao, Shuhua Tao, et al.
Biological Roles of Lipids in Rice
Reprinted from: *Int. J. Mol. Sci.* **2024**, *25*, 9046, https://doi.org/10.3390/ijms25169046 . . . . . . **176**

About the Editors

Yuxuan Hou

Yuxuan Hou is an associate professor at the State Key Laboratory of Rice Biology and Breeding, China National Rice Research Institute. Her research focuses on key aspects of rice bacterial diseases, including disease epidemiology, host–pathogen interactions, and the mechanisms of host resistance. By integrating approaches from genetics, cell biology, molecular biology, physiology, molecular pathology, transcriptomics, and proteomics, she and her team aim to elucidate the dynamics of rice bacterial disease occurrence, identify and characterize virulence genes in bacterial pathogens as well as resistance genes in rice, and dissect the underlying genetic regulatory pathways. This research provides valuable genetic resources and theoretical support for breeding rice varieties with enhanced resistance to bacterial diseases.

Zhiyong Li

Zhiyong Li is an associate professor at the State Key Laboratory of Rice Biology, China National Rice Research Institute. His main research directions include the cloning and mechanism analysis of key genes for carbon and nitrogen utilization in rice, and the construction and application of a rice gain-of-function mutant databases.

Preface

This Reprint collects pivotal research from the Special Issue "Genetic Regulation of Plant Growth and Protection" published in the *International Journal of Molecular Sciences*. The Special Issue focuses on elucidating the complex genetic and molecular mechanisms that govern plant development, adaptation, and resilience. It aims to synthesize recent advancements in understanding how key genes, signaling pathways, and regulatory networks—spanning transcription factors, hormone signaling, epigenetic modifications, and non-coding RNAs—coordinate plant growth while responding to abiotic and biotic stresses. The motivation for this Special Issue arises from the urgent need to translate foundational discoveries into strategies for sustainable agriculture, particularly in the face of climate change and environmental pressures. This work is addressed to researchers, plant scientists, and breeders seeking to integrate genetic insights into the development of stress-tolerant, high-yielding crops, and to advance the frontier of precision molecular breeding.

Yuxuan Hou and Zhiyong Li
Guest Editors

International Journal of
Molecular Sciences

MDPI

Editorial

Special Issue "Genetic Regulation of Plant Growth and Protection"

Jia Li, Xiaohong Tong, Zhiyong Li * and Yuxuan Hou *

State Key Laboratory of Rice Biology and Breeding, China National Rice Research Institute, Hangzhou 311400, China; 2022011010015@stu.hznu.cn (J.L.); tongxiaohong@caas.cn (X.T.)
* Correspondence: lizhiyong@caas.cn (Z.L.); houyuxuan@caas.cn (Y.H.)

Plants in natural environments are subject to diverse biotic and abiotic stresses, such as drought, low temperatures, and pests, which significantly impact their growth, development, and yield. To address these challenges, plants have evolved complex genetic regulatory networks that precisely control gene expression to balance growth and stress responses. In recent years, advancements in molecular biology and genomics have progressively unraveled the genetic mechanisms underlying plant growth and defense, including the roles of transcription factors, epigenetic modifications, hormone signaling pathways, and non-coding RNAs. The emergence of gene-editing technologies such as CRISPR-Cas9 has provided revolutionary tools for precisely modifying these regulatory genes and breeding stress-resistant crops. In this review, we summarize recent progress in the genetic regulation of plant growth and defense, focusing on key genes and their molecular mechanisms reported in *IJMS* (*International Journal of Molecular Sciences*), and discuss their potential applications in sustainable agriculture.

1. Genetic Regulation of Plant Growth and Development

In recent years, research in the field of plant functional genomics has progressively shifted from characterizing genes controlling individual traits to elucidating the complex regulatory networks underlying the coordinated development of multiple traits. The discovery and functional validation of key genes—ranging from transcription factors regulating abscission layer formation to phytohormone biosynthesis genes conserved across species and organelle-specific factors essential for photosynthetic efficiency—collectively delineate a multi-layered regulatory blueprint that governs plant development from organelle function to organ morphology, and from vegetative growth to reproductive success.

Xie et al. identified that in the BELL transcription factor family of *Zizania latifolia*, the genes *ZlqSH1a* and *ZlqSH1b* are homologs of the key shattering gene in rice. Their overexpression in rice promotes complete abscission zone development by regulating hormone signaling and cell wall metabolism-related genes (upregulation of *ERF* and downregulation of *PG1/PG2*), elucidating the genetic basis of strong seed shattering in *Zizania latifolia* and providing target genes for the domestication of varieties with moderate shattering (contribution 1). Liu et al. constructed a high spatiotemporal resolution multi-omics atlas of wheat spike development, revealing that cytokinin treatment significantly reduces basal spikelet abortion rates and increases fertile spikelet numbers, while abscisic acid inhibits overall spike development [1]. Regarding plant growth and biomass enhancement, the heterologous overexpression of the rice brassinosteroid (BR) biosynthesis key gene *OsCYP85A1* in poplar activates growth signaling pathways and secondary cell wall synthesis genes, significantly increasing plant height, stem diameter, and root system development, while also enhancing carbon assimilation capacity and disease resistance. These findings validate the cross-species conservation of the gene's growth-promoting function (contribution 2). Liu et al. employed interdisciplinary techniques including 3D imaging, genome-wide association study (GWAS), single-cell transcriptomics, and molecular biology to identify 72 key genes influencing root architecture for the first time. They further elucidated that the *PsiSKP2B* gene positively regulates poplar lateral root development by ubiquitin-mediated degradation of PsiZHD9 and PsiWOX4 proteins, thereby increasing root auxin content, providing a key genetic target for forest tree root improvement [2]. Rice lipid synthesis and regulatory genes are involved not only in fundamental processes such as membrane construction and energy storage but also influence plant fertility, nutritional quality, and stress resistance by regulating anther cuticle formation, grain composition, and wax accumulation (contribution 3). The *OsRNE* gene identified by Fang et al. ensures normal chloroplast development and photosynthesis by regulating chloroplast RNA metabolism, making it critical for rice seedling survival and growth (contribution 4).

These findings collectively outline a multi-layered regulatory network governing plant development, from organelle biogenesis to organ formation, and from vegetative growth to reproductive success, providing critical theoretical foundations and genetic resources for crop genetic improvement. Further elucidation of the interactions among these genes and their environmental response mechanisms will help reveal the coordinated regulatory principles underlying plant growth, development, and stress adaptation. By integrating multi-omics data mining with gene-editing technologies, the precise modification of functional combinations of key hub genes is expected to enable synergistic improvements in crop yield, quality, and stress resilience, thereby advancing the development of next-generation intelligent design breeding systems.

2. Genetic Regulation of Plant Stress Tolerance

During growth and development, plants inevitably face various abiotic stresses, such as drought, low temperature, salinity, and nutrient deficiency. A profound understanding of the molecular regulatory networks underlying plant responses to these adversities is crucial for genetic improvement of stress resistance in crops. The authors of recent studies across multiple species have achieved breakthrough advancements in this field.

In tobacco, *NtAITRs* function as transcriptional repressors in the ABA signaling pathway, negatively regulating drought tolerance. Gene editing-mediated knockout of this gene family significantly enhanced drought resistance without impairing normal growth (contribution 5). Previously, Yue et al. demonstrated that heterologous expression of the potato *NAC1* gene in *Arabidopsis* not only improved seed germination rates and green leaf retention but also reduced ROS accumulation while simultaneously increasing proline levels, thereby enhancing salt tolerance [3]. In rice, the "SAPK10-ABF1-*TPS2*" pathway elucidates a novel mechanism whereby the ABA signaling component OsABF1, upon phosphorylation and activation by SAPK10, directly promotes the expression of the trehalose biosynthesis gene *TPS2* to enhance cold tolerance (contribution 6). Another bZIP transcription factor, *OsbZIP72*, positively regulates cold tolerance at both seedling and booting stages. Knockout mutants exhibited significantly reduced seedling survival rates under 4 °C stress, whereas overexpression lines showed improved survival and higher seed-setting rates under low-temperature conditions during booting [4]. Furthermore, *OsbZIP72* promotes jasmonic acid biosynthesis via the "SAPK10-bZIP72-*AOC*" pathway while concurrently suppressing seed germination [5]. In oil-tea camellia, the SPX family gene *CoSPX-MFS3* acts as a key positive regulator of plant adaptation to low-phosphorus conditions by modulating organic acid secretion, phosphate transporter gene expression, and vacuolar phosphate homeostasis (contribution 7). Fang et al. revealed the molecular mechanism of phosphate anion efflux mediated by *Arabidopsis PHO1;H1* and its regulation by inositol polyphosphates. Both *Arabidopsis* PHO1 and its homolog PHO1;H1 are typical SPX-EXS family phosphate transporters involved in loading inorganic phosphate into the xylem vessel for root-to-shoot translocation [6]. Collectively, these findings substantially advance our understanding of plant adaptation mechanisms to drought, cold, and nutrient deficiencies at multiple levels, providing diverse genetic targets and strategic approaches for the molecular breeding of stress-resilient crops.

In summary, the discovery and functional characterization of key genes—from transcription factors in tobacco and rice (*NtAITRs*, *OsABF1*, and *OsbZIP72*) to transporter proteins in oil-tea camellia and Arabidopsis (*CoSPX-MFS3* and *PHO1;H1*)—systematically reveal multi-layered strategies employed by plants to combat environmental stresses through integrated hormone signaling, reprogrammed metabolic pathways, and optimized nutrient transport. These advances not only deepen our understanding of plant stress adaptation mechanisms but also mark a transition in crop breeding from traditional phenotype-based selection toward precision molecular design breeding.

3. Genetic Regulation of Plant Biotic Stress Defense

The authors of recent studies have multilaterally elucidated the complex dynamics of plant–pathogen interactions, spanning from pathogenic virulence strategies to plant immune networks. On the pathogen side, the novel rice bacterial blight strain LA20 utilizes its unique repertoire of TAL effectors (*Tal1b* and *Tal4*) to precisely target rice susceptibility genes *OsSWEET11* and *OsTFX1*, thereby overcoming resistance mediated by major R genes including *xa5*, *xa13*, and *Xa7* (contribution 8). This finding reveals how pathogen structural variation in effectors enables breakthrough of host resistance barriers, informing new strategies for developing broad-spectrum resistant cultivars. On the plant side, multiple key regulatory genes enhance resistance through distinct pathways. In tomato, *SlWRKY80* functions as a positive regulatory transcription factor that establishes a multi-layered defense network against root-knot nematodes by coordinately activating salicylic acid, jasmonic acid, and ethylene signaling pathways, maintaining ROS metabolic homeostasis, and enhancing root vitality (contribution 9). Furthermore, *SlWRKY80* participates in MeJA response and positively regulates salt-alkali stress tolerance. The SlWRKY80 protein directly binds to promoters of *SlSPDS2* and *SlNHX4*, activating their transcription to promote spermidine synthesis and Na+/K+ homeostasis, providing new theoretical foundations for understanding MeJA-mediated stress resistance and tomato production. In rice, *OsSPL42* acts as a putative negative regulator whose loss-of-function leads to ROS accumulation, programmed cell death, and lignin deposition, consequently enhancing resistance to specific *Xoo* strains [7]. Simultaneously, OsSPL42 interacts with multiple organellar RNA editing factors (*OsMORF8-1*/*OsMORF8-2*), affecting RNA editing processes (contribution 10). The *spl42* mutant exhibits impaired chloroplast development with abnormal thylakoid morphology, accumulated O^{2-} and H_2O_2, and enhanced activities of ROS-scavenging enzymes (CATB, AOX1a, and AOX1b). Transcriptional analysis results showed significant downregulation of chloroplast development and chlorophyll biosynthesis genes; in comparison, senescence-associated, ROS-producing, and defense-related genes were upregulated [8]. In potato, *StPAM16-1* was identified as a negative immune regulator that suppresses the response to common scab toxin through interaction with *COP9* signalosome subunit *StCSN5*. It has been demonstrated that silencing this gene enhances disease resistance (contribution 11).

These findings, spanning from pathogen virulence evolution to multi-layered plant immune regulation, provide crucial targets and theoretical foundations for designing durable and broad-spectrum disease resistance breeding strategies. The authors of future studies should focus on integrating multi-omics data to systematically elucidate key nodes within disease resistance signaling networks and their dynamic regulatory mechanisms, while simultaneously developing novel gene editing-based breeding technologies that enhance crop resistance without compromising agricultural product quality and safety. Through cross-species functional validation of candidate genes, we can accelerate the exploration and utilization of disease resistance genetic resources, thereby establishing scientific and technological support for building sustainable plant disease management systems.

4. Conclusions and Perspectives

In this review, we summarize the genetic regulatory mechanisms underlying plant growth and defense, covering the latest research advances in three major aspects: plant growth and development, abiotic stress tolerance, and biotic stress resistance. Through molecular biology and genomics technologies, scientists have revealed the crucial roles of multiple key genes and their regulatory networks in plant responses to environmental stresses and maintenance of growth and development. Researchers have found that BELL transcription factor family members (such as *ZlqSH1a* and *ZlqSH1b*) play a pivotal role in seed abscission by regulating phytohormone signaling and cell wall metabolic pathways to influence abscission layer development. Overexpression of the brassinosteroid (BR) biosynthesis gene *OsCYP85A1* significantly promotes plant growth while concurrently enhancing xylem formation and disease resistance. Lipid biosynthesis-related genes (*KCS* and *OsFBN1*) play multiple roles in plant growth and development, stress responses, and quality formation. The chloroplast development gene *OsRNE* affects photosynthesis by regulating chloroplast RNA metabolism, which is crucial for plant growth. In tobacco, the *NtAITR* gene negatively regulates drought tolerance through the ABA signaling pathway, and its knockout enhances drought resistance. The bZIP transcription factor *OsABF1* in rice participates in cold stress response by regulating trehalose synthesis. In *Camellia oleifera*, the *CoSPX-MFS3* gene enhances low-phosphorus adaptability by regulating phosphorus signaling pathways and organic acid accumulation. Genetic variation in the *TAL* effector gene of rice bacterial blight strain LA20 is the key factor enabling it to overcome rice resistance. In tomato, *SlWRKY80* enhances resistance to root-knot nematodes by activating SA/JA/ETH signaling and ROS metabolism. The *OsSPL42* gene in rice spotted-leaf mutant *spl42* affects disease resistance by regulating lignin biosynthesis and ROS balance. In potato, the *StPAM16-1* gene negatively regulates immune responses to common scab through interaction with the COP9 signalosome complex.

The authors of future studies should further elucidate the molecular mechanisms of key genes (*OsSPL42* and *StPAM16-1*) and their interactions with other pathways to refine our understanding of plant genetic regulatory networks. The application of gene-editing technologies such as CRISPR-Cas9 will enable precise modification of regulatory genes, facilitating the development of stress-resistant, high-yield, and high-quality crop varieties. Additionally, exploring the universality and potential of cross-species gene function validation (applying *OsCYP85A1* in woody plants) could broaden the scope of genetic improvement. Translating research findings into agricultural practice is crucial for breeding new crop varieties with enhanced resistance to diseases, drought, and low-phosphorus stress, ultimately improving crop yield and quality. Meanwhile, ecological balance must be prioritized to ensure the environmental safety of genetically improved crops and mitigate potential ecological risks. Strengthening interdisciplinary collaboration among the fields of molecular biology, genetics, ecology, and agronomy will drive comprehensive advancements in plant genetic regulation research. These efforts will provide scientific support for global food security and sustainable agriculture. In conclusion, continued in-depth research on plant genetic regulation will offer novel insights and tools to address agricultural challenges, contributing to the realization of sustainable agricultural development.

Author Contributions: All of the authors participated in the editing of this research topic. Y.H. and Z.L. designed the research topic, Z.L., X.T. and J.L. wrote the draft, and Y.H. provided suggestions and comments on the editorial. All authors have read and agreed to the published version of the manuscript.

Funding: This project was funded by the National Key R&D Program of China 2024YFD1200600, the National Natural Science Foundation of China (grant No. 32472115), and the Zhejiang Provincial Natural Science Foundation of China (LY22C130004 and LMS25C130007).

Conflicts of Interest: The authors declare that the research was conducted in the absence of any commercial or financial relationships that could be construed as a potential conflict of interest.

List of Contributions:

1. Xie, Y.-N.; Yang, T.; Zhang, B.-T.; Qi, Q.-Q.; Ding, A.-M.; Shang, L.-G.; Zhang, Y.; Qian, Q.; Zhang, Z.-F.; Yan, N. Systematic Analysis of *BELL* Family Genes in *Zizania latifolia* and Functional Identification of *ZlqSH1a/b* in Rice Seed Shattering. *Int. J. Mol. Sci.* **2022**, *23*, 15939.
2. Li, G.; Yao, X.; Chen, Z.; Tian, X.; Lu, L. The Overexpression of *Oryza sativa* L. *CYP85A1* Promotes Growth and Biomass Production in Transgenic Trees. *Int. J. Mol. Sci.* **2023**, *24*, 6480.
3. Zhou, K.; Luo, Z.; Huang, W.; Liu, Z.; Miao, X.; Tao, S.; Wang, J.; Zhang, J.; Wang, S.; Zeng, X. Biological Roles of Lipids in Rice. *Int. J. Mol. Sci.* **2024**, *25*, 9046.
4. Fang, H.; Song, L.; Liu, K.; Gu, Y.; Guo, Y.; Zhang, C.; Zhang, L. OsRNE Encodes an RNase E/G-Type Endoribonuclease Required for Chloroplast Development and Seedling Growth in Rice. *Int. J. Mol. Sci.* **2025**, *26*, 2375.
5. Li, G.; Ma, Y.; Wang, X.; Cheng, N.; Meng, D.; Chen, S.; Wang, W.; Wang, X.; Hu, X.; Yan, L.; et al. CRISPR/Cas9 Gene Editing of *NtAITRs*, a Family of Transcription Repressor Genes, Leads to Enhanced Drought Tolerance in Tobacco. *Int. J. Mol. Sci.* **2022**, *23*, 15268.
6. Shu, Y.; Zhang, W.; Tang, L.; Li, Z.; Liu, X.; Liu, X.; Liu, W.; Li, G.; Ying, J.; Huang, J.; et al. *ABF1* Positively Regulates Rice Chilling Tolerance via Inducing Trehalose Biosynthesis. *Int. J. Mol. Sci.* **2023**, *24*, 11082.
7. Chen, J.; Han, X.; Liu, L.; Yang, B.; Zhuo, R.; Yao, X. Genome-Wide Detection of SPX Family and Profiling of CoSPX-MFS3 in Regulating Low-Phosphate Stress in Tea-Oil Camellia. *Int. J. Mol. Sci.* **2023**, *24*, 11552.
8. Hou, Y.; Liang, Y.; Yang, C.; Ji, Z.; Zeng, Y.; Li, G.; E, Z. Complete Genomic Sequence of *Xanthomonas oryzae* pv. *oryzae* Strain, LA20, for Studying Resurgence of Rice Bacterial Blight in the Yangtze River Region, China. *Int. J. Mol. Sci.* **2023**, *24*, 8132.
9. Chen, Y.; Wang, Z.; Nie, W.; Zhao, T.; Dang, Y.; Feng, C.; Liu, L.; Wang, C.; Du, C. Study on the Function of *SlWRKY80* in Tomato Defense against Meloidogyne incognita. *Int. J. Mol. Sci.* **2024**, *25*, 8892.
10. Li, P.; Shang, H.; Xu, X.; Gong, J.; Wu, J.-L.; Zhang, X. A Novel Single Base Mutation in *OsSPL42* Leads to the Formation of Leaf Lesions in Rice. *Int. J. Mol. Sci.* **2024**, *25*, 11871.
11. Liu, L.; Song, S.; Liu, N.; Wang, Z.; Zhao, Y.; Zhong, N.; Zhao, P.; Wang, H. The Silencing of the *StPAM16-1* Gene Enhanced the Resistance of Potato Plants to the Phytotoxin Thaxtomin A. *Int. J. Mol. Sci.* **2025**, *26*, 1361.

References

1. Liu, Y.-Y.; Zhang, L.-L.; Zhu, a.-T.; Shen, L.-P.; Zhang, J.-Q.; Chen, J.; Lu, F.; Hao, Y.-F.; Chen, W.; Guo, Z.-F. Multi-omics Identifies Key Genetic and Metabolic Networks Regulating Spike Organ Development in Wheat. *Plant Cell* **2025**, koaf250. [CrossRef] [PubMed]
2. Li, J.; Bo, W.; Bu, C.; Zhou, J.; Li, P.; Wang, M.; Song, Y.; Liu, Q.; El-Kassaby, Y.-A.; Zhang, D. Integrating 3D Imaging, GWAS, and Single-cell Transcriptome Approaches to Elucidate Root System Architecture in Populus. *Plant Physiol.* **2025**, *199*, 432. [CrossRef] [PubMed]
3. Yue, L.; Zhuang, Y.; Gu, Y.; Li, H.; Tu, S.; Yang, X.; Huang, W. Heterologous Expression of *Solanum tuberosum NAC1* Gene Confers Enhanced Tolerance to Salt Stress in Transgenic *Nicotiana benthamiana*. *J. Plant Biol.* **2021**, *64*, 531–542. [CrossRef]
4. Gu, Y.-S.; Guo, H.-F.; Li, H.-H.; Su, R.-B.; Najeeb, U.-K.; Li, Z.-C. QTL Mapping by GWAS and Functional Analysis of *OsbZIP72* for Cold Tolerance at Rice Seedling Stage. *Crop J.* **2024**, *12*, 1697–1708. [CrossRef]

5. Wang, Y.; Hou, Y.; Qiu, J.; Wang, H.; Wang, S.; Tang, L.; Tong, X.; Zhang, J. Abscisic Acid Promotes Jasmonic Acid Biosynthesis via a 'SAPK10-bZIP72-*AOC*' Pathway to Synergistically Inhibit Seed Germination in Rice (*Oryza sativa*). *New Phytol.* **2020**, *228*, 1336–1353. [CrossRef] [PubMed]
6. Fang, S.; Yang, Y.; Zhang, X.; Yang, Z.; Zhang, M.; Zhao, Y.; Zhang, C.; Yu, F.; Wang, Y.F.; Zhang, P. Structural mechanism underlying PHO1;H1-mediated phosphate transport in Arabidopsis. *Nat. Plants* **2025**, *11*, 309–320. [CrossRef] [PubMed]
7. Shang, C.; Liu, X.; Chen, G.; Zheng, H.; Khan, A.; Li, G.; Hu, X. *SlWRKY80*-mediated Jasmonic Acid Pathway Positively Regulates Tomato Resistance to Saline-alkali Stress by Enhancing Spermidine Content and Stabilizing Na^+/K^+ Homeostasis. *Hortic. Res.* **2024**, *11*, uhae028. [CrossRef] [PubMed]
8. Liu, L.; Wang, Y.; Tian, Y.; Song, S.; Wu, Z.; Ding, X.; Zheng, H.; Huang, Y.; Liu, S.; Dong, X.; et al. Isolation and Characterization of SPOTTED LEAF42 Encoding a Porphobilinogen Deaminase in Rice. *Plants* **2023**, *12*, 403. [CrossRef] [PubMed]

International Journal of
Molecular Sciences

MDPI

Article

OsRNE Encodes an RNase E/G-Type Endoribonuclease Required for Chloroplast Development and Seedling Growth in Rice

Huimin Fang [1], Lili Song [2,3], Kangwei Liu [2,3], Yishu Gu [2,3], Yao Guo [2,3], Chao Zhang [2,3,*] and Long Zhang [2,3,*]

1 Guangling College, Yangzhou University, Yangzhou 225000, China; hmfang@yzu.edu.cn
2 Key Laboratory of Crop Genetics and Physiology of Jiangsu Province, Key Laboratory of Plant Functional Genomics of the Ministry of Education, Yangzhou University, Yangzhou 225009, China; 18300765057@163.com (L.S.); lkw18352768872@163.com (K.L.); yi09402024@163.com (Y.G.); 15554430252@163.com (Y.G.)
3 Co-Innovation Center for Modern Production Technology of Grain Crops of Jiangsu Province, Joint International Research Laboratory of Agriculture & Agri-Product Safety of the Ministry of Education, Yangzhou University, Yangzhou 225009, China
* Correspondence: chaozhang@yzu.edu.cn (C.Z.); zhanglong@yzu.edu.cn (L.Z.); Tel.: +86-514-8799-7217 (L.Z.); Fax: +86-514-8797-1747 (L.Z.)

Abstract: Chloroplast biogenesis is a crucial biological process in plants. Endoribonuclease E (RNase E) functions in the RNA metabolism of chloroplast and plays a vital role for chloroplast development in *Arabidopsis*. However, despite sharing 44.7% of its amino acid sequence identity with *Arabidopsis* RNase E, the biological function of rice OsRNE (*Oryza sativa* RNase E) remains unknown. Here, we identified a *white leaf and lethal 1* (*wll1*) mutant that displayed white leaves and died at the seedling stage. The causal gene *OsRNE* was isolated by MutMap+ method. CRISPR/Cas9-mediated knockout of *OsRNE* resulted in white leaves and seedling lethality, confirming *OsRNE* as the causal gene for the *wll1* phenotype. The albino phenotype of *osrne* mutant was associated with decreased chlorophyll content and abnormal thylakoid morphology in the chloroplast. The absence of *OsRNE* led to a significant reduction in the Rubisco large subunit (RbcL), and the 23S and 16S chloroplast rRNAs were nearly undetectable in the *osrne* mutant. *OsRNE* transcripts were highly expressed in green tissues, and the protein was localized to chloroplasts, indicating its essential role in photosynthetic organs. Furthermore, transcriptome analysis showed that most of the genes associated with photosynthesis and carbohydrate metabolism pathways in the *osrne* mutant were significantly down-regulated compared with those in WT. Chlorophyll- and other pigment-related genes were also differentially expressed in the *osrne* mutant. Our findings demonstrated that OsRNE plays an important role in chloroplast development and chlorophyll biosynthesis in rice.

Keywords: chlorophyll biosynthesis; chloroplast development; rice; RNA metabolism; RNase E/G; thylakoid

1. Introduction

Chloroplasts are essential for photosynthesis in higher plants, converting light energy into chemical energy [1]. Chlorophyll, synthesized within chloroplasts, determines leaf color and influences seedling growth. Any deficiency in chloroplast development leads to leaf color mutants, and many mutants have been identified, e.g., pale green, yellowing, white-striped, variegated, and albino mutants [2–6]. All of these mutants, especially albino mutants, are ideal specimens for studying potential molecular mechanisms of chloroplast and leaf development [7–11].

Chloroplast development occurs in three stages: (1) plastid DNA replication and synthesis, (2) chloroplast assembly, and (3) photosynthetic apparatus formation [12,13]. In the second stage, nuclear-encoded RNA polymerase (NEP) preferentially transcribes plastid housekeeping genes to promote transcription and translation in chloroplasts. In the third stage, plastid-encoded RNA polymerase (PEP) combines with the nuclear encoded protein to form a photosynthetic and metabolic system that controls chloroplast development. Chloroplasts are semi-autonomous organelles and have independent small genomes, but most of the proteins are encoded in the nucleus and transferred to chloroplasts after translation. Taking rice as an example, the chloroplast genome is about 135 kb, containing 34 RNA-coding genes and 120 protein-coding genes [14]. Approximately 3000 proteins play an important role in chloroplast functions, of which over 95% are encoded by nuclear genes [15]. In summary, chloroplast development is coordinately regulated by both plastid and nuclear-encoded genes. Therefore, cloning and identifying these nuclear genes will help elucidate the complex regulatory mechanisms of plant chloroplast development.

Post-transcriptional regulation plays a major role in chloroplast biogenesis, with nucleus-encoded ribonucleases facilitating RNA processing and degradation [16]. Endonucleases are divided into CSP41, ribonuclease E/G (RNase E/G), RNase III, RNase H, RNase J, RNase P, and RNase Z, and they have various functions in RNA metabolism, including maturation of rRNA and tRNA precursors, processing of polycistronic mRNAs, and RNA decay [16]. RNase E/G proteins contain RNase E and its truncated form, RNase G, and have been reported to play an important role in RNA metabolism in bacteria, cyanobacteria, and *Arabidopsis* [17–20]. In *Escherichia coli*, RNase E encoded by the *RNE* gene is a ribosomal RNA (rRNA)-processing enzyme. RNase E has RNA-binding activity, cleaves RNA endoribonucleolytically in AU-rich single-stranded regions, and interacts with polynucleotide phosphorylase (PNPase) and other proteins implicated in RNA processing and degradation [21–27]. *Anabaena* 3′–5′ exoribonuclease RNase II binds to RNase E and modulates its endoribonucleolytic activity [18]. *Arabidopsis* RNase E is located in the chloroplast as a high-molecular-weight complex [19]. Absence of RNase E results in multiple defects in chloroplast RNA metabolism in *Arabidopsis* [19,20,28]. Most importantly, polycistronic precursor transcripts over-accumulate in the knockout plants of *Arabidopsis*, while several mature monocistronic mRNAs are strongly reduced [20]. The mutant also exhibits defective transcript processing of mRNAs for ribosomal proteins, which results in plastid ribosome deficiency and impaired plant growth [20,28]. There is only a single RNase E/G gene in the nuclear genomes of higher plants and not in eukaryotes lacking chloroplasts, which implies that RNase E may play an important role in the chloroplast development of higher plants. However, little is known about the role of RNase E in rice.

In this study, we cloned the *OsRNE* gene through MutMap+ method, confirmed the function of *OsRNE* by CRISPR/Cas9 gene editing approach, and conducted a detailed study of seedling phenotypes. Meanwhile, a functional analysis of OsRNE, including expression pattern and subcellular localization, was carried out. Finally, RNA-seq was used to analyze the expression levels of genes related to chlorophyll and other pigments as well as genes involved in photosynthesis and carbohydrate metabolism pathways in *osrne* mutants. Our findings demonstrate that OsRNE plays an important role in chloroplast development, chlorophyll synthesis, and photosynthesis in rice.

2. Results

2.1. Genetic Mapping of the White Leaf Locus Using the MutMap+ Method

The *wll1* was isolated from a ^{60}Co-irradiated mutant pool of japonica variety Yandao 8 (YD8). Under paddy conditions, the seedlings of *wll1* exhibited an obvious white leaf phenotype from the emergence of the first leaf compared with the green leaves of YD8 plants and died within three weeks (Figure 1A,B). Because *wll1* is a lethal mutant, heterozygous plants were used to generate progeny for genetic analysis and mapping. The segregation of green and white phenotypes (73 green: 19 white; $X^2 = 0.93 < X^2_{0.05,1} = 3.84$) confirmed that the mutation is controlled by a recessive nuclear gene. MutMap+, a method suitable for identifying mutations that cause early development lethality or generally hamper crossing, was adopted to identify the causal gene [29]. The genomic DNA of 30 green leaf progenies and 30 white leaf progenies of the heterozygous plants were mixed in equal proportion to produce DNA pool 1 and pool 2, respectively. The DNA samples of pool 1 and pool 2 were separately sequenced to generate Illumina HiSeq paired-end reads that were mapped to reference genomes of Nipponbare (http://rice.uga.edu (accessed on 3 December 2023)). The numbers of clean reads obtained from DNA pool 1 (clean base 12.14 G) and pool 2 (clean base 16.03 G) were 80,915,726 and 106,896,834, respectively. The sequence depth of two DNA pools was 33 x and 43 x, respectively, corresponding to the rice genome (approximately 370 Mb). SNP/Indel index was calculated as the ratio between the number of reads of a mutant SNP/Indel and the total number of reads corresponding to the SNP/Indel. Drawing upon the fundamental principles of genetics, we established specific criteria to pinpoint the causal mutation underlying the *wll1* phenotype (Supplementary Figure S1A). These criteria are outlined as follows: (1) an SNP/Indel-index of 1 in the mutant pool, as the phenotype of albino seeding is homozygous mutation (*wll1/wll1*), and (2) an SNP/Indel-index of less than 0.5 in the wild-type pool, accounting for the presence of homozygous wild-type sequences (*WLL1/WLL1* with SNP/Indel-index = 0) and heterozygotes (*WLL1/wll1* with SNP/Indel-index of 0.5). After this filtration, only one Indel, with Indel-index = 1 (23 mutation reads/23 total reads) and 0.30 (3 mutation reads/10 total reads) in the mutant pool and wild-type pool, respectively, was identified (Supplementary Figure S1B). This Indel represents a single base G deletion in 14,169,090th bp of chromosome 8.

To verify the results of MutMap+, we randomly selected 10 *wll1* seedlings and amplified the Indel using a specific pair of primers for Sanger sequencing. All mutant seedings contained the homozygous mutation of the candidate Indel, suggesting that the mutant phenotype co-segregated with the genotype.

Based on the rice genome annotation project (https://rapdb.dna.affrc.go.jp, accessed on 1 January 2024), the candidate Indel, 14,169,090th bp of chromosome 8, is located at the 12th exon of *LOC_Os08g23430* gene (Figure 1C,D). The rice genome database annotates that *LOC_Os08g23430* encodes the rice OsRNE protein, belonging to the RNase E/G endonuclease family. The *OsRNE* gene comprises 14 exons and 13 introns (Figure 1C). The full length of its CDS is 3258 bp and encodes a protein of 1085 amino acids. Mutation of the candidate Indel in *wll1* resulted in frame shifts with premature termination (Figure 1C). If translated, the transcript in the mutant would encode a protein with 967 amino acids (Figure 1D).

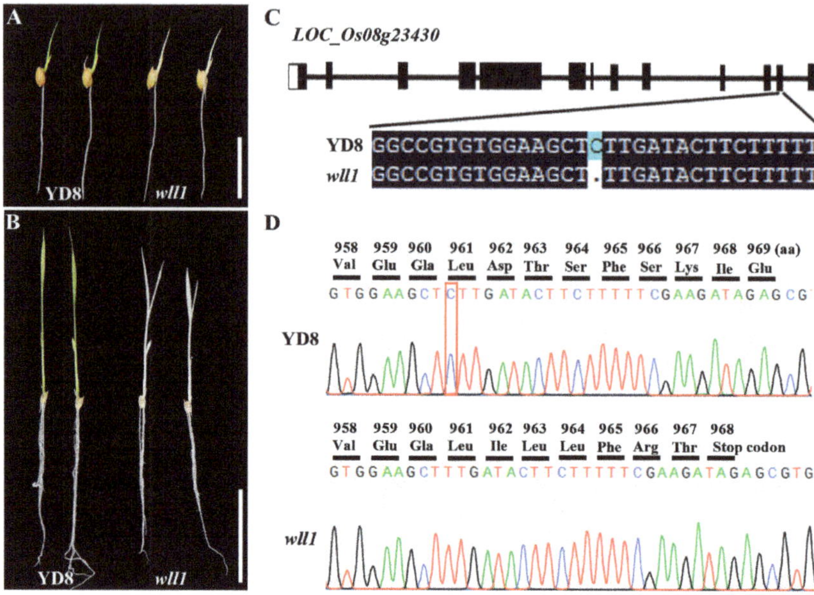

Figure 1. Isolation of causal mutation in white leaf phenotype through MutMap+ method. (**A**,**B**) Comparison of the four-day-old (**A**) and ten-day-old (**B**) seedlings of YD8 and *wll1*. Scale bars = 2 cm (**A**) and 5 cm (**B**). (**C**) Gene structure of *LOC_Os08g23430*. The candidate Indel (a single base C deletion) locates at the 12th exon of *LOC_Os08g23430* gene. (**D**) Sequence chromatograms of *wll1* mutation. The transcript of the mutant would encode a protein with 967 amino acids. Red box represents the missing single base C.

2.2. Functional Validation of OsRNE Gene Using CRISPR-Cas9 System

To further investigate the biological function of OsRNE in rice, a clustered, regularly interspaced, short palindromic repeats (CRISPR)/CRISPR-associated protein 9 (Cas9)-mediated gene editing system was used to generate osrne mutants. The guide RNA (gRNA) expression cassettes that specifically target the first exon of OsRNE were generated and transformed into rice variety Zhonghua 11 (wild-type, WT) (Figure 2A). Two independent homozygous mutant lines *osrne-1* and *osrne-2*, derived from the self-progeny population of heterozygous plants, were identified with an adenosine (A) and thymine (T) insertion in the 108th bp of the first extron, respectively (Figure 2B), which both resulted in frame-shifts with premature termination (Figure 2C). If translated, this transcript would encode a protein with 68 amino acids and does not include most of the domains in OsRNE (Figure 2C). As the two homozygous mutants caused the same type of premature termination, the mutant line *osrne-1* was used for further analysis.

The seedlings of WT grew normally, while the heterozygous mutant seedlings showed phenotypic separation of green and white (Figure 2D,E). Sanger sequencing showed that the *OsRNE* gene in the white seedlings underwent homozygous mutations, whereas normal green seedlings were either homozygote or heterozygote for the *OsRNE* allele. All the green seedlings grew normally to maturity, similar to the recipient parent Zhonghua 11, while the white seedlings did not survive past the three-leaf stage (Figure 2F,G). The knockout lines exhibited the same seedling phenotype as the *wll1* mutant, further demonstrating that *LOC_Os08g23430* is the gene responsible for the *wll1* phenotype.

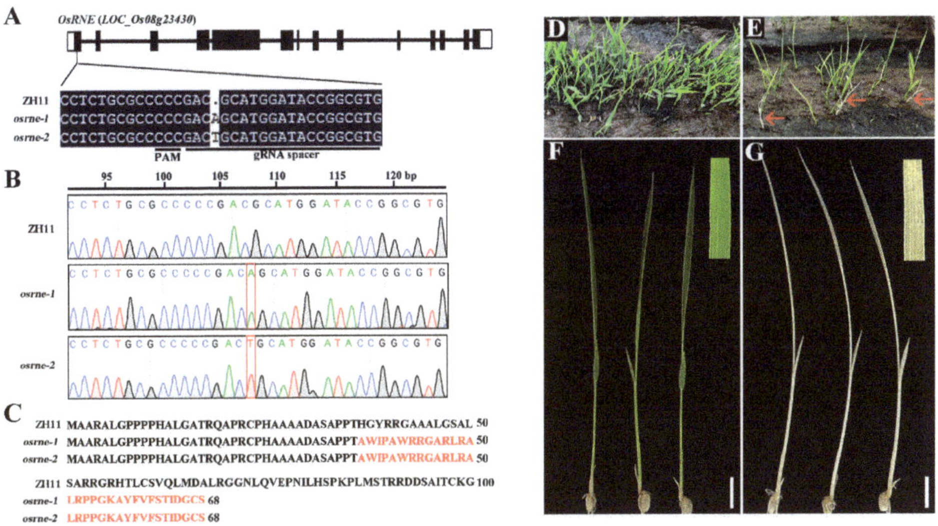

Figure 2. Knockout of *OsRNE* leads to albino phenotype. (**A**) Two independent lines were generated by CRISPR/Cas9 in the first exon of the *OsRNE* gene. (**B**) Mutation events were confirmed by Sanger sequencing. Red box represents the insertion of single base. (**C**) The transcript of *osrne* mutants would encode a protein with 68 amino acids. Red fonts show the amino acids formed by the frameshift mutation. The full-length CDS of *OsRNE* encodes a protein of 1085 amino acids, and Figure 2C shows the first 100 amino acids in ZH11. (**D–G**) Phenotypic analyses of WT and *osrne-1* seedlings. (**D**,**E**) Seedlings of WT (**D**) and heterozygous mutant (**E**) at the two-leaf stage in the field. The red arrows represent white seedlings separated from the heterozygous mutant. (**F**,**G**) The leaves of normal green (**F**) and white seedlings (**G**) at the two-leaf stage. Scale bars = 10 mm (**F**,**G**).

2.3. Phenotypic Characterization of Osrne Mutant

To assess the impact of the *OsRNE* mutation on pigment biosynthesis, we measured chlorophyll and carotenoid levels in WT and *osrne-1* leaves at the two-leaf stage. The levels of chlorophyll a (Chl a), chlorophyll b (Chl b), and carotenoids (Car) were significantly lower in *osrne-1* than in WT (Figure 3A), indicating a critical role for OsRNE in chlorophyll biosynthesis. Moreover, limited Rubisco large subunit RbcL proteins were found in the *osrne-1* leaves at the two-leaf stage (Figure 3B), which revealed that OsRNE might also play an important role in regulating Rubisco formation.

Eukaryotic ribosomes consist of a 60S large subunit and a 40S small subunit. The 60S and 40S subunits comprise rRNAs (28S, 5.8S, 5S, and 18S) and ribosomal proteins, while chloroplast ribosomes consisting of a 50S large subunit and a 30S small subunit comprise rRNAs (23S, 5S, and 16S) and ribosomal proteins. We then analyzed the composition and content of rRNAs using an Agilent 2100 (Agilent Technologies, Santa Clara, CA, USA) in WT and *osrne-1* leaves at the two-leaf seedling stage. The 28S and 18S rRNAs were almost the same in WT and *osrne-1* leaves (Figure 3C,D). However, the chloroplast rRNAs (23S and 16S) could not be detected in *osrne-1* leaves (Figure 3D). These results indicate that *osrne-1* was defective in the biogenesis of chloroplast ribosomes.

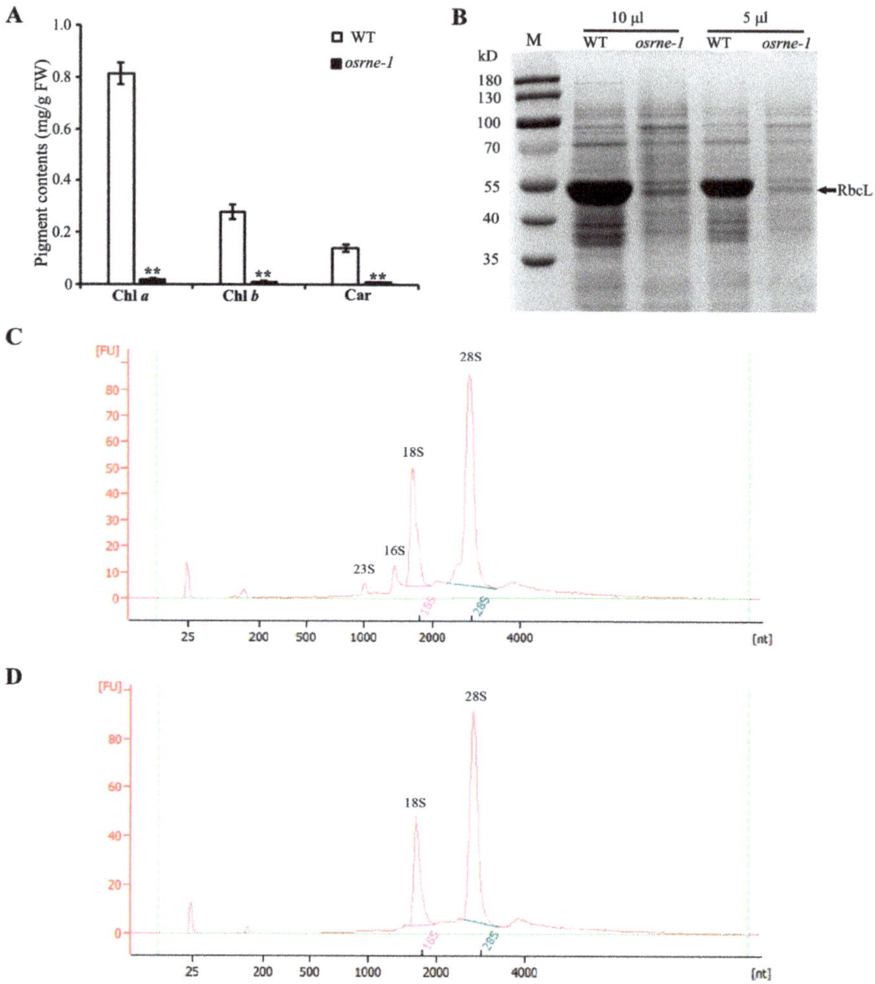

Figure 3. Analyses of pigment contents and rRNA from WT and *osrne-1* mutant. (**A**) Pigment contents in leaves at the two-leaf stage. Chl a, chlorophyll a; Chl b, chlorophyll b; Car, carotenoid. Error bars represent SD (standard deviation) of three biological replicates. Asterisks indicate a significant difference between the WT and *osrne-1* by Student's *t*-test; ** $p < 0.01$. (**B**) SDS-PAGE analysis of total proteins in leaves WT and *osrne-1* mutant. (**C**,**D**) rRNA analysis using Agilent 2100 in WT (**C**) and *osrne-1* (**D**) leaves at the two-leaf stage.

2.4. Chloroplast Development Was Impaired in Osrne Mutant

As a consequence of the deficiency in chlorophyll biogenesis in the mutant, we further investigated the ultrastructure of the chloroplast using transmission electron microscopy (TEM). WT chloroplasts exhibited well-developed grana thylakoids and stroma thylakoids (Figure 4A–C), whereas *osrne-1* mutants displayed two distinct phenotypes: one with short grana thylakoids and almost unobservable stroma thylakoids (Figure 4D–F) and another with a nearly complete absence of thylakoids (Figure 4G–I). These results imply that OsRNE is essential for chloroplast development from the beginning of seedling establishment.

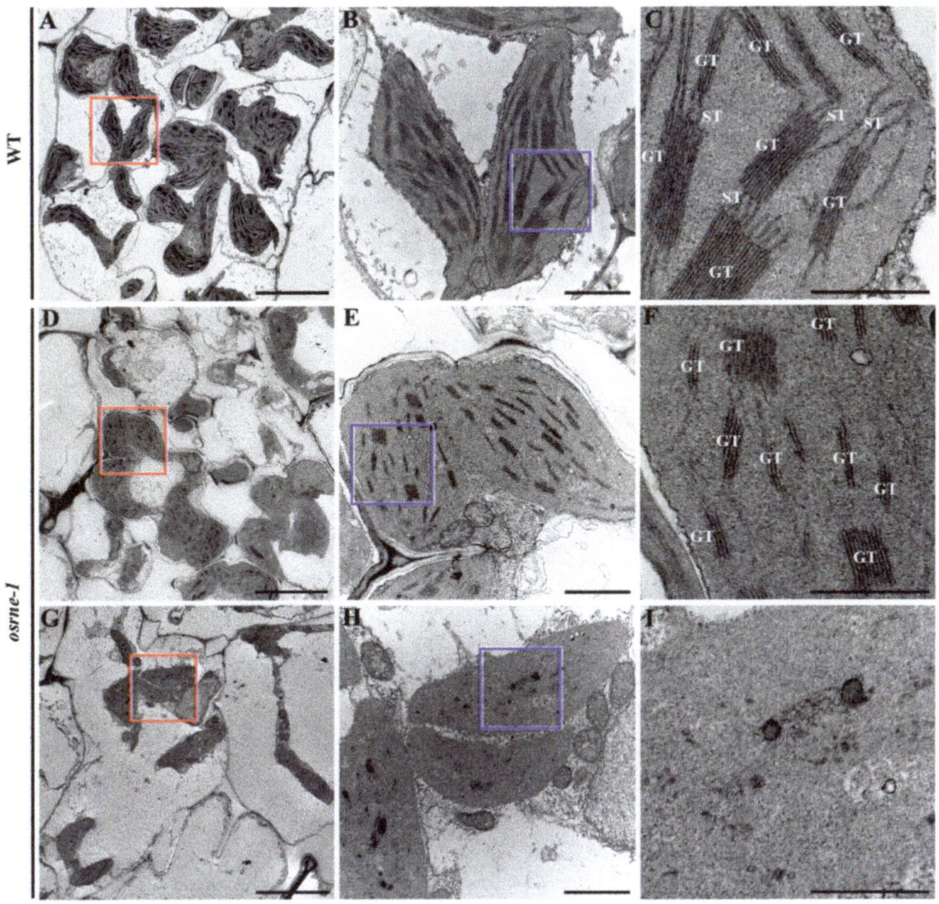

Figure 4. Chloroplasts ultrastructure in WT and *osrne-1* mesophyll cells at the two-leaf stage. (**B**,**E**,**H**) represent the magnified regions indicated by red outline in (**A**,**C**,**E**), respectively. (**C**,**F**,**I**) represent the magnified regions indicated by blue outline in (**B**,**E**,**H**), respectively. GT, grana thylakoid; ST, stroma thylakoid. Scale bars = 5 μm (**A**,**D**,**G**), 1 μm (**B**,**E**,**H**) and 0.5 μm (**C**,**F**,**I**).

2.5. Expression Analysis and Subcellular Localization of OsRNE

Predicting through the Rice eFP Browser (http://bar.utoronto.ca/efprice/, accessed on 6 March 2024), the transcripts of *OsRNE* were highly expressed in young leaf, young inflorescence, mature leaf, and shoot apical meristem (SAM) (Figure 5A). To confirm the expression pattern of *OsRNE*, total RNA was extracted from different organs of WT, including the roots, culms, leaves, leaf sheaths, and panicles at the heading stage. Quantitative RT-PCR (qRT-PCR) analysis showed that *OsRNE* exhibited constitutive expression (Figure 5B). The highest expression was observed in the leaves, while relatively lower expression was detected in the culms, panicles, and roots (Figure 5B). These data suggest that OsRNE plays an important role in photosynthetic organs.

The OsRNE contains a carbohydrate-binding module 2 (CBM2) domain at the N-terminal and a RNaseE/G domain at the C-terminal (Supplementary Figure S2). It has been reported that proteins containing CBM domains are localized in chloroplasts [30,31]. According to the prediction in WoLF PSORT (https://wolfpsort.hgc.jp/, accessed on 12 May 2024), the OsRNE is a chloroplast localization protein

(Supplementary Figure S3). To explore the subcellular localization of the OsRNE protein, the N-terminal 1-206 amino acids of OsRNE containing a chloroplast transit peptide (cTP) was fused to the green fluorescent protein (GFP) and transformed into *N. benthamiana* leaves and rice protoplasts. Fluorescent signals of the OsRNE-GFP fusion protein merged with chloroplast autofluorescence (Figure 5C,D), indicating that the OsRNE protein is a chloroplast-targeted protein.

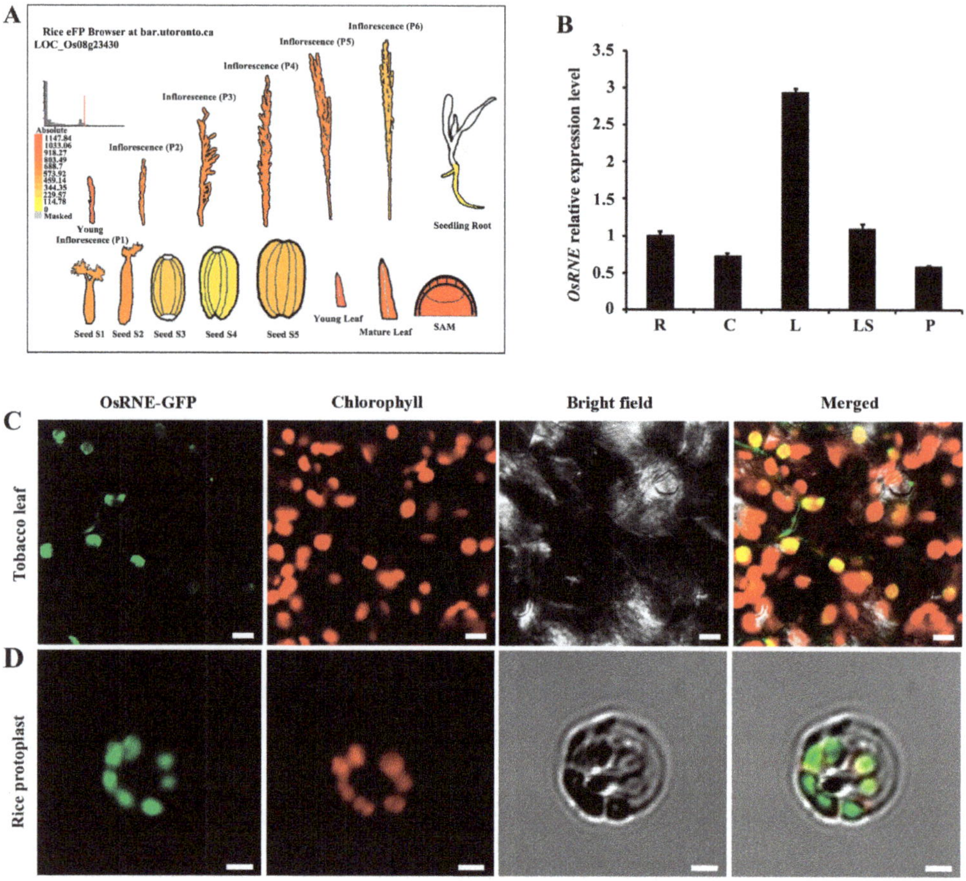

Figure 5. Expression pattern of *OsRNE* and subcellular localization of OsRNE. (**A**) Expression patterns of *OsRNE* based on the rice eFP browser. (**B**) qRT-PCR analysis of *OsRNE* relative expression level in different tissues at the heading stage. R, roots; C, culms; L, flag leaves; LS, leaf sheaths; P, panicles before heading. Error bars represent ± SD (n = 3). Confocal microscope images showing the subcellular localization of OsRNE-GFP in tobacco leaf (**C**) and rice protoplast (**D**). Signals from GFP fluorescence, chlorophyll autofluorescence, bright field, and merged images are shown. Scale bars = 10 μm (**C**) and 5 μm (**D**).

2.6. The Mutation of OsRNE Altered the Expression of Photosynthesis and Carbohydrate Metabolism Pathway Genes

To analyze the effect of the *OsRNE* mutation on gene expression, we compared the transcriptional profiles of nuclear and plastidic genes in *osrne-1* and WT seedlings using RNA-seq. A total of 4674 differentially expressed genes (DEGs, $q < 0.05$; $|\log_2(\text{FoldChange})| \geq 1$) were identified between *osrne-1* and WT. Among the 4674 DEGs, 2074 and 2600 genes were up-regulated and down-regulated, respectively (Figure 6A, Supplementary Tables S1 and S2).

The volcano plot shows distribution of a statistical significance and the altered expression level of the DEGs between *osrne-1* and WT (Figure 6B).

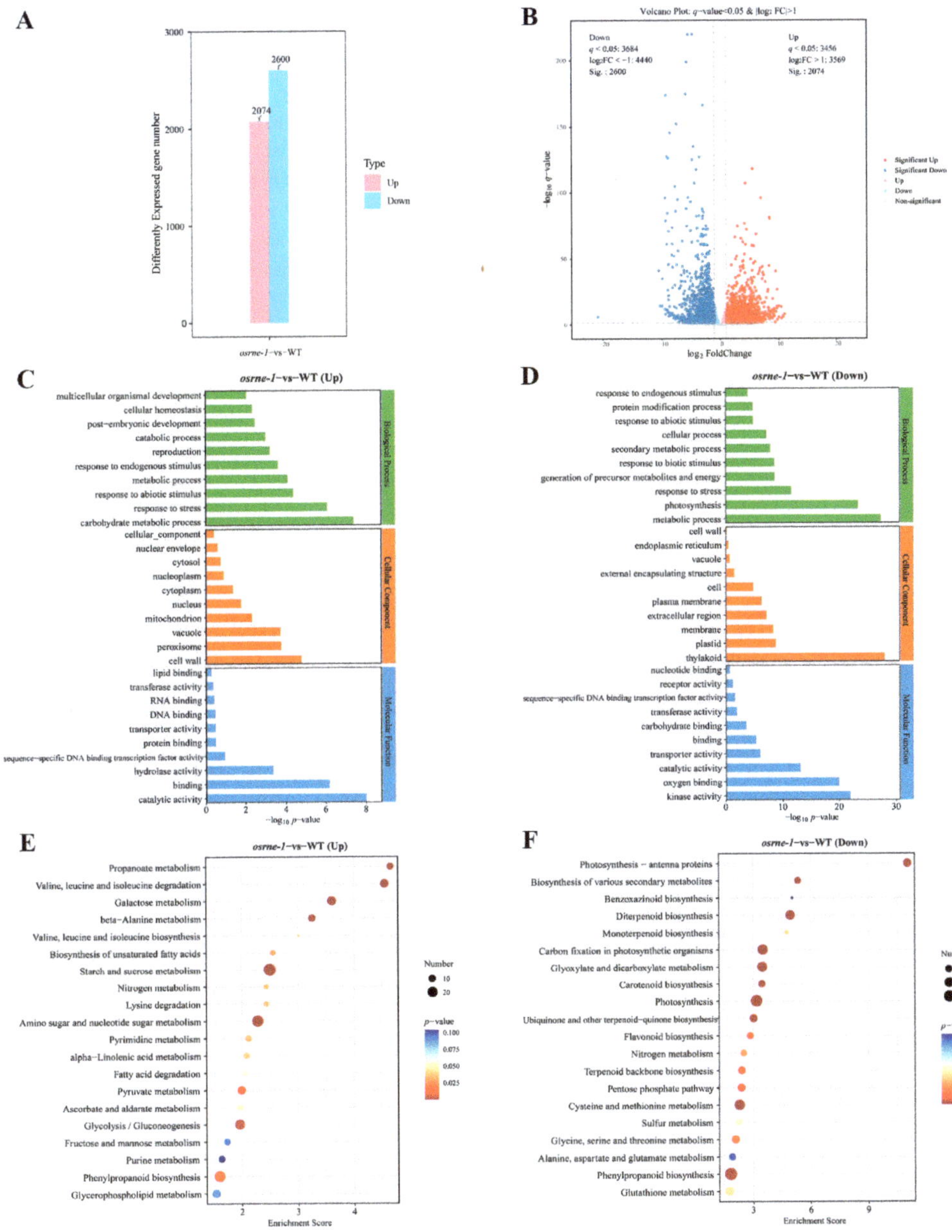

Figure 6. Transcriptome analysis of WT and *osrne-1* seedlings. (**A**) Number of DEGs from comparison of *osrne-1* and WT seedling transcriptomes. (**B**) Volcano plot of significantly up- or down-regulated DEGs. (**C**) GO enrichment analysis of the up-regulated DEGs. (**D**) GO enrichment analysis of the down-regulated DEGs. (**E**) KEGG pathway enrichment analysis of the up-regulated DEGs. (**F**) KEGG pathway enrichment analysis of the down-regulated DEGs.

Gene ontology (GO) analysis of the 4674 DEGs identified carbohydrate metabolic process (96 DEGs) for biological processes, cell wall (84 DEGs) for cellular components, and catalytic activity (282 DEGs) for molecular function as the most significant up-regulated DEGs (Figure 6C) in *osrne-1* vs. WT. Among the down-regulated DEGs, metabolic process (842 DEGs) and photosynthesis (73 DEGs) for biological processes, thylakoid (132 DEGs) for cellular components, and kinase activity (295 DEGs) for molecular function were the most significant (Figure 6D). In the significantly enriched GO, terms of down-regulated DEGs, metabolic process, photosynthesis, thylakoid, plastid, kinase activity, and oxygen binding, which are associated with photosynthesis and carbon fixation, were obviously identified (Figure 6D).

For KEGG pathway analysis, the top 20 significant categories were identified for up- and down-regulated DEGs. The up-regulated DEGs included propanoate metabolism, valine and leucine and isoleucine degradation, galactose metabolism, starch and sucrose metabolism, amino sugar and nucleotide sugar metabolism, and glycolysis/gluconeogenesis (Figure 6E). On the contrary, photosynthesis-antenna proteins, diterpenoid biosynthesis, photosynthesis, carbon fixation in photosynthetic organisms, biosynthesis of various secondary metabolites, carotenoid biosynthesis, ubiquinone and other terpenoid-quinone biosynthesis, and flavonoid biosynthesis were identified in the down-regulated DEGs (Figure 6F).

Further analysis showed that the expression levels of forty-five genes related to the photosynthesis pathway, twenty-five genes related to carbon fixation, ten genes related to the carotenoid biosynthesis, and ten genes related to the flavonoid biosynthesis, flavone, and flavonol biosynthesis pathway were down-regulated (Figure 7A–F, Supplementary Tables S3–S7). Moreover, the RNA-seq results indicate that most of the genes associated with the tricarboxylic acid cycle (TCA cycle), galactose metabolism, fatty acid degradation, and valine, leucine, and isoleucine degradation were up-regulated in *osrne-1* (Supplementary Figure S4). These results indicate that loss-of-function mutation of OsRNE mainly affects the expression of photosynthesis, carbohydrate metabolism, and pigment-related genes.

2.7. *OsRNE Mutation Affects the Expression of the Genes Involved in Chlorophyll Metabolism and Chloroplast Development*

Due to *OsRNE* mutation leading to white leaf phenotype, the expression levels of chlorophyll metabolism genes in *osrne*-1 and WT were compared from the transcriptome data. The expression of many chlorophyll metabolism genes showed significant differences between *osrne*-1 and WT (Figure 8A,B, Supplementary Tables S8 and S9). Common pathway genes of chlorophyll synthesis, including *OsCHLH* (*chlorophyllide a oxygenase H subunit*), *OsGUN4* (*genomes uncoupled 4*), *OsPORA* (*protochlorophyllide oxidoreductase A*), *OsPORB* (*protochlorophyllide oxidoreductase B*), *OsCRD1* (*Oryza sativa dicarboxylate diiron protein 1*), and *OsCAO1* (*chlorophyllide-a oxygenase 1*), were strongly decreased in *osrne-1* (Figure 8A). In the chlorophyll degradation pathway, *OsNYC1* (*non-yellow coloring 1*) and *OsRCCR1* (*red chlorophyll catabolite reductase 1*) showed an up-regulated transcript level in *osrne-1*, while the expression levels of *OsRCCR2* (*red chlorophyll catabolite reductase 2*), *OsPAO1* (*polyamine oxidase 1*), and *OsPAO5* (*polyamine oxidase 5*) were significantly decreased in *osrne-1* (Figure 8B). To further study the effects of the *OsRNE* mutation on the expression of other genes associated with chlorophyll biosynthesis and chloroplast development, four chloroplast development genes (*rpoA*, *rpoB*, *rpoC1*, and *rpoC2* (four subunits of the plastid-encoded polymerase, PEP)), seven photosynthetic-related genes (*atpA*, *atpB*, *psaA1*, *psaA2*, *psbD1* (three reaction-center polypeptides), *rbcL*, and *RCA* (*Rubisco activase*)), and six chlorophyll biosynthesis genes (*OsCAO1*, *OsCHLH*, *OsCHLI*, *OsCRD1*, *OsHEML* (*glutamyl tRNA reductase*), and *OsPORA*) were selected for qRT-PCR assay (Figure 8C–E). The transcript abundance of *rpoA*, *rpoB*, *rpoC1*, and *rpoC2* involved in chloroplast development significantly increased in *osrne-1* (Figure 8C), while the expression of photosynthetic-related genes, including nuclear-encoding gene (*RCA*) and plastid-encoding genes (*atpA*, *atpB*, *psaA1*, *psaA2*, *psbD1*, and *rbcL*), significantly decreased in *osrne-1* (Figure 8D). In addition, the transcription level of chlorophyll biosynthesis genes (*OsCAO1*, *OsCHLH*, *OsCHLI*, *OsCRD1*, *OsHEML*, and *OsPORA*) decreased significantly in *osrne-1* (Figure 8E).

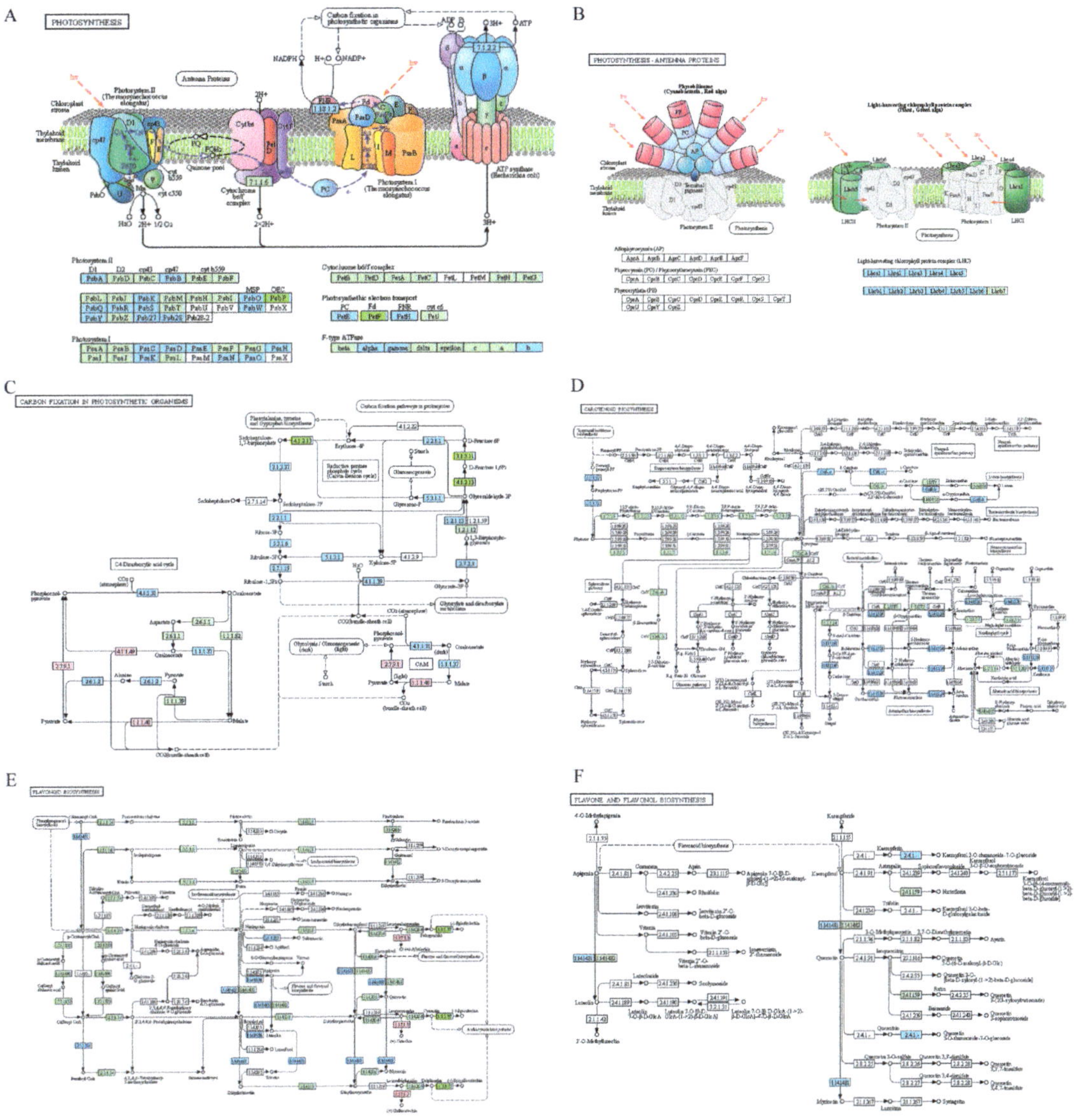

Figure 7. Differential map of genes in photosynthesis and carbohydrate metabolism pathways between WT and *osrne-1*. (**A**) Differential genes in photosynthesis pathway. (**B**) Differential genes of photosynthesis-antenna proteins. (**C**) Differential genes of carbon fixation in photosynthetic organisms. (**D**) Differential genes of carotenoid biosynthesis. (**E**) Differential genes of flavonoid biosynthesis. (**F**) Differential genes of flavone and flavonol biosynthesis. Blue boxes represent down-regulated genes, red boxes represent up-regulated genes, green boxes represent plant-specific genes, and dark green boxes represent both up-regulated genes and down-regulated genes.

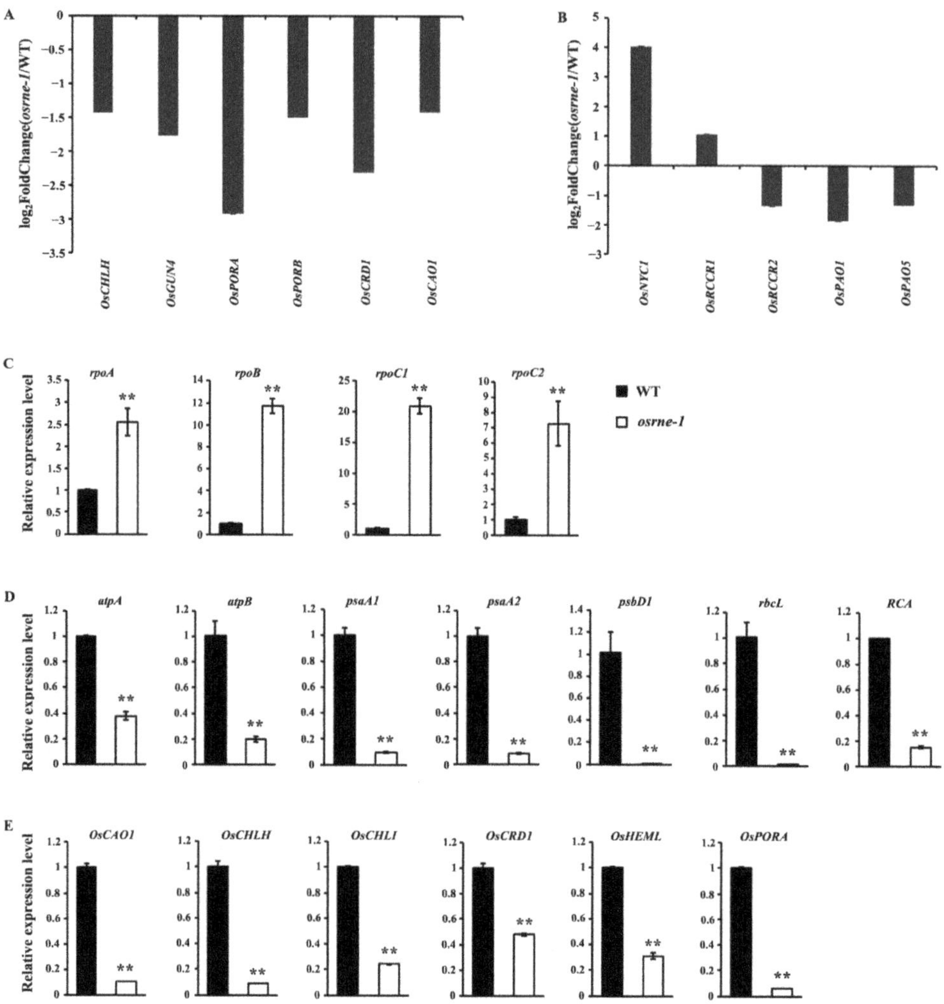

Figure 8. Differential expression of chlorophyll metabolism and chloroplast development genes in WT and *osrne-1*. The graphs show the $\log_2$ ratio of transcript levels involved in chlorophyll biosynthesis (**A**) and chlorophyll degradation pathway (**B**) in *osrne-1* compared with WT. Raw data are shown in Supplementary Tables S8 and S9. (**C**) Expression level of genes related to chloroplast biogenesis using qRT-PCR. (**D**) Expression level of genes related to photosynthesis using qRT-PCR. (**E**) Expression level of genes related to chlorophyll synthesis using qRT-PCR. The relative expression level of each gene was normalized using *Actin* as the internal control. Error bars are based on three independent biological replicates. Asterisks indicate statistically significant differences compared to the WT (** $p < 0.01$).

3. Discussion

Normal chloroplast development and chlorophyll biosynthesis are vital for the growth and development of green plants. RNase E is a key regulator of post-transcriptional gene expression, facilitating RNA degradation and the maturation of rRNAs and tRNAs in prokaryotes [16]. In *Arabidopsis*, loss of RNase E results in pale-green mutants with reduced plastid size, fewer thylakoids, and shorter granal stacks [28]. Processing of chloroplast transcripts still occurs in the *Arabidopsis* mutants, but some unprocessed high-molecular-weight transcripts accumulate [20]. Unlike in bacteria, *Arabidopsis* RNase E is not essential for survival [20,32]. Loss of RNE in *Arabidopsis* is perhaps partially compensated either by endonuclease RNase J or by other RNases rendering the mutant viable. As in the nuclear genome of *Arabidopsis*, there is only one RNase E/G gene in the nuclear genome of rice and not in the eukaryotes lacking chloroplasts [19,28]. Therefore, it is intriguing to determine the role of RNase E in rice. In this study, we proved that OsRNE is required for chloroplast development in rice by creating the *RNase E* mutants *wll1* and *osrne* in rice, which both display a lethal, chlorophyll-deficient, and white seedling phenotype (Figures 1 and 2). The *osrne* mutant harbors a single base insertion (an A or T) that causes the frame shifts with premature termination of OsRNE (Figure 2). This might disable the enzymatic activity of OsRNE protein and lead to a defect in RNA processing in the chloroplasts. Interestingly, *Arabidopsis* RNase E mutant plants require sucrose at all stages of growth and flowering in vitro, allowing for seed harvest [19,28]. However, the rice *osrne* mutant still displayed an albino phenotype on MS medium containing 2% sucrose and failed to survive beyond the three-leaf stage, possibly because *Arabidopsis atrne* mutants on sucrose-containing media develop pale-green leaves and can synthesize some pigments, whereas rice mutants with more severe chloroplast phenotypes remain albino and are unable to synthesize pigments. In future work, metabolomic analysis will be employed to screen for metabolic pathways that are particularly affected during chloroplast development.

Temporal and spatial expression of genes is in accordance to their functions. Our expression analysis showed that the transcript of *OsRNE* was abundantly found in green leaves in rice (Figure 5B), indicating that *OsRNE* has a predominant expression in photosynthetic organs. The *OsRNE* expression pattern suggests its important function in chloroplast development. TEM analysis of the chloroplast ultrastructure revealed that the *osrne* mutant had developmentally defective chloroplasts with a small amount of thylakoids or no thylakoids (Figure 4C–F). Thus, OsRNE may be involved in chloroplast biogenesis and participate in the formation of thylakoids during chloroplast build-up.

Transcriptome analysis using *osrne-1* and WT seedlings in rice showed that DEGs related to photosynthesis, thylakoid, carbon fixation, carotenoid biosynthesis, and flavonoid biosynthesis were enriched in GO and KEGG analysis (Figure 6C–F). These enrichment patterns of DEGs are consistent with the chlorophyll and photosynthesis deficiency in *osrne-1*. Moreover, the expression levels of both the genes for synthesis and degradation of valine, leucine, and isoleucine were increased (Figure 6E). Reports have shown that branched-chain amino acids are synthesized in chloroplasts [33–36]. We infer that the mutation in *OsRNE* may lead to defects in amino acid metabolism. The increased biosynthesis and degradation of valine, leucine, and isoleucine could maintain the valine, leucine, and isoleucine flow at a normal level for plant growth.

Chloroplast genes are transcribed by NEP and PEP [12,13]. Ordination of two-way transcription is essential for chloroplast development. The status of the chloroplast could influence the transcription of nuclear genes through retrograde signaling. In *osrne-1* plants, expression of many nuclear genes changed (Supplementary Tables S1 and S2). The RNA-seq results show that the expression of the most photosynthetic genes was reduced (Supplementary Table S3). Conversely, the expression of nuclear-encoded chloroplast development genes (*rpoA*, *rpoB*, *rpoC1*, and *rpoC2*) was increased (Figure 8C), indicating a feedback regulatory mechanism for chloroplast biogenesis-related genes. Some mRNA transcripts require ribosomal translation for efficient stabilization [37–39]. The reduced transcripts of *atpA*, *atpB*, *psaA1*, *psaA2*, *psbD1*, and *rbcL* might result from the inefficient PEP transcription activity (Figure 8D). Furthermore, the reduced transcript of *rbcL* and the reduction in RbcL protein level suggested that the *osrne* mutant was most likely defective in biogenesis of plastid ribosomes (Figures 3B and 8D). Moreover, the decreased expression of genes in the chlorophyll biosynthesis pathway (*OsCAO1, OsCHLH, OsCHLI, OsCRD1, OsHEML*, and *OsPORA*) and photosynthesis (*atpA, atpB, psaA1, psaA2, psbD1, rbcL*, and *RCA*) was likely due to the disruption of chloroplast development. Consequently, chlorophyll *a* and *b* accumulation decreased (Figure 3A). Therefore, the white leaf phenotype may be attributable to a defect in chloroplast development, resulting in the inability to carry out photosynthesis.

In *Escherichia coli*, RNase E forms a multiprotein complex called degradosome with PNPase, RNA-helicase, and enolase, involved in the degradation of bacterial mRNAs [16]. *Arabidopsis* RNase E forms a high-molecular-weight complex and interacts with an RNA-binding protein RHON1, which supports RNase E function in the chloroplast [32,40]. Therefore, in future work, we will use protein–protein interactions assays to screen the interacting proteins of OsRNE to gain a more comprehensive understanding of the regulatory network involved in chloroplast development.

In conclusion, we report herein that OsRNE is crucial for plastid gene expression, chlorophyll biosynthesis, and chloroplast development in rice.

4. Materials and Methods

4.1. Plant Materials and Growth Conditions

The *wll1* mutant with white leaf phenotype was initially identified from a ^{60}Co-irradiated mutant pool of the *japonica* rice cultivar Yandao 8. The *osrne* mutants were generated using the CRISPR/Cas9 system. A target sequence located within the first exon of *OsRNE* was selected and cloned into CRISPR/Cas9 vector. The CRISPR/Cas9 vector was then transferred into the rice *japonica* variety Zhonghua 11 *calli* via *A. tumefaciens*-mediated transformation [41]. Among the plants of the T_0 generation, the various types of *OsRNE* were confirmed by Sanger sequencing. As the *wll1* and homozygous *osrne* mutation was seedling-lethal, heterozygous plants were grown for reproduction. All rice materials were planted in the experimental field of Yangzhou University in May and harvested in October under natural conditions.

4.2. MutMap+

MutMap+ was performed as described previously [29,42]. Briefly, 30 green seedlings and 30 white seedlings from the heterozygous plant population were selected for DNA extraction. The DNA from individual plants was equally mixed to construct two DNA pools: pool 1 and pool 2. The DNA samples of pool 1 and pool 2 were randomly broken into fragments with 350 bp for the library construction, and then, two libraries were re-sequenced using an Illumina HiSequation 2500 (Novegene, Tianjin, China) to product the short sequence reads as raw data. After quality control of raw data and genome mapping, SNP/Indel were detected and annotated by GATK3.8 [43].

4.3. Determination of Pigment Content

Equal weights of fresh leaves of WT and *osrne-1* plants from two-leaf stage seedlings were soaked in 95% ethanol for 48 h in darkness and used to determine chlorophyll contents according to a described method [44]. Supernatant absorbances at 665, 649, and 470 nm were measured using a Hitachi U-1800 Spectrophotometer (Tokyo, Japan). Three independent biological repeats were measured for each sample.

4.4. Transmission Electron Microscopy (TEM) Analysis

Transverse sections of WT and *osrne-1* leaves from two-leaf stage seedlings were fixed in 2.5% glutaraldehyde in a phosphate buffer and further fixed in 1% OsO_4 three hours at 4 °C. Samples were further dehydrated through a series of ethanol (20%, 40%, 60%, 70%,

80%, 90%, 95%, and 100% ethanol, 10 min for each concentration) and then embedded in Spurr's medium. Ultrathin sections (approximately 70 nm) were produced using Leica EM UC7 ultramicrotome (Wetzlar, Germany). Sections were double-stained with uranyl acetate and lead citrate for 20 min and viewed with a Hitachi H-7650 transmission electron microscope (Hitachi High-Technologies Corp, Tokyo, Japan).

4.5. RNA Isolation and qRT-PCR Analysis

The RNA for analyzing the expression pattern of *OsRNE* was sampled from WT roots, culms, leaves, leaf sheaths, and panicles at the heading stage, respectively, through an RNAprep pure plant kit (Tiangen, Beijing, China). For studying the effects of the *OsRNE* mutation on the expression of genes associated with chlorophyll metabolism and chloroplast development, total RNA was extracted from two-leaf stage seedlings of WT and *osrne-1*. Then, 2 μg of total RNA was reverse-transcribed by NovoScript®Plus All-in-one 1st Strand cDNA Synthesis SuperMix (gDNA Purge) (Novoprotein, Suzhou, China). qRT-PCR was performed according to a previous description [45]. The primer sequences of *OsRNE*, internal control gene *Actin*, chlorophyll metabolism, and chloroplast development genes are listed in Supplementary Table S10.

4.6. Subcellular Localization of OsRNE

The OsRNE$^{1-206}$ coding sequence was cloned into pCAMBIA1305-GFP and pAN580-GFP to generate recombinant plasmid p35S:OsRNE$^{1-206}$-GFP and pAN580:OsRNE$^{1-206}$-GFP, respectively. The primer sequences used for OsRNE subcellular localization are shown in Supplementary Table S11. The recombinant plasmids p35S:OsRNE$^{1-206}$-GFP and pAN580:OsRNE$^{1-206}$-GFP were expressed in *Nicotiana benthamiana* leaves and rice protoplasts, respectively, as described previously [46]. The GFP signals were observed using a Zeiss LSM880 confocal laser microscope (Carl Zeiss, Oberkochen, Germany).

4.7. Protein Extraction and SDS-PAGE

WT and *osrne-1* leaves (100 mg) from two-leaf stage seedlings were separately ground into fine powder with liquid nitrogen, and the total proteins were extracted in 500 μL of extraction buffer (10% SDS, 1 M Tris-HCl, pH 8.0). SDS-PAGE detection was performed using the standard procedures. The proteins of WT and *osrne-1* were resolved in 10% SDS–polyacrylamide gel electrophoresis (PAGE) gels and stained with Coomassie Brilliant Blue R250 to examine the protein bands. To better present the results, 10 μL and 5 μL of total proteins from WT and *osrne-1* were loaded individually.

4.8. RNA-Seq Analysis

Total RNA from two-leaf stage seedlings of WT and *osrne-1* was separately extracted using the TRIzol reagent (Invitrogen, CA, USA) according to the manufacturer's protocol. Three biological replicates were used. The purity and quantification of RNA were evaluated using NanoDrop2000 (Thermo Scientific, Waltham, MA, USA). The integrity of RNA was assessed using Agilent2100 Bioanalyzer (Agilent Technologies, Santa Clara, CA, USA). The transcriptome sequencing and analysis were conducted by OE Biotech Co., Ltd. (Shanghai, China). Q value < 0.05 and $|\log_2(\text{FoldChange})| \geq 1$ were set as the threshold for significantly differential expression genes (DEGs). The GO and KEGG pathway analysis were performed using R (v3.2.0), respectively.

Supplementary Materials: The following supporting information can be downloaded at https://www.mdpi.com/article/10.3390/ijms26052375/s1.

Author Contributions: Conceptualization, methodology, software, and formal analysis, L.Z., C.Z. and H.F.; investigation, H.F., L.S., K.L. and Y.G. (Yishu Gu); resources, H.F., L.S., K.L. and Y.G.

(Yao Guo); data curation, H.F. and L.S.; writing—original draft preparation, H.F. and L.S.; writing—review and editing, L.Z.; supervision, C.Z. and L.Z.; project administration, L.Z.; funding acquisition, H.F. and L.Z. All authors have read and agreed to the published version of the manuscript.

Funding: This study was financially supported by grants from the Natural Science Foundation of Jiangsu Province (BK20221283), the Qinglan Project of Jiangsu (Su (2022) no.2) to Huimin Fang, the Qinglan Project of Yangzhou University to Long Zhang, and the Innovation Program for College Students of Yangzhou University (C202311117015Y).

Institutional Review Board Statement: Not applicable.

Informed Consent Statement: Not applicable.

Data Availability Statement: The raw transcriptome sequence data have been submitted to the NCBI Short Read Archive (SRA) with accession number <PRJNA1103341>.

Acknowledgments: We are grateful to Hui Zhang (Shanghai Normal University) for kindly providing the CRISPR/Cas9 vector. The RNA libraries were sequenced on the illumina NovaseqTM 6000 platform by OE Biotech, Inc., Shanghai, China. We are grateful to OE Biotech, Inc. (Shanghai, China), for assisting in high-through sequencing and bioinformatics analysis.

Conflicts of Interest: The authors declare no conflicts of interest. The funders had no role in the design of the study; in the collection, analyses, or interpretation of data; in the writing of the manuscript; and in the decision to publish the results.

References

1. Suomivuori, C.M.; Fliegl, H.; Starikov, E.B.; Balaban, T.S.; Kaila, V.R.I.; Sundholm, D. Absorption shifts of diastereotopically ligated chlorophyll dimers of photosystem I. *Phys. Chem. Chem. Phys.* **2019**, *21*, 6851–6858. [CrossRef] [PubMed]
2. Pogson, B.J.; Albrecht, V. Genetic dissection of chloroplast biogenesis and development: An Overview. *Plant Physiol.* **2011**, *155*, 1545–1551. [CrossRef] [PubMed]
3. Zhao, D.S.; Zhang, C.Q.; Li, Q.F.; Yang, Q.Q.; Gu, M.H.; Liu, Q.Q. A residue substitution in the plastid ribosomal protein L12/AL1 produces defective plastid ribosome and causes early seedling lethality in rice. *Plant Mol. Biol.* **2016**, *91*, 161–177. [CrossRef] [PubMed]
4. Wang, Y.; Ren, Y.L.; Zhou, K.N.; Liu, L.L.; Wang, J.L.; Xu, Y.; Zhang, H.; Zhang, L.; Feng, Z.M.; Wang, L.W.; et al. *WHITE STRIPE LEAF4* encodes a novel P-type PPR protein required for chloroplast biogenesis during early leaf development. *Front. Plant Sci.* **2017**, *8*, 1116. [CrossRef]
5. Zhou, K.N.; Ren, Y.L.; Zhou, F.; Wang, Y.; Zhang, L.; Lyu, J.; Wang, Y.H.; Zhao, S.L.; Ma, W.W.; Zhang, H.; et al. *Young Seedling Stripe1* encodes a chloroplast nucleoid-associated protein required for chloroplast development in rice seedlings. *Planta* **2017**, *245*, 45–60. [CrossRef]
6. Shim, K.C.; Kang, Y.A.; Song, J.H.; Kim, Y.J.; Kim, J.K.; Kim, C.; Tai, T.H.; Park, I.; Ahn, S.N. A frameshift mutation in the Mg-chelatase I subunit gene *OsCHLI* is associated with a lethal chlorophyll-deficient, yellow seedling phenotype in rice. *Plants* **2023**, *12*, 2831. [CrossRef]
7. García-Alcázar, M.; Giménez, E.; Pineda, B.; Capel, C.; García-Sogo, B.; Sánchez, S.; Yuste-Lisbona, F.J.; Angosto, T.; Capel, J.; Moreno, V.; et al. Albino T-DNA tomato mutant reveals a key function of 1-deoxy-D-xylulose-5-phosphate synthase (DXS1) in plant development and survival. *Sci. Rep.* **2017**, *7*, 45333. [CrossRef]
8. Yan, J.Q.; Liu, B.; Cao, Z.Q.; Chen, L.; Liang, Z.J.; Wang, M.; Liu, W.R.; Lin, Y.E.; Jiang, B. Cytological, genetic and transcriptomic characterization of a cucumber albino mutant. *Front. Plant Sci.* **2022**, *13*, 1047090. [CrossRef]
9. Du, J.J.; Wang, J.W.; Shan, S.C.; Mi, T.; Song, Y.L.; Xia, Y.; Ma, S.C.; Zhang, G.S.; Ma, L.J.; Niu, N. Low-temperature-mediated promoter methylation relates to the expression of *TaPOR2D*, affecting the level of chlorophyll accumulation in albino wheat (*Triticum aestivum* L.). *Int. J. Mol. Sci.* **2023**, *24*, 14697. [CrossRef]
10. Xu, Y.B.; Wu, Z.S.; Shen, W.; Zhou, H.Y.; Li, H.; He, X.H.; Li, R.B.; Qin, B.X. Disruption of the rice ALS1 localized in chloroplast causes seedling-lethal albino phenotype. *Plant Sci.* **2024**, *338*, 111925. [CrossRef]
11. An, H.Q.; Ke, X.L.; Li, L.; Liu, Y.T.; Yuan, S.H.; Wang, Q.Y.; Hou, X.; Zhao, J. ALBINO EMBRYO AND SEEDLING is required for RNA splicing and chloroplast homeostasis in *Arabidopsis*. *Plant Physiol.* **2023**, *193*, 483–501. [CrossRef] [PubMed]
12. Jarvis, P.; López-Juez, E. Biogenesis and homeostasis of chloroplasts and other plastids. *Nat. Rev. Mol. Cell Biol.* **2013**, *14*, 787–802. [CrossRef] [PubMed]

13. Pogson, B.J.; Ganguly, D.; Albrecht-Borth, V. Insights into chloroplast biogenesis and development. *Biochim. Biophys. Acta* **2015**, *1847*, 1017–1024. [CrossRef] [PubMed]
14. Pfalz, J.; Pfannschmidt, T. Essential nucleoid proteins in early chloroplast development. *Trends Plant Sci.* **2013**, *18*, 186–194. [CrossRef]
15. Reumann, S.; Inoue, K.; Keegstra, K. Evolution of the general protein import pathway of plastids (Review). *Mol. Membr. Biol.* **2005**, *22*, 73–86. [CrossRef]
16. Stoppel, R.; Meurer, J. The cutting crew—Ribonucleases are key players in the control of plastid gene expression. *J. Exp. Bot.* **2012**, *63*, 1663–1673. [CrossRef]
17. Lee, K.; Cohen, S.N. A functional orthologue of RNase E shows shuffling of catalytic and PNPase-binding domains. *Mol. Microbiol.* **2003**, *48*, 349–360. [CrossRef]
18. Zhou, C.; Zhang, J.Y.; Hu, X.Y.; Li, C.C.; Wang, L.; Huang, Q.Y.; Chen, W.L. RNase II binds to RNase E and modulates its endoribonucleolytic activity in the cyanobacterium *Anabaena* PCC 7120. *Nucleic Acids Res.* **2020**, *48*, 3922–3934. [CrossRef]
19. Schein, A.; Sheffy-Levin, S.; Glaser, F.; Schuster, G. The RNase E/G-type endoribonuclease of higher plants is located in the chloroplast and cleaves RNA similarly to the *E. coli* enzyme. *RNA* **2008**, *14*, 1057–1068. [CrossRef]
20. Walter, M.; Piepenburg, K.; Schöttler, M.A.; Petersen, K.; Kahlau, S.; Tiller, N.; Drechsel, O.; Weingartner, M.; Kudla, J.; Bock, R. Knockout of the plastid RNase E leads to defective RNA processing and chloroplast ribosome deficiency. *Plant J.* **2010**, *64*, 851–863. [CrossRef]
21. Bouvet, P.; Belasco, J.G. Control of RNase E-mediated RNA degradation by 5′-terminal base pairing in *E. coli*. *Nature* **1992**, *360*, 488–491. [CrossRef] [PubMed]
22. McDowall, K.J.; Kaberdin, V.R.; Wu, S.W.; Cohen, S.N.; Lin-Chao, S. Site-specific RNase E cleavage of oligonucleotides and inhibition by stem-loops. *Nature* **1995**, *374*, 287–290. [CrossRef] [PubMed]
23. Xu, F.F.; Cohen, S.N. RNA degradation in *Escherichia coli* regulated by 3′ adenylation and 5′ phosphorylation. *Nature* **1995**, *374*, 180–183. [CrossRef] [PubMed]
24. Alifano, P.; Rivellini, F.; Piscitelli, C.; Carlomagno, M. Ribonuclease E provides substrates for Ribonuclease P-dependent processing of a polycistronic mRNA. *Genes Dev.* **1995**, *8*, 3021–3031. [CrossRef]
25. Huang, H.; Liao, J.; Cohen, S. Poly(A)- and poly(U)-specific RNA 3′ tail shortening by *E. coli* ribonuclease E. *Nature* **1998**, *391*, 99–102. [CrossRef]
26. Mackie, G.A. Ribonuclease E is a 5′-end-dependent endonuclease. *Nature* **1998**, *395*, 720–723. [CrossRef]
27. Morita, T.; Maki, K.; Aiba, H. RNase E-based ribonucleoprotein complexes: Mechanical basis of mRNA destabilization mediated by bacterial noncoding RNAs. *Genes Dev.* **2005**, *19*, 2176–2186. [CrossRef]
28. Mudd, E.A.; Sullivan, S.; Gisby, M.F.; Mironov, A.; Kwon, C.S.; Chung, W.I.; Day, A. A 125 kDa RNase E/G-like protein is present in plastids and is essential for chloroplast development and autotrophic growth in *Arabidopsis*. *J. Exp. Bot.* **2008**, *59*, 2597–2610. [CrossRef]
29. Fekih, R.; Takagi, H.; Tamiru, M.; Abe, A.; Natsume, S.; Yaegashi, H.; Sharma, S.; Sharma, S.; Kanzaki, H.; Matsumura, H.; et al. MutMap plus: Genetic Mapping and Mutant Identification Without Crossing in Rice. *PLoS ONE* **2013**, *8*, e68529. [CrossRef]
30. Zhang, L.; Li, N.; Zhang, J.; Zhao, L.L.; Qiu, J.J.; Wei, C.X. The CBM48 domain-containing protein FLO6 regulates starch synthesis by interacting with SSIVb and GBSS in rice. *Plant Mol. Biol.* **2022**, *108*, 343–361. [CrossRef]
31. Wang, W.; Wei, X.J.; Jiao, G.A.; Chen, W.Q.; Wu, Y.W.; Sheng, Z.H.; Hu, S.K.; Xie, L.H.; Wang, J.Y.; Tang, S.Q.; et al. *GBSS-BINDING PROTEIN*, encoding a CBM48 domain-containing protein, affects rice quality and yield. *J. Integr. Plant Biol.* **2020**, *62*, 948–966. [CrossRef] [PubMed]
32. Stoppel, R.; Manavski, N.; Schein, A.; Schuster, G.; Teubner, M.; Schmitz-Linneweber, C.; Meurer, J. RHON1 is a novel ribonucleic acid-binding protein that supports RNase E function in the chloroplast. *Nucleic Acids Res.* **2012**, *40*, 8593–8606. [CrossRef] [PubMed]
33. Ellerström, M.; Josefsson, L.G.; Rask, L.; Ronne, H. Cloning of a cDNA for rape chloroplast 3-isopropylmalate dehydrogenase by genetic complementation in yeast. *Plant Mol. Biol.* **1992**, *18*, 557–566. [CrossRef] [PubMed]
34. Diebold, R.; Schuster, J.; Däschner, K.; Binder, S. The branched-chain amino acid transaminase gene family in *Arabidopsis* encodes plastid and mitochondrial proteins. *Plant Physiol.* **2002**, *129*, 540–550. [CrossRef]
35. Zybailov, B.; Rutschow, H.; Friso, G.; Rudella, A.; Emanuelsson, O.; Sun, Q.; van Wijk, K.J. Sorting signals, N-terminal modifications and abundance of the chloroplast proteome. *PLoS ONE* **2008**, *3*, e1994. [CrossRef]
36. Binder, S. Branched-chain amino acid metabolism in *Arabidopsis thaliana*. *Arab. Book* **2010**, *8*, e0137. [CrossRef]
37. Barkan, A. Nuclear mutants of maize with defects in chloroplast polysome assembly have altered chloroplast RNA metabolism. *Plant Cell* **1993**, *5*, 389–402. [CrossRef]
38. Bisanz, C.; Bégot, L.; Carol, P.; Perez, P.; Bligny, M.; Pesey, H.; Gallois, J.L.; Silva, L.M.; Mache, R. The Arabidopsis nuclear *DAL* gene encodes a chloroplast protein which is required for the maturation of the plastid ribosomal RNAs and is essential for chloroplast differentiation. *Plant Mol. Biol.* **2003**, *51*, 651–663. [CrossRef]

39. Kleinknecht, L.; Wang, F.; Stube, R.; Philippar, K.; Nickelsen, J.; Bohne, A.V. RAP, the sole octotricopeptide repeat protein in *Arabidopsis*, is required for chloroplast 16S rRNA maturation. *Plant Cell* **2014**, *26*, 777–787. [CrossRef]
40. Yang, Z.; Li, M.M.; Sun, Q.W. RHON1 co-transcriptionally resolves R-loops for *Arabidopsis* chloroplast genome maintenance. *Cell Rep.* **2020**, *30*, 243–256. [CrossRef]
41. Hiei, Y.; Ohta, S.; Komari, T.; Kumashiro, T. Efficient transformation of rice (*Oryza sativa* L.) mediated by *Agrobacterium* and sequence analysis of the boundaries of the T-DNA. *Plant J.* **1994**, *6*, 271–282. [CrossRef] [PubMed]
42. Zhang, C.; Mao, X.C.; Feng, X.X.; Sun, Y.L.; Wang, Z.R.; Tang, J.Q.; Yu, H.X. OsALB3 is required for chloroplast development by promoting the accumulation of light-harvesting chlorophyll-binding proteins in rice. *Plants* **2023**, *12*, 4003. [CrossRef] [PubMed]
43. Heldenbrand, J.R.; Baheti, S.; Bockol, M.A.; Drucker, T.M.; Hart, S.N.; Hudson, M.E.; Iyer, R.K.; Kalmbach, M.T.; Kendig, K.I.; Klee, E.W.; et al. Recommendations for performance optimizations when using GATK3.8 and GATK4. *BMC Bioinform.* **2019**, *20*, 557. [CrossRef]
44. Arnon, D.I. Copper enzymes in isolated chloroplast. *Plant Physiol.* **1949**, *24*, 1–15. [CrossRef]
45. Zhang, L.; Zhao, L.L.; Lin, L.S.; Zhao, L.X.; Liu, Q.Q.; Wei, C.X. A novel mutation of *OsPPDKB*, encoding pyruvate orthophosphate dikinase, affects metabolism and structure of starch in the rice endosperm. *Int. J. Mol. Sci.* **2018**, *19*, 2268. [CrossRef]
46. Fang, H.M.; Meng, Q.L.; Xu, J.W.; Tang, H.J.; Tang, S.Y.; Zhang, H.S.; Huang, J. Knock-down of stress inducible *OsSRFP1* encoding an E3 ubiquitin ligase with transcriptional activation activity confers abiotic stress tolerance through enhancing antioxidant protection in rice. *Plant Mol. Biol.* **2015**, *87*, 441–458. [CrossRef]

International Journal of
Molecular Sciences

MDPI

Article

The Silencing of the *StPAM16-1* Gene Enhanced the Resistance of Potato Plants to the Phytotoxin Thaxtomin A

Lu Liu [1,2,†], Shuangwei Song [1,2,†], Ning Liu [3,4], Zhiqin Wang [1,2], Yonglong Zhao [1,2], Naiqin Zhong [1,2,5], Pan Zhao [1,2,5,*] and Haiyun Wang [1,*]

1 State Key Laboratory of Plant Genomics, Institute of Microbiology, Chinese Academy of Sciences, Beijing 100101, China; liulu202206@163.com (L.L.); song_shuangwei@163.com (S.S.); wangzq1210@163.com (Z.W.); ylzhao97101@163.com (Y.Z.); nqzhong@im.ac.cn (N.Z.)

2 Engineering Laboratory for Advanced Microbial Technology of Agriculture, Chinese Academy of Sciences, Beijing 100101, China

3 Beijing Vegetable Research Center (BVRC), Beijing Academy of Agriculture and Forestry Science, Beijing 100097, China; liuning@nercv.org

4 National Engineering Research Center for Vegetables (NERCV), State Key Laboratory of Vegetable Biobreeding, Beijing Academy of Agriculture and Forestry Sciences, Beijing 100097, China

5 Key Laboratory of Potato Industry Integration and Development Enterprises in Inner Mongolia Autonomous Region, Hulunbuir 021000, China

* Correspondence: zhaop@im.ac.cn (P.Z.); wanghy@im.ac.cn (H.W.)

† These authors contributed equally to this work.

Abstract: Potato common scab (CS) caused by *Streptomyces scabiei* is a severe disease that threatens tuber quality and its market value. To date, little is known about the mechanism regulating the resistance of potato to CS. In this study, we identified a presequence translocase-associated motor 16 gene from potato (designated *StPAM16-1*) that is involved in the response to the phytotoxin thaxtomin A (TA) secreted by *S. scabiei*. The StPAM16-1 protein was localized in the mitochondria, and the expression of the gene was upregulated in potato leaves treated with TA. The suppression of *StPAM16-1* in potato led to enhanced resistance to TA and *S. scabiei*. Protein interaction analyses revealed that StPAM16-1 interacted with the subunit 5b of the COP9 signalosome complex (StCSN5). Similar to that of *StPAM16-1*, the expression levels of *StCSN5* significantly increased in potato leaves treated with TA. These results indicated that StPAM16-1 acted as a negative regulator and was functionally associated with StCSN5 in the immune response of potato plants against CS. Our study sheds light on the molecular mechanism by which PAM16 participates in the plant immune response. Furthermore, both *StPAM16-1* and *StCSN5* could be potential target genes in the molecular breeding of potato cultivars with increased resistance to CS.

Keywords: *Solanum tuberosum*; *Streptomyces scabiei*; thaxtomin A; StPAM16-1; StCSN5

1. Introduction

Potato common scab (CS), caused by pathogenic *Streptomyces* spp., is a severe soil-borne disease worldwide [1]. The symptoms of CS include scab-like superficial, raised and/or pitted lesions at infection sites, which results in a decrease in the quality and marketability of potato tubers [2]. Previous studies revealed that the phytotoxin thaxtomin, which is secreted by pathogenic *Streptomyces* spp., including *S. scabiei*, is the key pathogenic factor of CS [3]. As the predominant pathogenic component of the thaxtomin family, thaxtomin A (TA) affects plant growth and development in various ways, such as by inhibiting plant cellulose biosynthesis [4] and seedling growth [5], causing dramatic cell hypertrophy [6], altering Ca^{2+} and H^{+} flow [7,8] and triggering programmed cell death

(PCD) [9]. Although several approaches such as crop rotation, chemical treatment and biological control have been shown to alleviate this disease, control efficiency is limited [10]. Given that disease resistance in potato cultivars is the foundation of the integrated method, studying the genetics of tolerance in breeding is regarded as one of the best options for controlling CS. To date, the strongest source of resistance has been a Phureja group clone, and a major dominant resistance gene is in the populations of hybrids between *Solanum tuberosum* dihaploids and Phureja group clones [11,12]. Research on the genome evolution and diversity of wild and cultivated potatoes, as well as the graph pangenome [13–15], can provide critical insights into the potato genome and guide breeders in potato breeding. At present, the identification and functional characterization of defense-related genes to generate CS-resistant cultivars via molecular breeding are some of the desirable strategies for controlling CS.

The presequence translocase-associated protein import motor (PAM) subunit PAM16 is required for protein precursors to be translocated into the mitochondrial matrix [16]. In *Arabidopsis*, AtPAM16 may play a negative role in plant immunity by importing a negative regulator of plant immunity to mitochondria to inhibit the overaccumulation of ROS, and a single *Atpam16* mutant plant exhibited increased resistance to virulent pathogens [17]. *Atpam16-1* and *Atpam16-2* double-mutant plants showed enhanced SUPPRESSOR OF NPR1-1, CONSTITUTIVE 1 (SNC1)-mediated immunity [17]. The mechanism of non-target site resistance to the herbicide thaxtomin A conferred by the gene *pam16* mutant in *Marchantia polymorpha* increases ROS levels or decreases thaxtomin A metabolism in weeds [18]. The homologous gene thaxtomin resistance 1 (*TXR1*) was first identified during the screening of a TA-resistant mutant of *Arabidopsis* [5]. The study revealed that the uptake of TA by *txr1*-transformed plants was lower than that by wild-type (WT) plants, but the metabolism of TA did not differ between the WT and *txr1* mutant plants. These results implied that PAM16 (TXR1) likely plays an important role in plant responses to biotic stresses. In human cells, a protein homologous to yeast PAM16 named Magmas (mitochondria-associated granulocyte-macrophage colony-stimulating factor signaling molecule) was found to be an essential part of the mitochondrial precursor protein import machinery [19]. In addition, it was reported that Magmas can regulate cellular ROS levels by controlling ROS production and scavenging [20]. Therefore, it is of great interest to study how PAM16 regulates the plant immune system as a mitochondrial intima input protein and to analyze its physiological functions and mechanisms, for example, in potato responses to *S. scabiei* infection and TA treatment.

COP9 signalosome (CSN) is a conserved and nucleus-localized protein complex that is fractionated as a 450–550 kDa complex in gel filtration columns [21–23]. Studies have shown that CSN regulates the activity of CULLIN-RING E3 ubiquitin ligases (CRLs) by removing the ubiquitin-like protein RUB1 (related to ubiquitin 1 in plants)/NEDD8 (neural precursor cell-expressed developmentally downregulated-8 in animals), which interact with CRLs and thus affect the development and defense against pathogens and herbivorous insects [24]. CSN consists of eight subunits, among which CSN5 is a metalloprotease with a conserved JAMM (JAB1 MPN domain metalloprotease) motif and requires zinc ions as an activator [21,25]. CSN is widely present in eukaryotes. In plants, CSN plays an important role in regulating development and stress responses [26–30] and acts as an inhibitor of photomorphogenesis in *Arabidopsis* [31].

The identification and functional characterization of defense-related genes not only provide information for the interaction between potato and CS but also help in identifying candidate genes for molecular breeding to generate CS-resistant cultivars. In this study, we identified a phytotoxin TA-responsive gene, *StPAM16-1*, from tetraploid potato (*Solanum tuberosum* L.) cv. Shepody and found that the relative expression of *StPAM16-1* was induced by TA treatment and that StPAM16-1 was localized in mitochondria. We characterized the function of StPAM16-1 in response to TA treatment by VIGS and RNAi. StPAM16-1 could interact with StCSN5 via yeast two-hybrid and split luciferase complementation assays. Our results demonstrated that StPAM16-1 acted as a negative regulator of the defense response to TA in potato. As the expression of *StCSN5* was also induced by TA treatment, we speculated that the interaction between StPAM16-1 and StCSN5 may reflect a functional link between the two proteins in the immune response of potato to CS.

2. Results

2.1. Identification of StPAM16-1

A previous study reported that *AtPAM16/AtTXR1* is a TA-responsive gene that plays a negative role in the defense process in *Arabidopsis* [5]. In the aim of improving the disease tolerance of potato plants through genetic manipulation, we selected *PAM16* as the candidate gene and identified AtPAM16/AtTXR1 orthologs in the potato proteome via sequence alignment via the Potato Genomics Resource (http://solanaceae.plantbiology.msu.edu/; accessed on 15 November 2021) and NCBI databases (https://blast.ncbi.nlm.nih.gov/; accessed on 15 November 2021). A total of four proteins with high homology to AtPAM16/AtTXR1 were identified, among which the protein with the highest homology was named StPAM16-1 (PGSC0003DMP400044341). The other three homologous proteins were named StPAM16-2 (PGSC0003DMP400051461), StPAM16-3 (PGSC0003DMP400051472) and StPAM16-4 (PGSC0003DMP400026161), respectively. The sequence alignment and domain analysis of StPAM16-1, StPAM16-2, StPAM16-3, StPAM16-4, AtPAM16, AtPAM16L and ScPAM16 were carried out using DNAMAN. StPAM16-1 contains a highly conserved DnaJ domain (Figure 1a), which mediates chaperone binding [32]. Furthermore, the phylogenetic analysis of StPAM16-1 and its homologous proteins was carried out by MEGA 7.0. As shown in Figure 1b, PAM16 is present in various eukaryotes, and StPAM16-1 shares high homology with the PAM16 of *S. lycopersicum*.

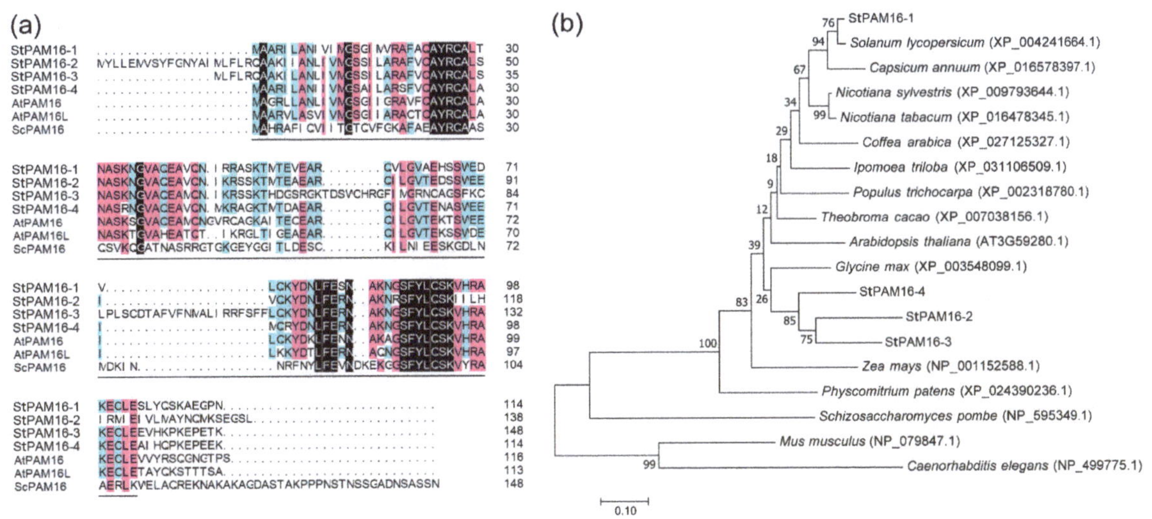

Figure 1. Homology analysis of StPAM16-1. (**a**,**b**) show multiple sequence alignment and phylogenetic analysis of StPAM16-1 and its homologous proteins, respectively. DnaJ domain is underlined by solid lines. Different colors of background represent similarity of amino acid sequences (black: 100%; pink: 75%; blue: 50%). Scale bar stands for evolutionary distance.

2.2. *StPAM16-1* Expression Is Induced by TA and Defense-Related Plant Hormones

To test whether StPAM16-1 participates in the response to TA in potato, *StPAM16-1* transcript levels were determined by quantitative reverse transcription PCR (RT-qPCR). The results showed that the relative expression of *StPAM16-1* was significantly upregulated at 6 h post-treatment and then gradually recovered to the background level (Figure 2a). Then, we examined the relative expression of *StPAM16-1* after treatment with several defense-related phytohormones, including salicylic acid (SA), jasmonic acid (JA) and ethylene (ETH). The results showed that the expression of *StPAM16-1* was upregulated after treatment with SA, JA and ETH (Figure 2b), with the highest enhancement occurring after SA treatment.

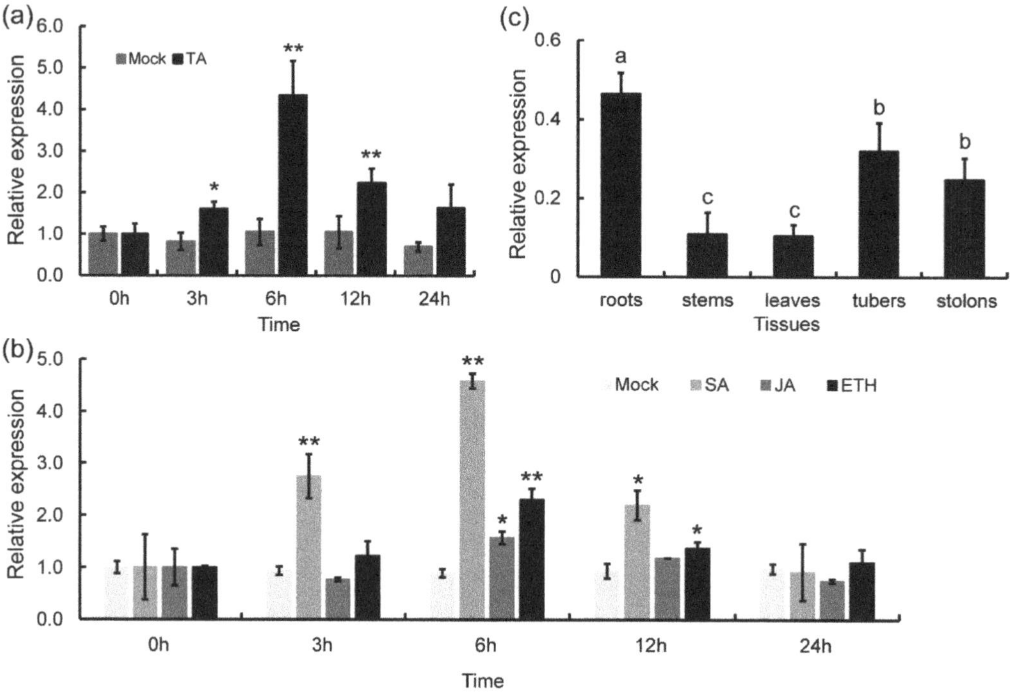

Figure 2. The expression pattern of *StPAM16-1* in potato plants. (**a**,**b**) The relative expression of *StPAM16-1* in potato leaves after TA (**a**), SA, JA and ETH (**b**) treatments. The asterisks denote statistically significant differences according to Student's *t* test; * $p < 0.05$, ** $p < 0.01$. (**c**) The relative expression of *StPAM16-1* in different organs of potato. Different letters indicate statistically significant differences as determined by Student's *t* test. *StActin* was used as a reference gene. The values represent the means ± standard deviation (SD). Three biological replicates were performed.

In addition, the relative expression of *StPAM16-1* in different tissues was detected by RT-qPCR. As shown in Figure 2c, *StPAM16-1* was expressed in all tissues but with relatively higher levels in the roots, tubers and stolons, which is closely related to its disease-related function.

2.3. *StPAM16-1* Localizes to Mitochondria

To further understand the biological function of StPAM16-1, the subcellular distribution of the proteins was analyzed using a tobacco transient expression system. An StPAM16-1-GFP fusion gene driven by the 35S promoter was constructed in the vector and transformed into competent Agrobacterium cells. The transformants were then injected into tobacco leaf cells. As shown in Figure 3, StPAM16-1-GFP fusion proteins were colocalized with mitochondrial markers, indicating that StPAM16-1 is localized to this organelle.

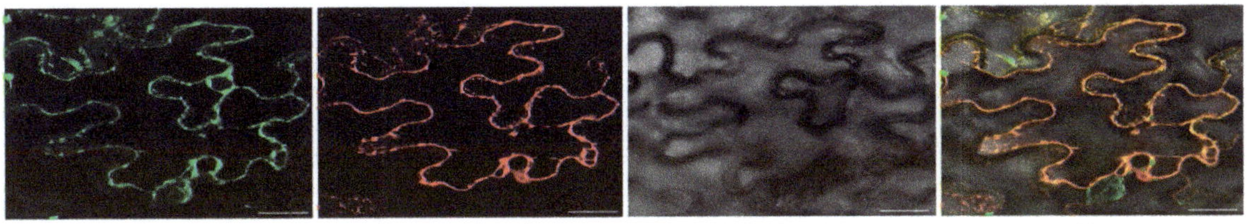

Figure 3. Subcellular localization of StPAM16-1 in *Nicotiana benthamiana* leaf cells. Bar = 25 μm.

2.4. Suppression of StPAM16-1 Expression Improves Tolerance of Potato Plants to Thaxtomin A and S. scabiei

To investigate the function of StPAM16-1 in response to TA, we first assessed the disease tolerance of potato plants with decreased *StPAM16-1* expression levels via virus-induced gene silencing (VIGS). To test the efficiency of the VIGS system, *StPDS*, which encodes a 15-cis-phytoene desaturase, was used as a marker. Approximately 40 days after injection, the albino phenotype appeared on *StPDS*-silenced potato leaves, and *StPDS* silencing was verified by RT-qPCR, which confirmed the efficiency of VIGS (Figure 4a,b). Then, we constructed the VIGS vector *pgR107-StPAM16-1* and transformed it into *Agrobacterium*. About 40 days after injection, total RNA was extracted from the leaves to determine the relative expression of *StPAM16-1* via RT-qPCR. The results showed that the relative expression of *StPAM16-1* was significantly lower in *StPAM16-1*-silenced potato leaves than in the control (Figure 4c). Subsequently, *StPAM16-1*-silenced and control potato plants were treated with 10 μM TA to test their tolerance. As shown in Figure 4d, there were more black necrotic spots on the control potato leaves than on the *StPAM16-1*-silenced potato leaves. Then, we used trypan blue to stain potato leaves treated with TA or sterile water and found that the necrosis of control potato leaves was more severe than that of *StPAM16-1*-silenced leaves. These results indicated that silencing *StPAM16-1* increased the tolerance of potato plants to TA.

To further explore the effects of StPAM16-1 in response to TA, the expression of *StPAM16-1* was suppressed by the RNAi approach, and an in-depth analysis was carried out. The *pANDA35K-StPAM16-1* RNAi vector was constructed and subsequently transformed into potato plants via *Agrobacterium*-mediated transformation. The relative expression of *StPAM16-1* in RNAi potato plants was measured by RT-qPCR. Three RNAi plants with different expression levels of *StPAM16-1* were selected and named RNAi1, RNAi2 and RNAi3. As shown in Figure 5a, the relative expression levels of *StPAM16-1* in the three *StPAM16-1*-RNAi plants were significantly lower than those in the WT plant, while the expression of *StPAM16-2*, *StPAM16-3* and *StPAM16-4* did not change in the WT and *StPAM16-1*-RNAi potato plants. Compared with the WT plants, the *StPAM16-1*-RNAi plants also exhibited a dwarf phenotype (Figure 5b). This finding is consistent with a previous report on *Arabidopsis* [5]. Therefore, *StPAM16-1*-RNAi plants were selected for further studies.

Figure 4. Increased tolerance of *StPAM16-1*-silenced potato plants to TA treatment. (**a**) Albino phenotype of control (*PVX:00*) and *StPDS*-silenced potato plants (*PVX:StPDS*) approximately 40 days after injection. (**b**) Relative expression of *StPDS* in *PVX:00* and *PVX:StPDS*. (**c**) Relative expression of *StPAM16-1* in PVX:00 and *StPAM16-1*-silenced potato plants (*PVX:StPAM16-1*). *StActin* was used as reference gene. Values represent means ± SD. Asterisks denote statistically significant differences as determined by Student's *t* test, ** $p < 0.01$. Three biological replicates were performed. (**d**) Phenotypes of *PVX:00* and *PVX:StPAM16-1* after TA treatment, which were stained with trypan blue. Mock represents *PVX:00* treated with sterile water.

We treated WT and three *StPAM16-1*-RNAi potato leaves with 10 μM TA. As shown in Figure 5c, the tolerance of *StPAM16-1*-RNAi potato leaves to TA was significantly enhanced as compared to the WT control. Further, WT and *StPAM16-1*-RNAi plants grown for 3–4 weeks were inoculated with *S. scabiei*. Potato tubers were harvested after 3 months, and the disease incidence of tubers was counted. The results showed that there were a large number of scab patches on the WT tubers, while there were fewer scab patches on the *StPAM16-1*-RNAi tubers (Figure 5d). Meanwhile, the incidence rate, disease index and relative control effect of the WT and *StPAM16-1*-RNAi tubers were also calculated. Compared with those of the WT tubers, the incidence rate (Figure 5e) and disease index (Figure 5f) of the *StPAM16-1*-RNAi tubers significantly decreased, while the relative control effect clearly increased (Figure 5g). These results indicated that the inhibition of *StPAM16-1* improved the tolerance of potato plants to *S. scabiei*.

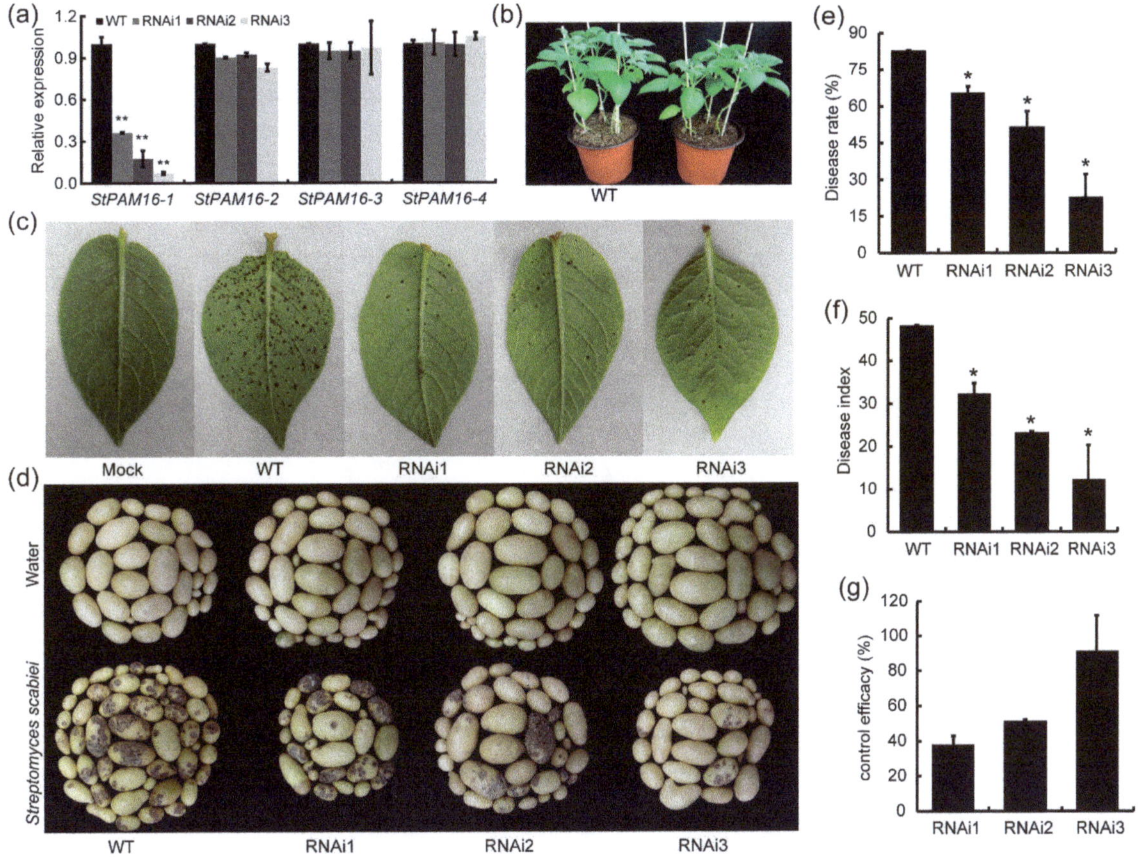

Figure 5. Repression of *StPAM16-1* expression enhanced tolerance of potato plants to TA and *S. scabiei* infection. (**a**) Relative expression levels of *StPAM16-1* and its homologous genes in WT and *StPAM16-1*-RNAi potato plants. (**b**) Phenotypes of WT and *StPAM16-1*-RNAi potato plants. (**c**) Repression of *StPAM16-1* expression increased tolerance of potato plants to TA treatment. (**d**) Phenotypes of WT and *StPAM16-1*-RNAi potato tubers after *S. scabiei* infection. Mock represents WT potato tubers treated with sterile water. (**e**) Rate of diseased tubers of WT and *StPAM16-1*-RNAi potato plants. (**f**) Disease index of WT and *StPAM16-1*-RNAi potato tubers. (**g**) Relative disease control effect of *StPAM16-1*-RNAi in potato tubers. Values represent means ± SD. Asterisks denote statistically significant differences according to Student's *t* test; * $p < 0.05$, ** $p < 0.01$. Three biological replicates were performed.

2.5. StPAM16-1 Interacts with StCSN5

To elucidate the functional mechanism of StPAM16-1, the cDNA of potato leaves treated with TA was used to construct a yeast two-hybrid library. The cDNA was cloned into the yeast expression vector pGADT7 for prey protein screening, and StPAM16-1 cDNA was inserted into the pGBKT7 vector as bait. Four candidate proteins that interact with StPAM16-1 were initially screened from the yeast two-hybrid library (Table 1), including COP9 signalosome complex subunit 5b-like (StCSN5), anthocyanin-3-O-glucosyltransferase 2-like and fructose-diphosphate aldolase-like protein and the uncharacterized protein LOC102596406.

Table 1. StPAM16-1-interacting proteins.

Number	Protein Description	NCBI Database Accession No.
1	PREDICTED: COP9 signalosome complex subunit 5b-like protein	XP_006351153.1
2	PREDICTED: anthocyanidin 3-O-glucosyltransferase 2-like protein	XP_006353729.1
3	fructose-bisphosphate aldolase-like protein	NP_001275379.1
4	PREDICTED: uncharacterized protein LOC102596406	XP_006344642.1

Among these four candidate proteins, CSN was reported to play an important role in regulating development and stress response [33]. Hence, we selected StCSN5 to verify the interaction between StPAM16-1. Yeast two-hybrid assays revealed a strong interaction between StPAM16-1 and StCSN5 (Figure 6a). Moreover, the interaction between StPAM16-1 and StCSN5 in plant cells was verified by the split luciferase complementation assay. As shown in Figure 6b, when an *Agrobacterium* culture containing cLUC-StPAM16-1/nLUC-StCSN5 vectors was injected into tobacco leaves, strong fluorescence was produced on the leaves. However, no fluorescence was detected in tobacco leaves injected with *Agrobacterium* cultures containing nLUC/cLUC-StPAM16-1, nLUC-StCSN5/cLUC or nLUC/cLUC. The results confirmed that StPAM16-1 and StCSN5 could interact in plant cells.

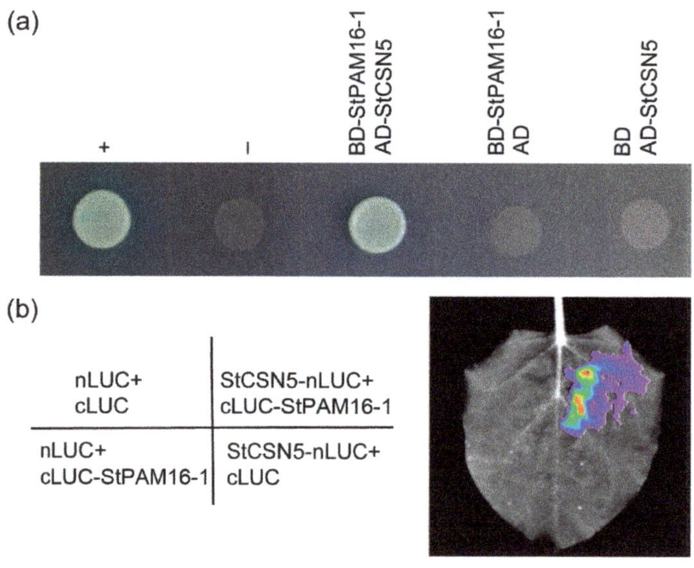

Figure 6. Interaction between StPAM16-1 and StCSN5. (**a**) Interaction between StPAM16-1 and StCSN5 was verified by yeast two-hybrid assay. +, positive control (BD-L53/AD-T); –, negative control (BD-Lam/AD-T). (**b**) Split luciferase complementation assay of interaction between StPAM16-1 and StCSN5 in plant cells.

2.6. Expression of StCSN5 Is Significantly Induced by TA and Several Defense-Related Plant Hormones

Given that the expression of *StPAM16-1* was induced by TA, we speculated that *StCSN5* expression might also respond to TA treatment. To verify this hypothesis, RT-qPCR was used to analyze the relative expression of *StCSN5* after TA treatment. Figure 6 shows that the transcription of *StCSN5* was indeed induced by TA treatment and peaked at 6 hpi. This expression pattern of *StCSN5* is similar to that of *StPAM16-1*. Hence, it can be speculated that the functions of these two genes are closely related to the responses of potato plants to TA and *S. scabiei* infection (Figure 7a).

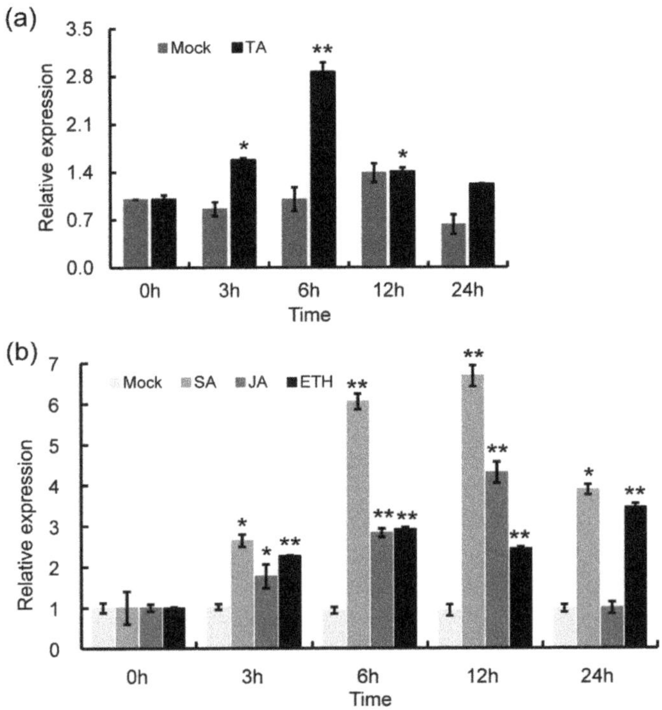

Figure 7. Relative expression of *StCSN5* in potato leaves after TA (**a**), SA, JA and ETH (**b**) treatments. *StActin* was used as reference gene. Values represent means ± SD. Asterisks denote statistically significant differences according to Student's *t* test; * $p < 0.05$, ** $p < 0.01$. Three biological replicates were performed.

Furthermore, we tested the expression of *StCSN5* in potato plants after defense-related hormone treatment by RT-qPCR. The results showed that the expression of *StCSN5* was induced by JA and ETH (Figure 7b) treatments.

3. Discussion

As a serious potato disease, CS has been one of the most prevalent concerns for farmers and industry [34]. In our previous study, we found that the phytotoxin TA secreted by pathogenic *S. scabiei* could trigger a series of host immune responses in potato [35], including changes in many metabolic pathways, enzymatic activities and ROS accumulation. Here, we identified a thaxtomin responsive gene from potato, *StPAM16-1*, that negatively regulates potato immune responses against TA and CS.

In this study, through a series of experiments, we showed that the relative expression of *StPAM16-1* in potato leaves was induced by TA (Figure 2a) and that *StPAM16-1*-inhibited potato plants generated by VIGS and RNAi acquired increased resistance to TA and *S. scabiei* (Figures 4 and 5). Plants are constantly confronted with the invasions of a variety of pathogens, including bacteria, fungi, viruses, nematodes and oomycetes [36]. To combat these attacks, plants have an evolved delicate and two-layered innate immune system, i.e., pattern-triggered immunity (PTI) and effector-triggered immunity (ETI) [37]. Many plant proteins participate and play either positive or negative roles in these immune processes in different plant-pathogen interaction systems [38,39]. During the interaction between potato and pathogens, the expression of the *StLysM-RLK05* gene was upregulated in tubers infected with *S. scabiei* [40], and the expression of *StLysM-RLK05* was significantly upregulated in resistant tubers compared with susceptible tubers infected with *S. scabiei*, suggesting a positive regulatory role of the protein in the immune response. In another study, Tai et al. reported that the expression of two *MYB* and three *bHLH* genes was correlated with CS resistance, indicating that these genes may contribute to the defense response of potato against CS [41]. Similarly, in *Arabidopsis thaliana*, AtPAM16/AtTXR1 participated in the response to TA and the pathogens *Pseudomonas syringae* and *Hyaloperonospora arabidopsidis* Noco2 [5,17]. Above all, we believe that StPAM16-1, which is similar to AtPAM16, is a negative regulator of the immune response of potato plants to TA and *S. scabiei* infection.

Similar to the AtPAM16 protein [12], the subcellular distribution analysis revealed that StPAM16-1 is a mitochondria-localized protein (Figure 3). The translocation of mitochondrial preproteins was significantly decreased in the yeast *pam16* mutant compared with the WT [16,42]. Similarly, a subunit of the *Arabidopsis* PAM complex, AtPAM16, was found to bind to the TIM23 complex and participate in translocating mitochondrial preproteins into the matrix [43,44]. Additionally, a previous study reported that TXR1 (PAM16) was a regulator of a transport system, and the increased resistance of the Arabidopsis mutant *txr1* to TA was due to a decrease in the rate of toxin uptake [5]. In our study, we observed that the expression of *StPAM16-1* was significantly induced by TA treatment, suggesting that upregulated *StPAM16-1* might play the same role as AtTXR1 in facilitating the entry of TA toxins into potato cells. Based on this information, we speculated that StPAM16-1 plays a role in response to TA and *S. scabiei* possibly by translocating a preprotein into mitochondria. This preprotein could be TA or a mitochondrial preprotein related to TA. As TA is only a cyclic dipeptide molecule containing tryptophan and phenylalanine residues [45,46], it is likely that TA binds to mitochondrial preproteins and is co-translocated into the mitochondrial matrix. However, such a role of StPAM16-1 in the translocation of mitochondrial preproteins needs to be further assessed.

To investigate the functional mechanism of StPAM16-1, we screened four proteins that interact with StPAM16-1 from a yeast two-hybrid library and verified the interaction between StPAM16-1 and StCSN5 via yeast two-hybrid and split luciferase complementation assays (Figure 6). As the fifth subunit of CSN [47], CSN5 plays an important role in regulating plant growth, development and various abiotic or biotic stress tolerances. It interacts with CRLs to regulate protein ubiquitination and degradation by regulating the ubiquitin/26S proteasome system [33]. It also interacts with CRLs to regulate various plant hormone-related signal transduction pathways, such as auxin, jasmonic acid and salicylic acid signals [26–29]. In *Arabidopsis*, CSN5 interacts with the ascorbic acid (AsA) biosynthetic enzyme GDP-Man pyrophosphorylase (VTC1) to inhibit the synthesis of AsA, thus negatively regulating the salt stress tolerance of plants [30]. It also interacts with CRLs to assist in the plant response to cold stress through ubiquitination [48]. Cui et al. reported that the silencing of the *CSN5* gene in susceptible grapevine enhanced resistance to powdery mildew [49]. The expression levels of the SA-associated marker genes *PR1*, *PR3* and *PAD4* were higher in the *CSN5*-RNAi plants and lower in the *CSN5*-overexpressing plants than in the WT plants.

Several studies have reported that CSN5 regulates plant resistance to pathogen and pest invasion by activating the JA pathway while inhibiting the SA pathway. Shang et al. [29] reported that tomato CSN4 and CSN5 play critical positive roles in resistance against root-knot nematodes (RKNs), which is positively related to JA content. The downregulation of tomato CSN5 resulted in reduced resistance to *Botrytis cinerea* and *Manduca sexta* larvae, which corresponded with reduced JA content [50]. Zhang et al. [51] reported that CSN5 negatively regulates the resistance of wheat to *Puccinia triticina*, as well as *Puccinia striiformis* f. sp. *tritici*, which is closely related to the SA pathway [28]. In our study, the expression of *StCSN5* was significantly induced by SA, JA and ETH (Figure 7), suggesting that StCSN5 is also involved in the response to phytohormone signaling. The StPAM16-1 protein contains a highly conserved DnaJ domain (Figure 1a) that binds to the heat shock protein for its proper folding, thus playing key roles in reactive oxygen species (ROS) scavenging and signaling [52]. When StPAM16-1 interacts with StCSN5, we hypothesize that the binding of StPAM16 to StCSN5 may assure the correct folding of StCSN5 and enable it to participate in the regulation of resistance pathways such as phytohormone signaling response. Moreover, as StPAM16-1 interacted with StCSN5 and the expression levels of both genes were induced by TA, we speculated that after TA treatment, an increase in the amount of the StCSN5 protein might be needed to degrade the accumulated StPAM16-1 protein to assist in plant defense. However, it is also possible that StPAM16-1 and StCSN5 form a complex to synergistically regulate the host immune response.

Further investigations will focus on the correlation between StPAM16-1 and defensive pathways, including phytohormone, metabolism and R-related genes. Generating *StCSN5*-silenced plants and investigating the biochemical activity of StCSN5 may provide more insights into the function of StPAM16-1-StCSN5 in the host immune response.

4. Materials and Methods

4.1. Plant Growth and Treatments

The tetraploid potato (*Solanum tuberosum* L.) cultivar Shepody was micropropagated on solid Murashige and Skoog (MS) medium and subcultured every four weeks at 23 °C with a 16 h light–8 h dark photoperiod. Three or four weeks later, potato plants with the same node segments and similar lengths were transplanted into pots containing nutrient soil and vermiculite (v/v = 3:1) and grown in a greenhouse under the same conditions as those used for tissue-cultured potato plants. Potato plants were watered with tap water once a week and Hogland nutrient solution once a month. *N. benthamiana* plants were grown in the greenhouse at 26 °C with the same photoperiod as that of the potato plants. For TA and hormone treatments, potato plants grown for three weeks were treated with 10 μM TA, 1 mM SA, 50 μM JA or 1 mM ETH and sampled at the indicated time points for further analysis. Leaves treated with sterile water were used as controls. A total of 3 pots (25 cm in diameter) with 5 seedlings per pot were used to replicate each treatment.

4.2. Pathogen Cultivation and Inoculation

The *S. scabiei* strain 4.1765, a typical pathogenic isolate from the China General Microbiological Culture Collection Center (CGMCC), was used in this study. *S. scabiei* was cultured on oatmeal agar medium (OMA) plates at 28 °C. Then, the spores were scraped off and dissolved in sterile water to make the spore suspension. The concentration of the spore suspension was adjusted to approximately 1×10^7 CFU/mL using the flat colony counting method. A diluted spore suspension was used to inoculate 3-week-old potato plants.

CS disease assessment was performed as previously described. Tubers > 2 g were selected and washed under running water, and then CS severity was evaluated. The disease index was calculated by the following equation: Disease index = $[\sum(n \times 1 + n \times 2 + n \times 3 + n \times 4 + n \times 5)/(N \times 5)] \times 100$. (n = number of tubers corresponding to the numerical grade. N = total number of potato tubers assessed. 5 = high score on the severity of scale). The percentage of tuber area covered was 0. No symptoms of scab; 1. 0–12.5%; 2. 12.6–25%;

3. 26–50%; 4. 51–75%; 5. 76–100%. Control efficacy = (disease index of control − disease index of treated)/disease index of control × 100%. Rate of disease = [number of infected tubers/total number of tubers] × 100× 100%.

4.3. Subcellular Localization of StPAM16-1

The coding region of *StPAM16-1* was amplified by PCR and fused with GFP under the control of the CaMV 35S promoter in the plant expression vector pGWB6. Plasmids containing GFP or mitochondrial markers were used as controls. The plasmids were transformed into *Agrobacterium tumefaciens* strain GV3101 competent cells, which were subsequently infiltrated into the leaves of 4-week-old *N. benthamiana* plants. After 48 h of incubation, fluorescent signals were visualized using a confocal laser scanning microscope (Leica TCS SP8, Leica Microsystems, Wetzlar, Germany).

4.4. Virus-Induced Gene Silencing

A specific cDNA fragment of *StPAM16-1* was amplified by PCR and inserted into the plant expression vector pgR107. pgR107 and pgR107-StPDS were used as negative and positive controls, respectively. Then, the plasmids were transformed into *A. tumefaciens* strain GV3101 competent cells (containing pJIC SA_Rep), which were subsequently infiltrated into 2-week-old potato leaves [53]. Potato plants were incubated in the dark for 24 h and then transferred to a greenhouse. Potato plants harboring pgR107-StPDS were used as a positive control.

4.5. Vector Construction and Potato Transformation

The specific cDNA fragment of *StPAM16-1* was amplified by PCR and inserted into the plant expression vector pANDA35K under the control of the CaMV 35S promoter. Then, the plasmid was transformed into GV3101 competent cells. For potato transformation, the stems of 3-week-old sterile potato plants were cut into 1 cm long pieces and immersed in A. tumefaciens resuspension solution for 10 min. After removing the excess solution, the stems were cultured on MS1 (MS + 4 mg/mL zeatin + 1 mg/mL indole-3-acetic acid) medium in the dark for 2 d. Then, the stems were transferred to MS2 (MS1 + 100 mg/L kanamycin + 200 mg/L cephalosporins) medium for 3–4 weeks to produce callus. The seedlings grown from callus were transplanted into MS3 (MS1 + 100 mg/L kanamycin) medium, and transgenic identification was carried out after the plants developed roots.

4.6. RNA Extraction and RT-qPCR Analysis

Total RNA was extracted from various potato tissues using a Plant Total RNA Extraction Kit (Tiangen, Beijing, China) according to the manufacturer's instructions. The same tissues from 5 plants in one pot were pooled together as one sample for RNA extraction. Three replicates were performed for each tissue sample. Then, cDNA was synthesized from 2 μg of RNA using TransScript® One-Step gDNA Removal and cDNA Synthesis SuperMix (Transgene, Beijing, China). RT-qPCR was performed using SYBR® Green Real-time PCR Master Mix (Toyobo, Osaka, Japan). All RT-qPCR experiments were performed using a CFX96 Touch instrument (Bio-Rad, Hercules, CA, USA) with three technical replicates and biological replicates. The relative expression of genes was calculated by the $2^{-\Delta\Delta Ct}$ method. *StActin* (PGSC0003DMG400027746) was used as the reference gene. The primers used for RT-qPCR are shown in Table S1.

4.7. Y2H Assay

The Y2H assay was performed according to the Matchmaker Gold Yeast Two-Hybrid System's instructions (Clontech, Mountain View, CA, USA). The ORFs of *StPAM16-1* and *StCSN5* were inserted into pGBKT7 and pGADT7, respectively. The plasmids BD-53/AD-T,

BD-StPAM16-1/AD, BD/AD-StCSN5, BD/AD and BD-StPAM16-1/AD-StCSN5 were co-transformed into yeast strain AH109 competent cells, which were subsequently cultured on DDO medium (SD medium/-Leu/-Trp) at 28–30 °C for 3–5 days. The yeast cells were subsequently resuspended and cultured in QOD medium (SD medium/-Leu/-Trp/-His/-Ade) supplemented with 40 μg/mL X-α-Gal at 30 °C. Then, the growth of colonies was observed and photographed.

4.8. Split Luciferase Complementation Assays

For the firefly luciferase complementation imaging assay [47], the ORF regions of *StPAM16-1* and *StCSN5* were inserted into the plant expression vectors pCAMBIA1300-CLuc and pCAMBIA1300-NLuc, respectively, and subsequently transformed into *Agrobacterium* cell strain GV3101. The Agrobacterium cultures were mixed and co-injected into the leaves of 4-week-old *N. benthamiana* plants. The plants were cultured in the dark for 24 h and then transferred to a greenhouse for 24 h. The leaves were sprayed with 0.5 mM D-luciferin potassium salt solution, and the Luc signals were captured using a plant in vivo imaging system (NightSHADE LB 985, Berthold Technologies, Baden, Germany).

4.9. Trypan Blue Staining

Trypan blue staining was performed as described previously [35]. Briefly, potato leaves were placed in a small beaker and washed with distilled water three times. The trypan blue staining solution (0.02% m/v trypan blue; ethanol-phenol-water-83% lactic acid = 2:1:1:1) was added to the beaker and boiled for 5–10 min. Then, 2.5 g/mL chloral hydrate solution was added to the beaker for decolorization, and the leaves were photographed.

4.10. Phylogenetic Analysis

PAM16-1 and relative homologous proteins sequences were downloaded from Potato Genomics Resource (http://solanaceae.plantbiology.msu.edu/, accessed on 15 November 2021) and NCBI (http://blast.ncbi.nlm.nih.gov/, accessed on 15 November 2021) databases. Multiple sequence alignment and phylogenetic analysis were performed using DNAMAN and MEGA7.0 software. Phylogenetic tree was constructed using Neighbor-joining method.

4.11. Statistical Analysis

Statistical analysis was performed using GraphPad Prism 9.3.0 (GraphPad Software, Boston, MA, USA). Three independent replications were performed for each test, and all values are reported as the mean with standard deviations (SDs). Student's t test (unpaired, two-tailed) was used to test for statistical significance for two groups compared at the 95% significance level ($p < 0.05$) or 99% significance level ($p < 0.01$).

5. Conclusions

Potato CS is a soil-borne disease that seriously threatens tuber quality and its market value. To date, an effective approach to controlling CS is lacking. In this study, we identified a gene designated *StPAM16-1* that responds to the phytotoxin TA and demonstrated that StPAM16-1 acted as a negative regulator of CS defense in potato. Importantly, we found that StPAM16-1 localized to the mitochondria and interacted with the COP9 signalosome subunit StCSN5, and the expression of StCSN5 significantly increased in potato leaves after they were treated with TA. These results provide important clues for understanding the functional mechanism of StPAM16-1. The identification of important regulators involved in the interaction between potato and *S. scabiei* could provide candidate genes for the genetic manipulation of plants with increased disease resistance. Based on our results, we propose that *StPAM16-1* can be used as a target gene for gene editing to develop new varieties of potato resistant to CS. Moreover, after the function of StCSN5 was verified, *StCSN5* could also be used as a target gene.

Supplementary Materials: The following supporting information can be downloaded at https://www.mdpi.com/article/10.3390/ijms26031361/s1.

Author Contributions: L.L., S.S., N.L. and H.W. conceived and designed this study; L.L., S.S., Z.W. and Y.Z. performed the experiments and data analyses; L.L. drafted the original manuscript; L.L., H.W., P.Z. and N.Z. reviewed and edited the manuscript. All authors have read and agreed to the published version of the manuscript.

Funding: This research was funded by Hebei Science and Technology Major Project, grant number 22287501Z; the Key Technologies R & D Program of Inner Mongolia, grant number 2021GG0300; and the Priority Research Program of Chinese Academy of Science, grant number XDA24020104.

Institutional Review Board Statement: Not applicable.

Informed Consent Statement: Not applicable.

Data Availability Statement: All data generated or analyzed during this study are included in the published article and Supplementary Materials.

Acknowledgments: We thank Guixian Xia at Institute of Microbiology, Chinese Academy of Sciences, for critically reading the manuscript and Lei Su for technical support with confocal microscopes. We also thank David Baulcombe (Sainsbury Laboratory, Norwich, UK) for kindly providing us with the PVX vector pGR107 and the helper plasmid pJIC SA_Rep.

Conflicts of Interest: The authors declare no conflicts of interest.

References

1. Lapaz, M.I.; Huguet-Tapia, J.C.; Siri, M.I.; Verdier, E.; Loria, R.; Pianzzola, M.J. Genotypic and Phenotypic Characterization of *Streptomyces* Species Causing Potato Common Scab in Uruguay. *Plant Dis.* **2017**, *101*, 1362–1372. [CrossRef] [PubMed]
2. Loria, R.; Bukhalid, R.A.; Fry, B.A.; King, R.R. Plant pathogenicity in the genus *Streptomyces*. *Plant Dis.* **1997**, *81*, 836–846. [CrossRef] [PubMed]
3. Loria, R.; Kers, J.; Joshi, M. Evolution of plant pathogenicity in *Streptomyces*. *Annu. Rev. Phytopathol.* **2006**, *44*, 469–487. [CrossRef] [PubMed]
4. Bischoff, V.; Cookson, S.J.; Wu, S.; Scheible, W.R. Thaxtomin A affects CESA-complex density, expression of cell wall genes, cell wall composition, and causes ectopic lignification in *Arabidopsis thaliana* seedlings. *J. Exp. Bot.* **2009**, *60*, 955–965. [CrossRef]
5. Scheible, W.R.; Fry, B.; Kochevenko, A.; Schindelasch, D.; Zimmerli, L.; Somerville, S.; Loria, R.; Somerville, C.R. An Arabidopsis mutant resistant to thaxtomin A, a cellulose synthesis inhibitor from *Streptomyces* species. *Plant Cell* **2003**, *15*, 1781–1794. [CrossRef]
6. Leiner, R.H.; Fry, B.A.; Carling, D.E.; Loria, R. Probable involvement of thaxtomin A in pathogenicity of *Streptomyces scabies* on seedlings. *Phytopathology* **1996**, *86*, 709–713. [CrossRef]
7. Tegg, R.S.; Melian, L.; Wilson, C.R.; Shabala, S. Plant cell growth and ion flux responses to the streptomycete phytotoxin thaxtomin A: Calcium and hydrogen flux patterns revealed by the non-invasive MIFE technique. *Plant Cell Physiol.* **2005**, *46*, 638–648. [CrossRef] [PubMed]
8. Errakhi, R.; Dauphin, A.; Meimoun, P.; Lehner, A.; Reboutier, D.; Vatsa, P.; Briand, J.; Madiona, K.; Rona, J.P.; Barakate, M.; et al. An early Ca^{2+} influx is a prerequisite to thaxtomin A-induced cell death in *Arabidopsis thaliana* cells. *J. Exp. Bot.* **2008**, *59*, 4259–4270. [CrossRef]
9. Duval, I.; Brochu, V.; Simard, M.; Beaulieu, C.; Beaudoin, N. Thaxtomin A induces programmed cell death in *Arabidopsis thaliana* suspension-cultured cells. *Planta* **2005**, *222*, 820–831. [CrossRef] [PubMed]
10. Dees, M.W.; Wanner, L.A. In Search of Better Management of Potato Common Scab. *Potato Res.* **2012**, *55*, 249–268. [CrossRef]
11. Murphy, A.M.; De Jong, H.; Tai, G.C.C. Transmission of resistance to common scab from the diploid to the tetraploid level via 4x-2x crosses in potatoes. *Euphytica* **1995**, *82*, 227–233. [CrossRef]
12. Alam, Z. Inheritance of Scab Resistance in 24-Chromosome Potatoes. Ph.D. Thesis, University of Wisconsin-Madison, Madison, WI, USA, 1972.
13. Tang, D.; Jia, Y.; Zhang, J.; Li, H.; Cheng, L.; Wang, P.; Bao, Z.; Liu, Z.; Feng, S.; Zhu, X.; et al. Genome evolution and diversity of wild and cultivated potatoes. *Nature* **2022**, *606*, 535–541. [CrossRef]

14. Zhou, Y.; Zhang, Z.; Bao, Z.; Li, H.; Lyu, Y.; Zan, Y.; Wu, Y.; Cheng, L.; Fang, Y.; Wu, K.; et al. Graph pangenome captures missing heritability and empowers tomato breeding. *Nature* **2022**, *606*, 527–534. [CrossRef]
15. Cheng, L.; Wang, N.; Bao, Z.; Zhou, Q.; Guarracino, A.; Yang, Y.; Wang, P.; Zhang, Z.; Tang, D.; Zhang, P.; et al. Leveraging a phased pangenome for haplotype design of hybrid potato. *Nature* **2025**. [CrossRef]
16. Frazier, A.E.; Dudek, J.; Guiard, B.; Voos, W.; Li, Y.; Lind, M.; Meisinger, C.; Geissler, A.; Sickmann, A.; Meyer, H.E.; et al. Pam16 has an essential role in the mitochondrial protein import motor. *Nat. Struct. Mol. Biol.* **2004**, *11*, 226–233. [CrossRef] [PubMed]
17. Huang, Y.; Chen, X.; Liu, Y.; Roth, C.; Copeland, C.; McFarlane, H.E.; Huang, S.; Lipka, V.; Wiermer, M.; Li, X. Mitochondrial AtPAM16 is required for plant survival and the negative regulation of plant immunity. *Nat. Commun.* **2013**, *4*, 2558. [CrossRef] [PubMed]
18. Köhler, C.; Casey, C.; Köcher, T.; Champion, C.; Jandrasits, K.; Mosiolek, M.; Bonnot, C.; Dolan, L. Reduced coenzyme Q synthesis confers non-target site resistance to the herbicide thaxtomin A. *PLoS Genet.* **2023**, *19*, e1010423. [CrossRef]
19. Waingankar, T.P.; D'Silva, P. Multiple variants of the human presequence translocase motor subunit Magmas govern the mitochondrial import. *J. Biol. Chem.* **2021**, *297*, 101349. [CrossRef]
20. Srivastava, S.; Sinha, D.; Saha, P.P.; Marthala, H.; D'Silva, P. Magmas functions as a ROS regulator and provides cytoprotection against oxidative stress-mediated damages. *Cell Death Dis.* **2014**, *5*, e1394. [CrossRef]
21. Qin, N.; Xu, D.; Li, J.; Deng, X.W. COP9 signalosome: Discovery, conservation, activity, and function. *J. Integr. Plant Biol.* **2020**, *62*, 90–103. [CrossRef] [PubMed]
22. Wei, N.; Deng, X.W. The COP9 signalosome. *Annu. Rev. Cell Dev. Biol.* **2003**, *19*, 261–286. [CrossRef] [PubMed]
23. Mikus, P.; Zundel, W. COPing with hypoxia. *Semin. Cell Dev. Biol.* **2005**, *16*, 462–473. [CrossRef]
24. Stratmann, J.W.; Gusmaroli, G. Many jobs for one good cop–The COP9 signalosome guards development and defense. *Plant Sci.* **2012**, *185–186*, 50–64. [CrossRef] [PubMed]
25. Serino, G.; Deng, X.W. The COP9 signalosome: Regulating plant development through the control of proteolysis. *Annu. Rev. Plant Biol.* **2003**, *54*, 165–182. [CrossRef]
26. Schwechheimer, C.; Serino, G.; Callis, J.; Crosby, W.L.; Lyapina, S.; Deshaies, R.J.; Gray, W.M.; Estelle, M.; Deng, X.W. Interactions of the COP9 signalosome with the E3 ubiquitin ligase SCFTIR1 in mediating auxin response. *Science* **2001**, *292*, 1379–1382. [CrossRef]
27. Feng, S.; Ma, L.; Wang, X.; Xie, D.; Dinesh-Kumar, S.P.; Wei, N.; Deng, X.W. The COP9 signalosome interacts physically with SCF COI1 and modulates jasmonate responses. *Plant Cell* **2003**, *15*, 1083–1094. [CrossRef]
28. Bai, X.; Huang, X.; Tian, S.; Peng, H.; Zhan, G.; Goher, F.; Guo, J.; Kang, Z.; Guo, J. RNAi-mediated stable silencing of *TaCSN5* confers broad-spectrum resistance to *Puccinia striiformis* f. sp. *tritici*. *Mol. Plant Pathol.* **2021**, *22*, 410–421. [CrossRef]
29. Shang, Y.; Wang, K.; Sun, S.; Zhou, J.; Yu, J.Q. COP9 Signalosome CSN4 and CSN5 Subunits Are Involved in Jasmonate-Dependent Defense Against Root-Knot Nematode in Tomato. *Front. Plant Sci.* **2019**, *10*, 1223. [CrossRef] [PubMed]
30. Wang, J.; Yu, Y.; Zhang, Z.; Quan, R.; Zhang, H.; Ma, L.; Deng, X.W.; Huang, R. *Arabidopsis* CSN5B interacts with VTC1 and modulates ascorbic acid synthesis. *Plant Cell* **2013**, *25*, 625–636. [CrossRef]
31. Wei, N.; Tsuge, T.; Serino, G.; Dohmae, N.; Takio, K.; Matsui, M.; Deng, X.W. The COP9 complex is conserved between plants and mammals and is related to the 26S proteasome regulatory complex. *Current Biol.* **1998**, *8*, 919–922. [CrossRef] [PubMed]
32. Girard, M.; Poupon, V.; Blondeau, F.; McPherson, P.S. The DnaJ-domain protein RME-8 functions in endosomal trafficking. *J. Biol. Chem.* **2005**, *280*, 40135–40143. [CrossRef]
33. Schwechheimer, C. The COP9 signalosome (CSN): An evolutionary conserved proteolysis regulator in eukaryotic development. *Biochim. Biophys. Acta* **2004**, *1695*, 45–54. [CrossRef]
34. Braun, S.; Gevens, A.; Charkowski, A.; Allen, C.; Jansky, S. Potato Common Scab: A Review of the Causal Pathogens, Management Practices, Varietal Resistance Screening Methods, and Host Resistance. *Am. J. Potato Res.* **2017**, *94*, 283–296. [CrossRef]
35. Liu, L.; Hao, L.; Liu, N.; Zhao, Y.; Zhong, N.; Zhao, P. iTRAQ-Based Proteomics Analysis of Response to *Solanum tuberosum* Leaves Treated with the Plant Phytotoxin Thaxtomin, A. *Int. J. Mol. Sci.* **2021**, *22*, 12036. [CrossRef]
36. Lapin, D.; Van den Ackerveken, G. Susceptibility to plant disease: More than a failure of host immunity. *Trends Plant Sci.* **2013**, *18*, 546–554. [CrossRef]
37. Yuan, M.; Jiang, Z.; Bi, G.; Nomura, K.; Liu, M.; Wang, Y.; Cai, B.; Zhou, J.M.; He, S.Y.; Xin, X.F. Pattern-recognition receptors are required for NLR-mediated plant immunity. *Nature* **2021**, *592*, 105–109. [CrossRef] [PubMed]
38. Saijo, Y.; Loo, E.P. Plant immunity in signal integration between biotic and abiotic stress responses. *New Phytol.* **2020**, *225*, 87–104. [CrossRef]
39. Zhou, J.M.; Zhang, Y. Plant Immunity: Danger Perception and Signaling. *Cell* **2020**, *181*, 978–989. [CrossRef]
40. Nazarian-Firouzabadi, F.; Joshi, S.; Xue, H.; Kushalappa, A.C. Genome-wide in silico identification of LysM-RLK genes in potato (*Solanum tuberosum* L.). *Mol. Biol. Rep.* **2019**, *46*, 5005–5017. [CrossRef]
41. Tai, H.H.; Goyer, C.; Murphy, A.M. Potato MYB and bHLH transcription factors associated with anthocyanin intensity and common scab resistance. *Botany* **2013**, *91*, 722–730. [CrossRef]

42. Kozany, C.; Mokranjac, D.; Sichting, M.; Neupert, W.; Hell, K. The J domain-related cochaperone Tim16 is a constituent of the mitochondrial TIM23 preprotein translocase. *Nat. Struct. Mol. Biol.* **2004**, *11*, 234–241. [CrossRef] [PubMed]
43. Li, Y.; Dudek, J.; Guiard, B.; Pfanner, N.; Rehling, P.; Voos, W. The presequence translocase-associated protein import motor of mitochondria: Pam16 functions in an antagonistic manner to Pam18. *J. Biol. Chem.* **2004**, *279*, 38047–38054. [CrossRef] [PubMed]
44. Chacinska, A.; van der Laan, M.; Mehnert, C.S.; Guiard, B.; Mick, D.U.; Hutu, D.P.; Truscott, K.N.; Wiedemann, N.; Meisinger, C.; Pfanner, N.; et al. Distinct forms of mitochondrial TOM-TIM supercomplexes define signal-dependent states of preprotein sorting. *Mol. Cell Biol.* **2010**, *30*, 307–318. [CrossRef]
45. Johnson, E.G.; Krasnoff, S.B.; Bignell, D.R.; Chung, W.C.; Tao, T.; Parry, R.J.; Loria, R.; Gibson, D.M. 4-Nitrotryptophan is a substrate for the non-ribosomal peptide synthetase TxtB in the thaxtomin A biosynthetic pathway. *Mol. Microbiol.* **2009**, *73*, 409–418. [CrossRef]
46. King, R.R.; Calhoun, L.A. The thaxtomin phytotoxins: Sources, synthesis, biosynthesis, biotransformation and biological activity. *Phytochemistry* **2009**, *70*, 833–841. [CrossRef] [PubMed]
47. Liu, Y.; Shah, S.V.; Xiang, X.; Wang, J.; Deng, Z.B.; Liu, C.; Zhang, L.; Wu, J.; Edmonds, T.; Jambor, C.; et al. COP9-associated CSN5 regulates exosomal protein deubiquitination and sorting. *Am. J. Pathol.* **2009**, *174*, 1415–1425. [CrossRef] [PubMed]
48. Schwechheimer, C.; Serino, G.; Deng, X.W. Multiple ubiquitin ligase-mediated processes require COP9 signalosome and AXR1 function. *Plant Cell* **2002**, *14*, 2553–2563. [CrossRef]
49. Cui, K.-C.; Liu, M.; Ke, G.-H.; Zhang, X.-Y.; Mu, B.; Zhou, M.; Hu, Y.; Wen, Y.-Q. Transient silencing of *VvCSN5* enhances powdery mildew resistance in grapevine (*Vitis vinifera*). *Plant Cell Tiss. Org. Cult.* **2021**, *146*, 621–633. [CrossRef]
50. Hind, S.R.; Pulliam, S.E.; Veronese, P.; Shantharaj, D.; Nazir, A.; Jacobs, N.S.; Stratmann, J.W. The COP9 signalosome controls jasmonic acid synthesis and plant responses to herbivory and pathogens. *Plant J.* **2011**, *65*, 480–491. [CrossRef]
51. Zhang, H.; Wang, X.; Giroux, M.J.; Huang, L. A wheat *COP9 subunit 5*-like gene is negatively involved in host response to leaf rust. *Mol. Plant Pathol.* **2017**, *18*, 125–133. [CrossRef]
52. Gray, J.; Rustgi, S.; von Wettstein, D.; Reinbothe, C.; Reinbothe, S. Common functions of the chloroplast and mitochondrial co-chaperones cpDnaJL (CDF1) and mtDnaJ (PAM16) in protein import and ROS scavenging in *Arabidopsis thaliana*. *Commun. Integr. Biol.* **2015**, *9*, e1119343. [CrossRef] [PubMed]
53. Xu, Y.; Liu, L.; Zhao, P.; Tong, J.; Zhong, N.; Zhang, H.; Liu, N. Genome-Wide Identification, Expression Profile and Evolution Analysis of Karyopherin beta Gene Family in *Solanum tuberosum* Group Phureja DM1-3 Reveals Its Roles in Abiotic Stresses. *Int. J. Mol. Sci.* **2020**, *21*, 931. [CrossRef]

International Journal of
Molecular Sciences

MDPI

Article

A Novel Single Base Mutation in *OsSPL42* Leads to the Formation of Leaf Lesions in Rice

Panpan Li, Huihui Shang, Xia Xu, Junyi Gong, Jian-Li Wu and Xiaobo Zhang *

State Key Laboratory of Rice Biology, China National Rice Research Institute, Hangzhou 310006, China; li_pp1203@163.com (P.L.); huihuishengji1101@163.com (H.S.); mailxuxia@163.com (X.X.); junyigong8107@163.com (J.G.); beishangd@163.com (J.-L.W.)
* Correspondence: zhangxiaobo@caas.cn; Tel.: +86-571-63370295

Abstract: Rice spotted-leaf mutants serve as valuable resources for studying plant programmed cell death (PCD) and disease resistance mechanisms, making them crucial for research on disease resistance in rice. Map-based cloning was used to identify and clone the spotted-leaf gene *OsSPL42*. Then, functional complementation and CRISPR/Cas9 techniques were also employed to further validate the function of this gene. By applying leaf clippings for bacterial blight (BB) inoculation, the BB resistance of different rice lines was assessed. The results in this study were as follows: The *OsSPL42* behaved as a recessive nuclear gene and was narrowed down to a 111 kb region on chromosome 8. All T_0 transgenic rice plants in the complementation experiments exhibited a wild-type phenotype, without any lesion spots on the rice leaves. This suggests that the *LOC_Os08g06100* encoding O-methyltransferase is the candidate gene for the mutant *spl42*. The *OsSpl42* is widely expressed and the OsSPL42-GFP protein is mainly localized in the cytoplasm. *OsSPL42* overexpression lines are more susceptible to BBs, which indicates that *OsSPL42* may act as a negative regulator of rice resistance to BB. In summary, we speculate that *OsSPL42* plays an important role in the regulation of pathogen response, providing new insights into plant defense mechanisms.

Keywords: rice; spotted-leaf mutant; map-based cloning; bacterial blight; lignin; defensive response

1. Introduction

The formation of leaf lesions that mimic the hypersensitive response (HR) is an abnormal phenomenon due to variations in internal genetic factors or environmental stresses in plants [1]. In rice, these mutants are typically referred to as spotted-leaf mutants, which are valuable models for studying chlorophyll metabolism, phytohormone signaling, reactive oxygen species (ROS) homeostasis, programmed cell death (PCD), and innate immunity [2–4].

The formation of spotted leaves in rice is largely associated with chlorophyll metabolism [5]. For instance, genes related to chlorophyll synthesis, such as *OsCHLH*, *OsPORB*, and *OsCAO1*, are key genes affecting chlorophyll synthesis [6–8]. The *OsCHLH* gene encodes a subunit of magnesium ion chelatase (Mg-chelatase), which is the first step in chlorophyll synthesis. Mg-chelatase is responsible for embedding magnesium ions (Mg^{2+}) into precursor molecules [5]. Mutations or deletions in *OsCHLH* result in the inability to successfully embed magnesium ions, leading to limited chlorophyll synthesis. This not only results in a drop in chlorophyll content but also triggers uneven pigmentation of the leaves, eventually forming spots. *OsPORB* encodes a subtype of protochlorophyllide oxidoreductase (POR), a light-dependent precursor of chlorophyllide. Under light conditions, the enzyme converts the protochlorophyllide precursor into chlorophyll. Mutations or downregulation of *OsPORB* expression lead to an incomplete step, resulting in a decrease in chlorophyll content and the appearance of blotches [9]. *OsCAO1* (chlorophyllide a oxygenase 1) is involved in the conversion of chlorophyllide b into chlorophyllide a. This step is critical for the stability of chlorophyll molecules in the chloroplast complex [6]. Mutation or loss of *OsCAO1* results in reduced chlorophyll stability, which may lead to the formation of spotted leaves.

In addition to genes related to chlorophyll synthesis and degradation, the spot formation in rice spotted-leaf mutants is also related to the plant hormone signaling pathway. Notably, plant hormones such as gibberellin (GA), indole acetic acid (IAA), and jasmonic acid (JA) play an important role in this process. The physiological and biochemical changes of the formation mechanism of rice mutant spotted leaves include many aspects involving gene expression and regulation, as well as a series of biochemical pathways and physiological processes [10].

The formation of spotted leaves is usually accompanied by the accumulation of ROS, leading to an imbalance in the REDOX reaction, and then causing a series of physiological and biochemical changes. Excessive ROS can cause PCD and HR [11]. To maintain the balance of ROS, multiple ROS scavenging enzymes are present in plants, such as catalase (CAT), superoxide dismutase (SOD), peroxidase (POD), ascorbate peroxidase (APX) and other enzymes [12]. Studies have shown that the leaf tissue of spotted-leaf mutants undergoes PCD mediated by genetic programming [13]. This phenomenon usually causes mutant plants to have high disease resistance, thus preventing the invasion and spread of pathogens in the environment [14–20]. Some studies have shown that PCD plays a certain role in preventing the invasion and spread of rice blast and BB. The spotted-leaf mutants generally showed increased resistance to fungi and bacteria [21]. This change makes spotted-leaf mutants an important resource for studying the innate immune mechanism of plants. Therefore, the identification and study of spotted-leaf mutants are of great significance for the analysis of the mechanism of action of spotted-leaf mutants and the innate immune mechanism of plants.

Rice spotted-leaf mutants are extensively studied due to their enhanced disease resistance against a range of pathogens, particularly BB caused by *Xanthomonas oryzae* pv. *Oryzae (Xoo)*. Notable mutants such as *spl11* [22], *spl30* [23], *spl24* [24], *spl40* [25], *Spl26* [26], and *spl41* [27] exhibit significant resistance to BB. *spl24* is a novel lesion mimic mutant with semi-dominant inheritance. This mutant enhances innate immunity and facilitates the identification of key genes involved in disease resistance, which can be applied in breeding strategies [24]. Similarly, *spl40* has been shown to confer strong resistance to BB by modulating the jasmonic acid (JA) and salicylic acid (SA) signaling pathways. The upregulation of these pathways in *spl40* triggers systemic acquired resistance (SAR), which is crucial for its defense response [25]. *spl11*, originally derived from mutagenized rice populations, shows non-specific resistance to both rice blast and BB, possibly through a broader defense activation mechanism. The non-specific nature of this resistance points to the potential involvement of generalized immune pathways rather than pathogen-specific responses [22]. However, despite forming leaf lesions, *spl30* does not exhibit typical signs of PCD. This unusual phenotype is hypothesized to result from the accumulation of as yet unidentified pigments rather than the activation of cell death-related pathways. The absence of PCD in *spl30* distinguishes it from other lesion mimic mutants and suggests an alternative resistance mechanism that may not rely on cell death for defense [23]. The *Spl26* gain-of-function mutation positively regulates plant immunity by enhancing ROS production and promoting cell death, which is central to its heightened resistance to BB [26]. Conversely, the *spl41* mutant acts as a negative regulator of plant immunity. This mutation impairs defense responses, providing insights into the fine-tuning of immune pathways and the potential for manipulating such genes in breeding programs [27]. These mutants have been incorporated into rice breeding programs, utilizing their enhanced resistance traits to improve the overall disease resistance of commercial varieties. The spotted-leaf mutants, particularly those like *spl24* and *spl40*, which activate key immune pathways, have been instrumental in creating rice lines with durable resistance. These mutants were originally identified through techniques such as EMS mutagenesis, and their integration into breeding schemes has led to the development of rice cultivars with robust, field-level resistance to BB. The mechanism of resistance in most of these mutants can be traced back to the spontaneous induction of programmed cell death, which plays a central role in limiting pathogen spread (except for *spl30*). Future studies should focus on environmental factors influencing the expression of resistance traits in these mutants. Understanding the environmental plasticity of their resistance mechanisms will be essential for optimizing their deployment in different agroecological zones. Such research will contribute to the enhancement of commercial rice varieties, ensuring greater adaptability and resilience in the face of pathogen pressure and changing climatic conditions.

In this study, we isolated a novel rice spotted-leaf mutant *spl42* from an ethyl methyl sulfonate (EMS)-induced mutant bank of the indica variety of ZJ100. A point mutation in *OsSPL42* was identified and was responsible for the lesion formation of *spl42*. *OsSPL42* was widely distributed and localized to cytoplasm and might negatively regulate disease resistance to *Xoo* in rice.

2. Results

2.1. OsSPL42 Controls the Spotted-Leaf Phenotype of spl42

In this study, the mutant *spl42* was obtained through EMS mutagenesis of ZJ100, and it was determined that the spotted-leaf phenotype of the mutant is controlled by a single recessive gene. At the same time, approximately 200 pairs of SSR markers and 10 pairs of InDel markers were used to map the candidate gene. The candidate gene was narrowed down to a 111 kb region between InDelD2 and InDelG8 (Table S1). Using the website Gramene (http://www.gramene.org/, accessed on 8 March 2021), we found ten open reading frames (ORFs) in the 111 kb region. The DNA sequences of ten ORFs were PCR amplified and sequenced, and then the DNA sequence from *spl42* and the wild control were carefully aligned. The results showed (Figure 1A) that the ORF10, *LOC_Os08g06100*, had a single base substitution (T247A) at the 247th base of the gene, resulting in the change of the 42nd amino acid (AA) (L42H). Therefore, the *LOC_Os08g06100*, which encodes O-methyltransferase involved in lignin monomer biosynthesis [28], was supposed to be the candidate gene.

To further determine whether the mutation of this gene leads to the growth and development defects of the mutant, we conducted functional complementation experiments. The binary vector pCAMBIA1300-OsSPL42 containing the *OsSPL42* promoter and genomic coding region was introduced into the mutant *spl42* callus. We obtained 16 positive transgenic plants confirmed by PCR assistant DNA sequencing. All the 16 *spl42* genetic background transgenic plants showed no lesion mimic phenotype as *spl42*. We selected three of them to collect plenty of seeds and they were named lines CP1, CP2, and CP3 for the next experiments (Figure 1B,C). By observing the phenotype of T_1, it was found that the T_1 complementary transgenic lines showed phenotypic separation, and all plants exhibiting a wild-type phenotype were transgenic positive. This indicated that the single T_0 positive transgenic strain was heterozygous, confirming that *OsSPL42* is successfully complementary to the mutant *spl42*, and further proving that *OsSPL42* is a single recessive gene controlling spotted leaves.

In addition, photosynthetic pigment and soluble protein contents in CP1, CP2, CP3, *spl42*, and wild-type were detected (Figure 1D–I). The results showed that the contents of chlorophyll a, chlorophyll b, chlorophyll a/b, carotenoids, total chlorophyll, and soluble protein in CP1, CP2, and CP3 were significantly higher compared to mutants. The levels of photosynthetic pigments and soluble protein content in the three complementary lines were partial recovery. CP1, CP2, and CP3 only recovered to 76.7%, 65.5%, and 82.6% of the ZJ100 of the chlorophyll a content, respectively, showing partial recovery but not a full recovery to the wild-type level. However, the chlorophyll a content of *spl42* recovered to 47.3% of ZJ100, which was significantly lower than the complementary lines. This is a very common phenomenon. The number of chloroplasts in living cells showed partial recovery, and necrotic spots did not appear when *OsSPL42* was introduced into the mutant. It was speculated that the appearance of spotted leaves during the growth and development of the mutant was caused by a mutation in the *OsSPL42* gene.

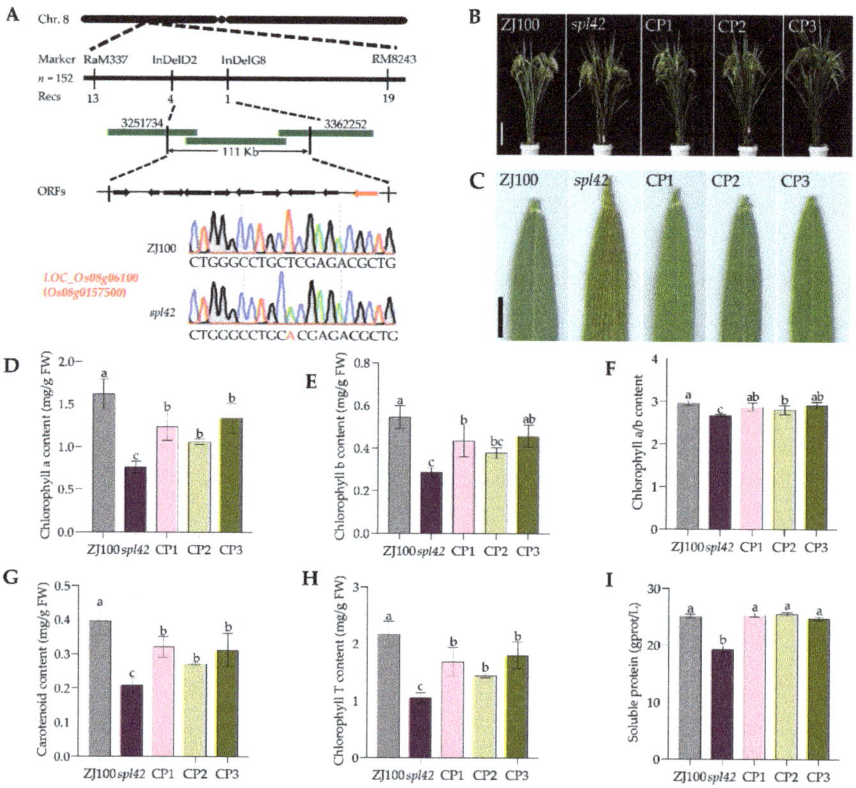

Figure 1. Identification and functional validation of *OsSPL42*. (**A**) The red letter A represents the mutation site, including T directly above A, showing the wild-type ZJ100 mutating from T to A. (**B**) Plant phenotype of wild-type ZJ100 and mutant *spl42*, and complementary lines at the filling stage. Bar = 17 cm. (**C**) Top second leaves of phenotype of wild-type ZJ100 and mutant *spl42*, and complementary lines at the tillering stage. Bar = 1 cm. (**D**) Chlorophyll a content. (**E**) Chlorophyll b content. (**F**) The ratio of chlorophyll a to chlorophyll b. (**G**) Carotenoid content. (**H**) Total chlorophyll content. (**I**) Soluble protein. CP, complementation; values are means ± SD, $n = 3$; and different letters indicate significant differences by one-way ANOVA and Duncan's test $p \leq 0.05$.

Agronomic traits of ZJ100, *spl42*, and three complementary lines were investigated (Table 1). Most agronomic traits were also recovered after functional complementation, indicating that the growth and development of the mutant was caused by a mutation in *OsSPL42*, resulting in alterations in agronomic traits. However, the 1000-grain weight of the complementary lines was lower than that of both mutant wild types. This indicates that the lower 1000-grain in *spl42* is not caused by the mutation in *OsSPL42*.

Table 1. Agronomic traits of wild type ZJ100, mutant *spl42*, and complementary lines.

Material	Plant Height (cm)	No. Tiller/Plant	Panicle Length (cm)	No. Filled Grain/Panicle	Seed Setting (%)	1000-Grain Weight (g)
ZJ100	98.0 ± 0.9^{b}	12.3 ± 1.5^{b}	23.2 ± 0.6^{b}	135.8 ± 7.7^{a}	79.8 ± 0.2^{a}	19.8 ± 0.3^{a}
spl42	100.9 ± 1.9^{b}	11.0 ± 1.0^{b}	24.4 ± 0.6^{ab}	104.9 ± 9.0^{b}	60.2 ± 1.5^{b}	18.2 ± 0.4^{b}
CP1	98.2 ± 1.1^{b}	15.3 ± 0.6^{a}	23.7 ± 0.7^{b}	130.1 ± 8.6^{a}	77.7 ± 1.7^{a}	17.5 ± 0.4^{bc}
CP2	98.3 ± 2.5^{b}	16.3 ± 0.6^{a}	26.1 ± 1.7^{a}	131.3 ± 27.0^{a}	78.5 ± 4.3^{a}	17.9 ± 0.5^{bc}
CP3	106.6 ± 2.1^{a}	15.3 ± 1.1^{a}	25.0 ± 0.9^{ab}	132.8 ± 14.9^{a}	78.8 ± 1.7^{a}	17.4 ± 0.5^{c}

Note: CP, Complementation; values are means ± SD, $n = 3$; and different letters indicate significant differences by one-way ANOVA and Duncan's test $p \leq 0.05$.

We also utilized CRISPR/cas9 technology for gene editing of *OsSPL42* in Nipponbare (NPB) callus. A total of 19 *OsSPL42* knockout lines were obtained. None of these 19 knockout lines showed leaf lesions in T_0. Only one homozygous knockout line (Cr1) in the T_1 generation exhibited leaf lesions. This line and the two other 2 homozygous knockout lines (Cr5 and Cr6) were selected for the next experiments (Figure 2) to explore the function of *OsSPL42* gene. The accumulation of soluble proteins and chlorophyll was detected in the wild-type NPB and three knockout lines. The results showed that the levels of chlorophyll a, chlorophyll b, chlorophyll a/b, carotenoids, total chlorophyll, and soluble proteins in Cr1 were significantly lower compared to the wild type. In contrast, the contents of Cr5 and Cr6 did not show significant differences compared to the wild type.

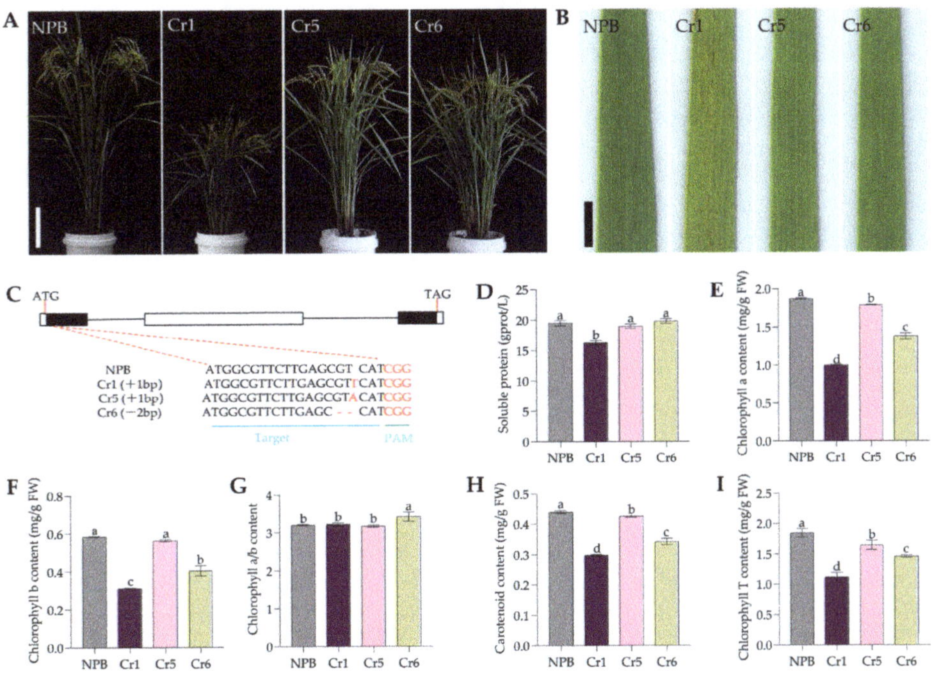

Figure 2. Phenotypic identification and functional verification of *OsSPL42* knockout strain. (**A**) Plant phenotype of NPB and knockout line at filling stage. Bar = 17 cm. (**B**) Leaves of phenotype of NPB and knockout line at filling stage. Bar = 1 cm. (**C**) Knockout target sequence. (**D**) Soluble protein. (**E**) Chlorophyll a content. (**F**) Chlorophyll b content. (**G**) The ratio of chlorophyll a to chlorophyll b. (**H**) Carotenoid content. (**I**) Total chlorophyll content. Cr, knockout; values are means ± SD, $n = 3$; and different letters indicate significant differences by one-way ANOVA and Duncan's test $p \leq 0.05$.

2.2. Leucine in OsSPL42 Is Highly Conserved

The CDS of *OsSPL42* gene has a total length of 1107 bases and encodes 368 AAs. The gene was predicted to contain three domains, lom complexity, dimerization, and O-methyltransferase, using http://smart.embl-heidelberg.de/ (accessed on 2 April 2021) (Figure 3A). lom complexity is a region with low sequence complexity, spanning from the 5th AA to the 20th AA. The dimerization domain is located at the N-terminus of many plant O-methyltransferases and has been shown to mediate the dimerization of several proteins. The O-methyltransferase domain includes a series of O-methyltransferases, some of which utilize S-adenosylmethionine as a substrate. The primary role of DNA methylation in prokaryotes is to protect host DNA from degradation by restriction enzymes.

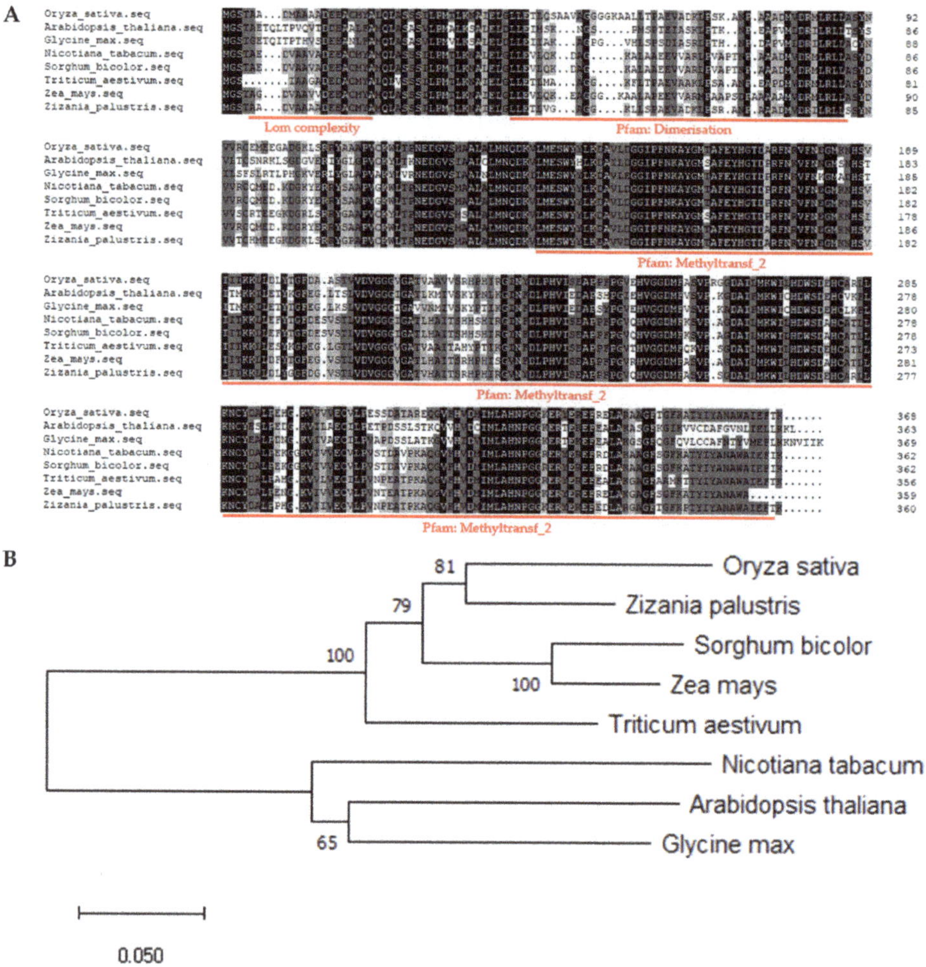

Figure 3. Homologous sequence alignment and evolutionary tree analysis of OsSPL42. (**A**) OsSPL42 homologous sequences were compared by DNAMAN v7 software (part), and the part underlined in red represents the domain; (**B**) construction of an evolutionary tree using the adjacency method through MEGA v6 software. All SPL42 proteins are numbered in NCBI as follows: *Oryza* OsSPL42 (KAB8107446.1), *Zizania palustris*. (KAG8086825.1), *Sorghum bicolor* SbSPL42 (AFO69477.1), *Zea* ZmSPL42 (AAQ24342.1), *Triticum aestivum* TaSPL42 (NP_001392772.1), Arabidopsis *thaliana* AtSPL42 (OAO95542.1), *Glycine* GmSPL42 (KAH1255570.1), and *Nicotiana tabacum* NtSPL42 (CAA52462.1).

Through http://www.ncbi.nlm.nih.gov/ (accessed on 18 November 2022), we performed OsSPL42 protein sequence alignment (Figure 3B). OsSPL42 orthologues were selected from four monocot plants and three dicotyledonous plants. Monocot plants besides rice includes *palustris, Zea may, bicolor,* and *Triticum aestivum,* and the accession numbers of their orthologues in NCBI are as follows, respectively: KAG8086825.1, AAQ24342.1, AFO69477.1, and NP_001392772.1. The accession numbers of the orthologs of *Arabidopsis thaliana, Nicotiana tabacum,* and *Glycine max* are OAO95542.1, CAA52462.1, and KAH1255570.1 in NCBI. The similarity between OsSPL42 and its orthologs varies, ranging from 59.13% in *soybean bicolor* to 87.23% in wild rice. The mutation site of the *OsSPL42* gene is in the second domain, and the mutation site was leucine in all the orthologs of these seven species. The AAs after mutation are consistent, indicating that the mutation site is highly conserved.

At the same time, we conducted an evolutionary tree analysis on these species. The results indicated that the OsSPL42 protein of rice and its homologous protein in *Zizania palustris* were located on the same branch of the evolutionary tree, indicating their close genetic relationship. Additionally, four monocotyledonous plants were clustered on one branch, while three dicotyledonous plants were grouped on another branch. The results indicated that the OsSPL42 protein is closely related to monocotyledonous plants, and its sequence is conserved.

2.3. *Altered Physiological/Biochemical Reactions Related to ROS Scavenging System in Spotted-Leaf Mutant spl42*

The formation of speckled leaf spots is mostly related to ROS accumulation. To investigate whether the cause of speckled leaf spots is related to ROS accumulation, 3,3′-diaminobenzene (DAB) and nitroblue tetrazolium (NBT) staining experiments were used. The accumulation of hydrogen peroxide (H_2O_2) and superoxide anion (O_2^-) in leaves was detected. The results of the DAB staining (Figure 4A) showed obvious reddish-brown material deposits in and around the lesion site of *spl42* leaves, but not in ZJ100. There were blue material deposits in and around the lesion site of *spl42* leaves. This indicates the presence of large deposits of H_2O_2 and O_2^- near *spl42* spots.

To further verify this result, the H_2O_2 content in mutant, wild-type, and complementary lines CP1, CP2, and CP3 was quantitatively determined (Figure 4B). It was observed that the H_2O_2 content of CP1, CP2, and CP3 was significantly lower than that of the mutant *spl42*. Additionally, the H_2O_2 content of ZJ100 was lower than that of the mutant, although it did not reach a statistically significant level. However, the trend was basically consistent with that of CP1, CP2, and CP3, all of which were lower than that of the mutants. The CAT activity of CP1, CP2, CP3, and ZJ100 plants was significantly higher than that of the mutant (Figure 4C). This indicates that the activity of the H_2O_2 scavenging enzyme was decreased in the mutant plants, resulting in the accumulation of H_2O_2, which caused cell death at the leaf spot. However, when the *OsSPL42* gene was introduced into this mutant, H_2O_2 deposition disappeared, and CAT activity recovered. This suggests that the *OsSPL42* gene might regulate the ROS outbreak in rice plants (Figure 4D–G).

In the same way, ROS accumulation in the wild-type NPB and corresponding knockout lines was detected. The results (Figure 5A–F) showed that compared with the wild type, the H_2O_2, APX, and MDA contents of the knockout lines were significantly higher. Additionally, the CAT and SOD contents were significantly lower in the knockout lines than in the wild type, while the POD contents did not show a significant difference. These results indicate that the loss of gene function following the knockout of *OsSPL42* disrupted the ROS scavenging system, resulting in the accumulation of ROS.

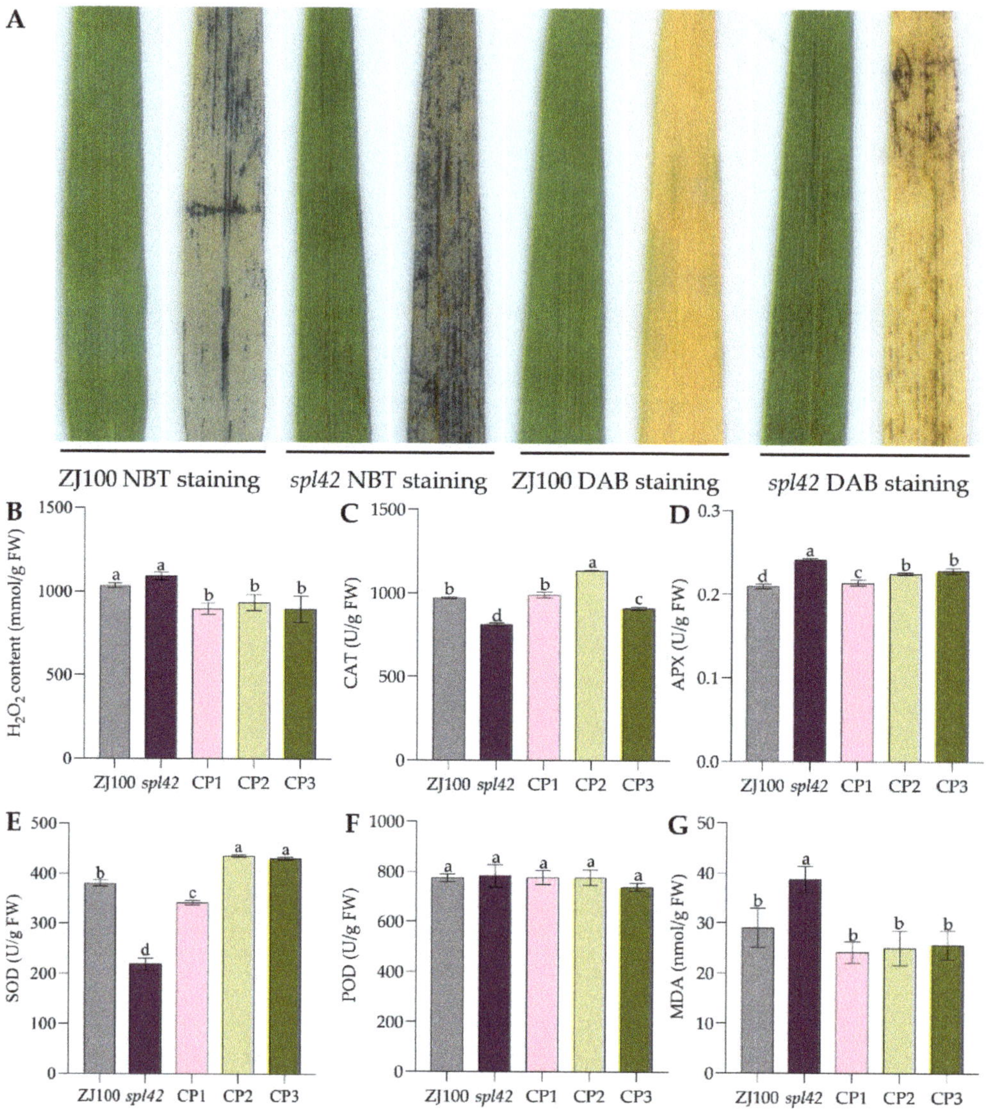

Figure 4. Determination of ROS accumulation of complementation lines of the *spl42* mutant at 10 weeks of age. (**A**) Wild-type ZJ100 and mutant *spl42* NBT and DAB staining (**B**) H_2O_2 content. (**C–F**) ROS scavenging enzyme activities (**G**) MDA content CP, complementation; values are means ± SD, $n = 3$; and different letters indicate significant differences by one-way ANOVA and Duncan's test $p \leq 0.05$.

After introducing a binary vector pCAMBIA1300-UBI-OsSPL42, in which *OsSPL42* was driven by a maize UBI promoter, into the callus of NPB, a total of 9 *OE-OsSPL42* overexpressing lines were obtained. The leaves of the overexpression plants exhibited normal green leaf phenotypes like those of wild-type NPB (Figure 6A,B). Expression lines OE2, OE3, and OE8 were selected to investigate *OsSPL42* gene expression, as shown in the figure. The gene expressions of OE2, OE3, and OE8 were significantly higher than those of the wild type by 3.5 times, 8.6 times, and 23.3 times, respectively (Figure 6C).

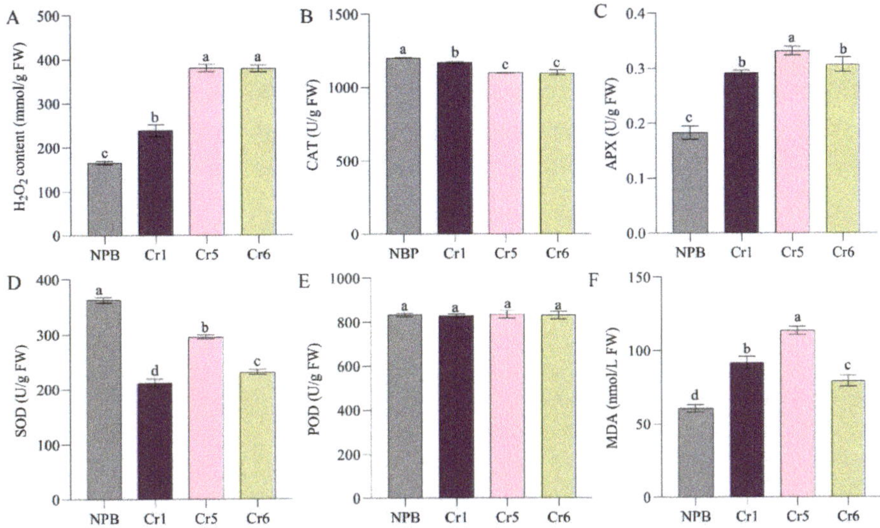

Figure 5. Determination of ROS accumulation of knockout lines. (**A**) H_2O_2 content. (**B–E**) ROS scavenging enzyme activities. (**F**) MDA content. Cr, knockout; values are means ± SD, $n = 3$; and different letters indicate significant differences by one-way ANOVA and Duncan's test $p \leq 0.05$.

Figure 6. Analysis expression and phenotypes of overexpression lines. (**A**) Plant phenotype of NPB and overexpression line at filling stage. Bar = 17 cm. (**B**) Leaves of phenotype of NPB and overexpression line at filling stage. Bar = 1 cm. (**C**) Expression of *OsSPL42* gene in leaves of wild-type NPB and overexpressed strains at filling stage. OE, overexpression; values are means ± SD, $n = 3$; and different letters indicate significant differences by one-way ANOVA and Duncan's test $p \leq 0.05$.

The analysis of soluble protein content and ROS accumulation in wild-type and overexpression lines is shown in Figure 7. The overexpression lines with higher expression levels, OE3 and OE8, exhibited significantly elevated total protein content compared to the wild-type NPB, whereas OE2, an overexpression line with lower expression levels, did not show a significant difference from NPB. The ROS accumulation patterns varied across the lines as well; compared to NPB, CAT activity was significantly reduced in OE2 and OE3, while OE8 showed a marked increase. H_2O_2 content in the three overexpression lines showed trends different to those of CAT activity: OE2 and OE3 displayed significantly higher H_2O_2 content than NPB, whereas OE8 exhibited a significant reduction. There were no significant differences in POD content among the lines; however, trends in APX, SOD, and MDA content paralleled those of soluble protein. Specifically, the overexpression lines OE3 and OE8 had significantly elevated levels compared to NPB, while OE2 showed no significant difference. These findings suggest that variations in the expression levels of the *OsSPL42* gene influence soluble protein content and ROS accumulation, indicating a potential regulatory role for *OsSPL42* in maintaining ROS balance and soluble protein content in rice. Moreover, these differences in gene expression appear to affect the activities of both oxidative and scavenging enzymes within the plant's redox enzyme system in a dose-dependent manner.

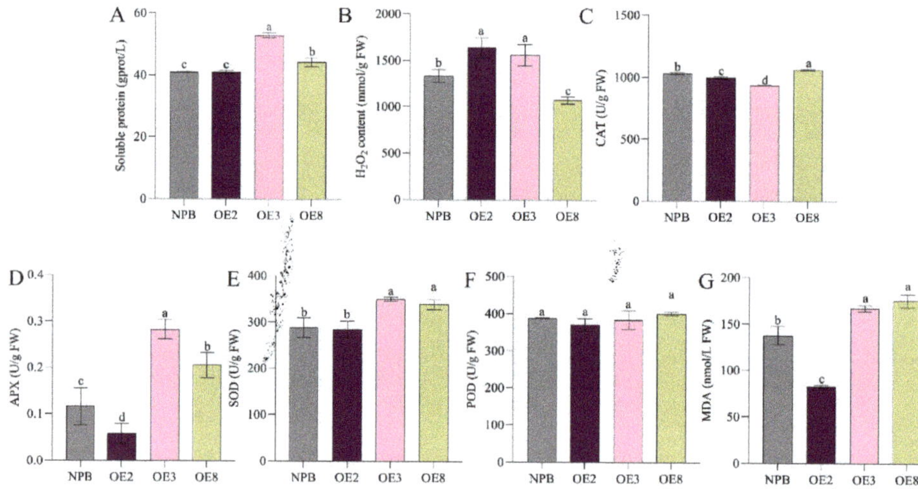

Figure 7. Determination of ROS accumulation of overexpression lines. (**A**) Soluble protein. (**B**) H_2O_2 content. (**C–F**) ROS scavenging enzyme activities. (**G**) MDA content. OE, overexpression; values are means ± SD, $n = 3$; and different letters indicate significant differences by one-way ANOVA and Duncan's test $p \leq 0.05$.

2.4. Enhanced Disease Resistance in Mutation spl42

Spotted-leaf mutants typically exhibit increased resistance to pathogens by inducing HR. In this study, 16 strains of *Xoo* were inoculated with the mutant *spl42*, ZJ100, and a susceptible control variety, Jinggang 30 (JG30), to assess their resistance to BB. The inoculation period occurred during the full tillering stage, and the lesion length was measured using the leaf clipping method 2 weeks after inoculation (Table S2). The disease responses of *spl24* to different strains varied significantly. Among them, resistance to race OS-225 was highly enhanced ($p \leq 0.01$). Therefore, OS-225 was chosen for the next experiment.

The resistance of rice mutants to BB is usually associated with the expression of defense response genes. Therefore, the relative expression levels of the defense response gene in ZJ100 and mutant *spl42* were detected. As shown in Figure 8C, compared with

ZJ100, the expression levels of 7 out of the 13 defense response genes (*OsPR1a*, *OsPR1b*, *OsPAL3*, *OsAOS2*, *OsWRKY45*, *OsJamyb*, and *OsPBZ1*) were significantly upregulated in the mutant, while the expression levels of the other 6 defense response genes were significantly downregulated. This result is consistent with both increased and decreased resistance in resistance evaluation experiments.

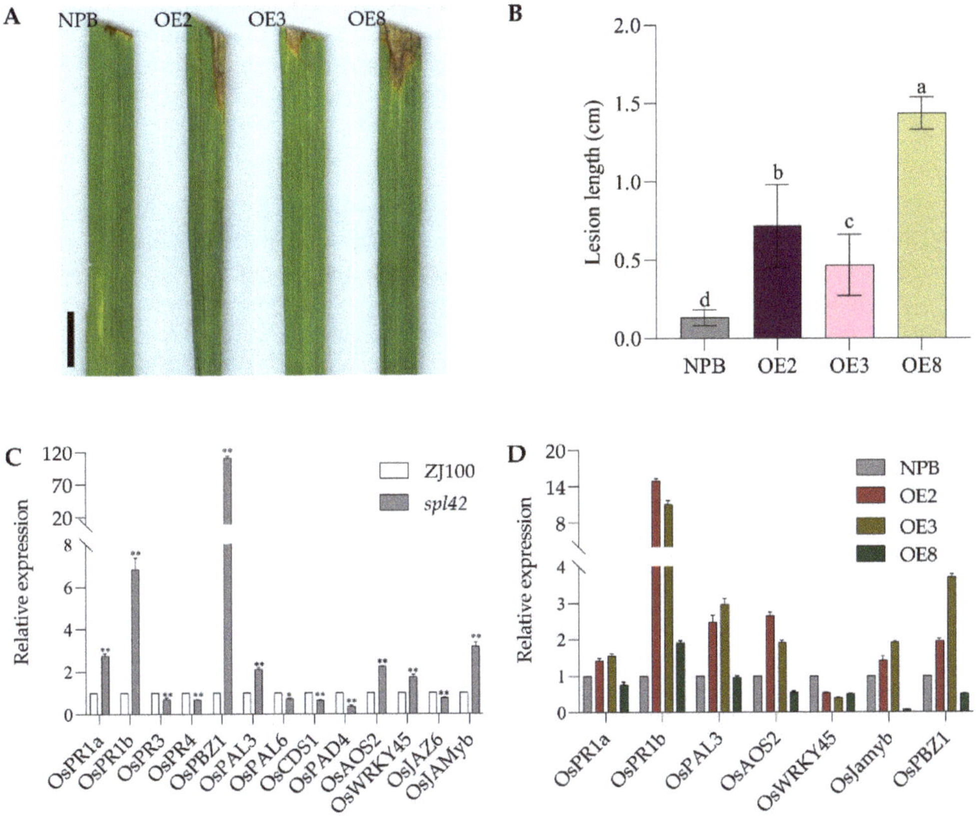

Figure 8. Evaluation of disease resistance in overexpressed lines and analysis of defense response gene expression. (**A**) Results of inoculation of wild-type NPB and overexpression lines with BB pathogen race Os-225 at the tillering stage. Bar = 1 cm. (**B**) Lesion length; values are means ± SD, n = 6; and different letters indicate significant differences by one-way ANOVA and Duncan's test $p \leq 0.05$. (**C**) Analysis of defense response gene expression of wild-type ZJ100 and mutant *spl42*; the ordinate indicates that ZJ100 is set to 1, and the remaining gene expressions are multiples of ZJ100; *, significance at $p \leq 0.05$; **, high significance at $p \leq 0.01$. (**D**) Analysis of defense response gene expression of wild-type NPB and overexpressed strains at the tillering stage. The ordinate indicates that NPB is set to 1, and the remaining gene expressions are multiples of NPB; OE, overexpression.

To verify the impact of overexpression on BB resistance, OS-225 was selected to inoculate both the wild-type NPB and the overexpressed plants. The results showed that the lesion length of overexpressed plants was significantly longer than that of the wild-type NPB (Figure 8A,B), indicating that plant resistance was significantly weakened after overexpressing the *OsSPL42* gene. Meanwhile, seven defense response genes (*OsPR1a*, *OsPR1b*, *OsPAL3*, *OsAOS2*, *OsWRKY45*, *OsJamyb*, and *OsPBZ1*), whose expression levels were upregulated in the mutant, were selected for expression analysis, as shown in Figure 8C,D. Overexpression of the *OsSPL42* gene resulted in the dysregulated expression of various defense response genes.

The relative expression levels of different genes were either upregulated or downregulated, and the expression levels varied significantly. Compared with the wild type, the expression of the *OsWRKY45* defense response gene was significantly downregulated in the overexpressed plants, while the expression of the *OsPR1b* defense response gene was significantly upregulated in the overexpressed plants. The expression levels of *OsPR1a*, *OsPAL3*, *OsAOS2*, *OsJamyb*, and *OsPBZ1* defense response genes in OE2 and OE3 were upregulated compared with those of the wild type, while those in OE8 were downregulated or showed no change. These results verified that the spotted-leaf mutant *spl42* induced increased resistance to some BB. The results showed that the overexpression plants were weakened to BB, suggesting that *OsSPL42* may negatively regulate the resistance of rice to BB.

2.5. OsSPL42 Regulates PCD in Rice

To verify whether the appearance of spots was accompanied by cell death, trypan blue staining was performed on ZJ100 and mutant *spl42* leaves at the late tillering stage. The results (Figure 9A) showed that blue precipitates appeared in both ZJ100 and mutant *spl42* leaves. However, fewer blue precipitates appeared in ZJ100 leaves, while more blue precipitates were observed in/around the lesions of the mutant *spl42* leaves. Therefore, we used the TUNEL (terminal-deoxynucleoitidyl transferase) method to detect the degree of DNA fragmentation, an apoptosis indicator (Figure 9C). The results showed that more DNA fragmentation (green, TUNEL-positive signals) was detected in mutant *spl42*, compared to that in wild-type ZJ100. This suggests that the *spl42* mutation induced large-scale DNA fragmentation in the cells. Therefore, we examined the relative expression levels of metacaspase (MC) family-related genes, cell death markers [29] in leaves of ZJ100 and mutant *spl42*. The results showed (Figure 9B) that compared with ZJ100, the relative expression levels of *OsMC2*, *OsMC4*, *OsMC5*, *OsMC6*, *OsMC7*, and *OsMC8* in the mutant *spl42* were upregulated and reached a statistically significant level. The relative expression level of *OsMC3* was upregulated but did not reach a statistically significant level. The relative expression level of *OsMC1* was significantly downregulated ($p < 0.01$). In summary, the experimental results here indicate that cell death occurs in the mutant *spl42*, and the expression of related genes involved in the regulation of PCD is altered.

2.6. OsSPL42 Is Widely Expressed and OsSPL42 Localizes to Cytoplasm

To study the expression pattern of the *OsSPL42* gene, we observed its expression in transgenic plant tissues using the GUS report system, where the GUS gene is under the control of the *OsSPL42* promoter. The results showed that blue color was present in the root, stem, leaf, sheath, and grain of transgenic plants (Figure 10B), indicating that the *OsSPL42* gene was expressed in these tissues. At the same time, the relative expression level of the *OsSPL42* gene in various tissues was also detected by qRT-PCR (Figure 10A). The results indicated that wild-type ZJ100 and mutant *spl42* plants exhibited differential gene expression patterns in the *OsSPL42* gene. Notably, in ZJ100, the expression of the *OsSPL42* gene peaked in stems at 10 weeks after germination. In the *spl42* mutant, the *Osspl42* gene was expressed at the highest level in the leaves at the filling stage. These results confirm that the *OsSPL42* is a widely expressed gene.

In this study, the full-length CDS of the *OsSPL42* gene was cloned into the GFP transient expression vector pAN580. The resulting PAN580-*OsSPL42* plasmid was used to identify the intracellular localization of the OsSPL42 protein. Rice protoplast transient transformation technology was used to observe the subcellular localization of the OsSPL42-GFP protein. The results showed that the green fluorescence signal of the OsSPL42-GFP protein was mainly located in the cytoplasm, the same as the signal of the GFP protein alone. This indicates that both OsSPL42 and GFP proteins were mainly present in the rice cytoplasm (Figure 11).

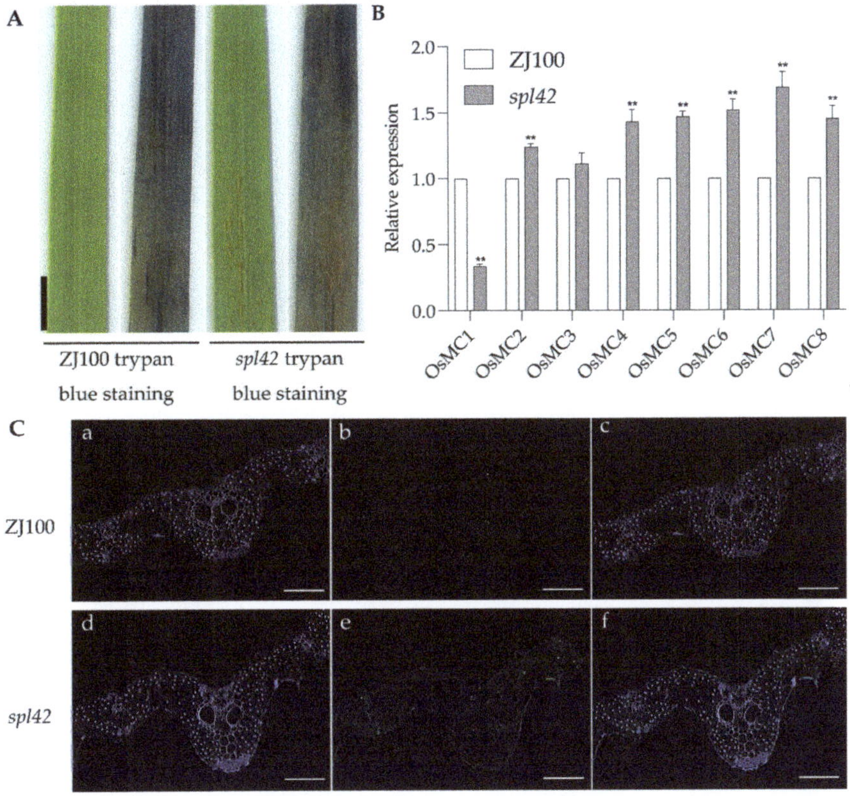

Figure 9. Cell death analysis of the mutant *spl42*. (**A**) Wild-type ZJ100 and mutant *spl42* trypan blue staining at 10 weeks of age. Bar = 1 cm. (**B**) Expression levels of ZJ100 and *spl42* PCD-related genes at 10 weeks of age. Values are means ± SD, $n = 3$; **, high significance at $p \leq 0.01$. (**C**) ZJ100 (**a–c**) and *spl42* (**d–f**) TUNEL experiments. The blue signal is DAPI signal, and the green signal is TUNEL signal, indicating the presence of fragmented DNA. Bar = 100 μm.

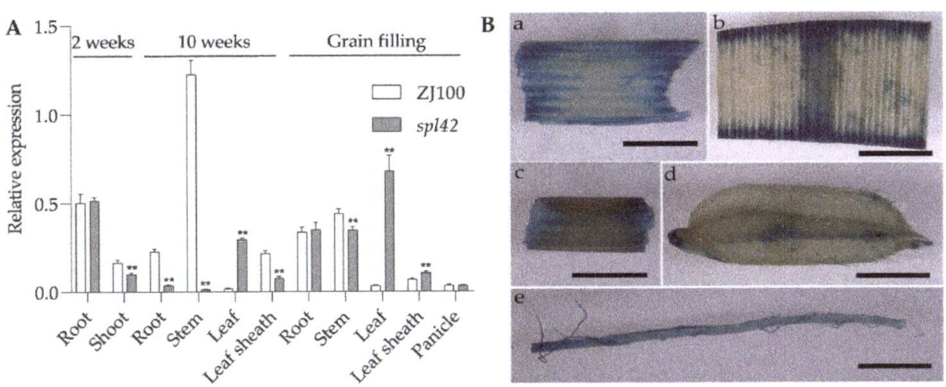

Figure 10. qRT-PCR and GUS analysis of *OsSPL42*. (**A**) Analysis of *OsSPL42* gene expression in wild-type ZJ100 and mutant *spl42*; **, high significance at $p \leq 0.01$. (**B**) Transgenic plants GUS staining analysis: (**Ba**) leaf sheath; (**Bb**) leaves; (**Bc**) third stem of the plant; (**Bd**) seed; (**Be**) young roots; Bar = 1 mm.

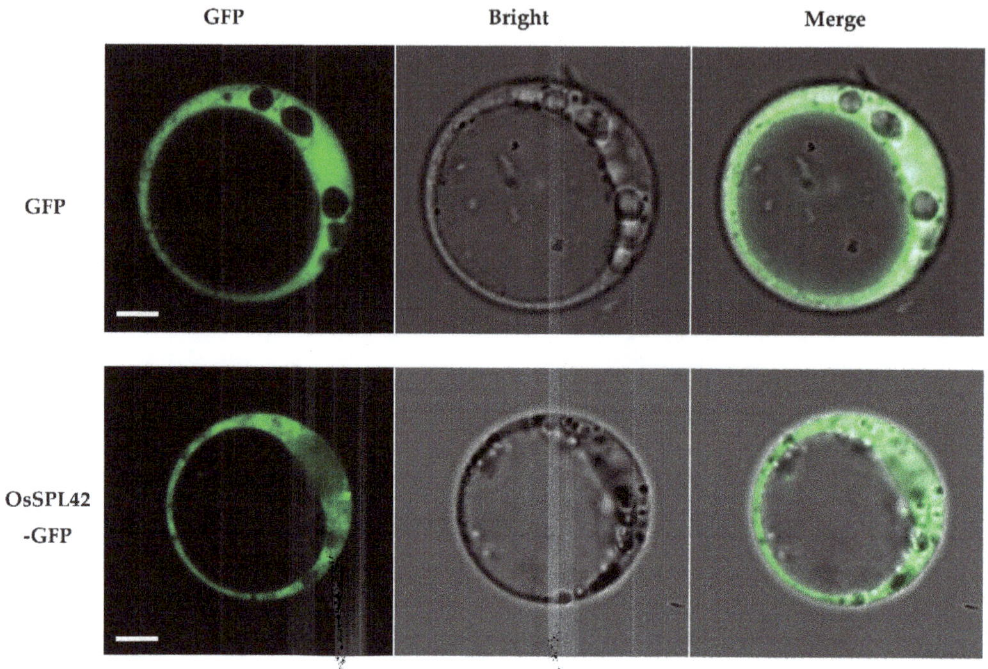

Figure 11. Sub-cellular localization of OsSPL42 protein. Note: GFP, green fluorescence protein; bar = 5 μm.

3. Discussion

The spotted-leaf mutants exhibit leaf lesions, usually a type of PCD induced by pathogen invasion [30]. These mutants are crucial for the investigation of the molecular mechanisms underlying PCD. Studies have shown that the development of lesions in these mutants often results in enhanced disease resistance, making them ideal subjects for studying the molecular mechanisms of disease resistance [31–33]. In this study, the mutant *spl42*, derived from EMS in indica rice, exhibits growth and development deficiencies like other spotted-leaf mutants. By 10 weeks of age, the leaf lesions in mutant *spl42* increase, accompanied by chlorophyll degradation and a decrease in soluble protein content. It is hypothesized that the formation of mutant leaf spots is related to chloroplasts. Many spotted-leaf mutants show a decrease in photosynthetic pigments, which is accompanied by changes in agronomic traits [24,34,35]. In this study, the main physiological and agronomic characters, such as the seed setting rate and 1000-grain weight of mutant *spl42*, were changed by photosynthetic pigment. Many leaf lesions are caused by light exposure [36,37]. The mutants *Spl24* and *HM47* showed light-induced lesion formation under natural light [24,38]. The experiment revealed no significant difference between ZJ100 and mutant *spl42* before and after shading, and spots still appeared in the shaded part of the mutant *spl42* leaves after shading. This suggests that the generation of mutant spots is not light-induced, and thus, the shading experiment is not mentioned in the paper.

In the trypan blue staining experiment, the blue color of the mutant *spl42* leaves was darker, and PCD detection found a large amount of fragmented DNA in mutant *spl42*, with upregulated expression levels of 6 MC family-related genes. The results of these three experiments confirmed the existence of cell death in mutant *spl42*, which may lead to chlorophyll degradation. ROS production is closely related to localized cell death [39]. ROS, a product of REDOX reactions, is removed by corresponding reactive oxygen scavenging systems, including CAT, SOD, and POD, to prevent the harmful effects on cells [40]. The accumulation of ROS in mutant *spl42*, particularly the accumulation of O_2^- and H_2O_2, may lead to the death and browning of leaf cells. The SOD activity of mutant *spl42* decreased, while the APX activity increased, suggesting that the damage to the ROS scavenging system might eventually lead to the generation of mutant *spl42* leaf spots.

Lignin can improve plant stress resistance, affecting metabolic pathways [41], salt tolerance [42], and drought tolerance of plants [43]. The COMT gene family can participate in plant abiotic stress responses [44]. In this study, genetic and localization analysis revealed that the mutant *spl42* phenotype was controlled by a recessive gene, *Osspl42*, which encodes an O-methyltransferase protein involved in lignin monomer biosynthesis [45], may be widely involved in various abiotic stress responses [46]. However, the involvement of *OsSPL42* in invasion, such as BB, requires further investigation. *OsSPL42* was found to be the same gene as *OsCOMT*, *OsCAldOMT1*, and *OsCAD2*, encoding caffeic acid 3-O-methyltransferase [28,45,47]. Rice caffeic acid *O*-methyltransferase (*OsCOMT1*) can catalyze the 5-Omethylation of 5-hydroxyferulate (5-HFA) and 5-hydroxyconiferaldehyde (5-HCAld). *OsCOMT1* can act as *5-HCAld OMT (OsCAldOMT1)* in the S-lignin biosynthesis pathway and participate in lignin biosynthesis [48]. Lam (2019) showed that the lignin monomers involved in the synthesis of OsCAldOMT1 include S-lignin, tricin-lignin, and selgin-lignin, with substrates including 5-hydroxyferulate (5-HFA) and 5-hydroxyconiferaldehyde (5-HCAld). The determination of lignin monomer content in rice is more complicated and requires chemical determination. We detected the lignin content of each rice line (Figure S1). The results showed that the total lignin content of the mutant was significantly higher than that of the wild-type and complementary plants (Figure S1A), and the total lignin content of the three knockout lines was significantly higher than that of the wild-type NPB. The total lignin content of overexpressed lines with high *OsSPL42* gene expression was significantly higher than that of wild-type NPB (Figure S1B). This suggests that the abnormal expression of *OsSPL42* might be associated with the accumulation of total lignin in rice leaves. The content of S-lignin monomer in the mutant was significantly higher than that of the wild type (Figure S1C), indicating that the abnormal *OsSPL42* gene might be accompanied by an increase in the content of S-lignin monomer, which would increase the total lignin content. However, this does not prove that *OsSPL42* is negatively regulated with lignin biosynthesis. Due to technical problems, we could not directly detect Tricin-lignin monomer content in rice plants. Therefore, lines with abnormal *OsSPL42* gene were associated with increased total lignin and S-lignin monomer content, but tricin-lignin monomer content may be decreased, which requires further research.

Studies have shown that *OsCOMT* can also inhibit the degradation of chlorophyll and chloroplasts, improve photosynthetic efficiency, and significantly delay the senescence of leaves at the filling stage. *OsCOMT* also positively regulates vascular bundle development [49]. Therefore, *OsCOMT* can improve rice yield through the dual regulation of leaf senescence and vascular development, and is a positive regulator of grain yield, significantly affecting plant height, panicle length, number of filled grains/panicle, and 1000-grain weight [23]. This study confirmed that *OsSPL42* regulates the formation of rice leaf spots, and whether lignin is related to the formation of rice leaf spots is very valuable for further study.

In the protein structure of OsSPL42, the domain of 132–366AA includes a series of *O*-methyltransferases, some of which utilize S-adenosylmethionine as a substrate. This is further evidence that *OsSPL42* encodes an *O*-methyltransferase. The mutant *spl42* was mutated at the 42nd base of *OsSPL42* gene, so the AAs of the 31–89AA domain were changed, but SMART predicted that the dimerization activity of the protein mediated by this structure did not show significant changes. Whether there is any effect on *O*-methyltransferase activity is worth researching further in the future.

ROS can function as a regulator of cell death [50]. DAB and NBT staining showed the accumulation of ROS like H_2O_2 and O_2^- in mutant *spl42*. Functional complementation experiments demonstrated that ROS accumulation was absent in the complementary plants, confirming that the mutant phenotype was restored after gene functional complementation, and the ROS clearance system was also restored. This further suggests that *OsSPL42* leads to cell death and regulates the production of leaf lesions in mutant *spl42*, influencing rice agronomic traits. Many genes in plants exhibit pleiotropic effects [51], and traits controlled by multiple genes or quantitative trait loci (QTLs), such as 1000-grain weight [52], can be influenced by other genes or QTLs. In the knockout experiment of the study, only Cr1 showed leaf lesions, while Cr5 and Cr6 did not leaf lesions. We performed off-target verification for the knockout experiments, confirming that off-target effects on other genes were excluded. So, we guess that other genes or QTLs may have compensatory effects on this trait, requiring further research.

The spotted-leaf mutants often enhance plant resistance to pathogens through hypersensitivity, which is an ideal material for studying hypersensitivity and PCD. In the early stage of this study, the resistance of the mutant *spl42* to BB was verified through inoculation with wild type ZJ100, mutant *spl42*, and a susceptible variety (JG30) of rice, *Xanthomonas oryzae* pv. *oryzae*. The results showed that the resistance of the mutant *spl42* to BB was different (Table S2). The resistance of OS-225 subspecies with significantly enhanced resistance was identified on overexpressed plants, and the results confirmed that overexpressed plants had reduced resistance to BB. Therefore, we speculated that *OsSPL42* gene may negatively regulate rice resistance to BB. This was also confirmed by the expression of defense response genes in ZJ100, *spl42*, and overexpressed plants.

The study raises an interesting question regarding the tradeoffs between enhanced BB resistance and other agronomic traits in rice, particularly in the context of *OsSPL42* regulation. *spl42* demonstrated heightened resistance to *Xoo*, with the overexpression of *OsSPL42* in transgenic lines leading to reduced BB resistance. However, *spl42* exhibited reduced agronomic performance and seed quality. Interestingly, these overexpressed lines exhibited improved agronomic traits, such as an increased tiller number (Figure 6). This suggests a potential tradeoff between disease resistance and growth [53]. The reduction in BB resistance in overexpressed plants could be attributed to a reallocation of resources from defense mechanisms to growth-promoting pathways, such as those involving auxins or gibberellins, which are often upregulated in response to overexpression of certain regulatory genes [54]. It is well established that plant immune responses, particularly hypersensitive responses and programmed cell death, are metabolically costly and can hinder growth and yield under certain conditions [55]. As such, the overexpression of *OsSPL42* may prioritize growth at the expense of immune defense, leading to an increase in agronomic performance but a corresponding decrease in disease resistance. To determine the practical implications of these tradeoffs, it will be necessary to conduct field trials that assess both disease resistance and yield-related traits under varying environmental conditions and pathogen pressures. Understanding the balance between growth and defense will be critical for optimizing the use of *OsSPL42* in breeding programs. But future research should also explore the molecular mechanisms underlying these tradeoffs, particularly the signaling pathways and transcription factors involved in the dual regulation of growth and defense. However, the underlying mechanism of this phenomenon requires further investigation.

4. Materials and Methods

4.1. Plant Materials

spl42 was obtained from an ethyl methanesulfonate (EMS)-induced mutant bank of ZJ100. The F_1 population was obtained by crossing *spl42* with a japonica rice, and F_2 from F_1 was used for mapping and genetic analysis. The transgenic plants of functional complementation, gene knockout, overexpression, and GUS staining were the experimental materials for functional study, including T_0 plants and T_1 and T_2 plants obtained from self-cross.

4.2. Growth Conditions and Agronomic Trait Evaluation

ZJ100, *spl42*, F_1, and F_2 populations were grown in the paddy field of the China Rice Research Institute in Hangzhou, Zhejiang Province. The transgenic plants were grown in glass greenhouses at the China National Rice Research Institute (CNRRI), Hangzhou, China. The row spacing of all plants was 16.7 cm $\times$ 26.7 cm, with normal fertilizer and water management. After ripening, 3 plants were randomly selected to evaluate their agronomic traits, including plant height, panicle length, number of tillers/plants, number of filled grains/panicle, seed setting rate, and 1000-grain weight. The means from three replicates were subjected to one-way analysis of variance and Duncan's multiple test ($p \leq 0.05$).

4.3. Determination of Photosynthetic Pigments

First, 0.1 g of fresh leaves were selected from the same part of the rice plants at about 10 weeks of age, and three plants were randomly selected from each group (excluding side-row plants). Then, the method of Arnon and Wellburn was used to determine the photosynthetic pigment content of rice plant leaves [56,57]. We repeated 2 times in the experiment and chose 1 time result to present in the paper.

4.4. Histochemical Analysis and PCD Detection

The leaves of ZJ100 and *spl42* at about 10 weeks of age were selected for histochemical staining, including DAB staining to detect the H_2O_2 content, NBT staining to detect the O_2^- content, and trypan blue staining to confirm cell death [58,59]. The test of PCD was carried out with the kit provided by using a Fluorescein In Situ Cell Death Detection Kit following the manufacturer's instructions (Roche, Basel, Switzerland). The principle is terminal-deoxynucleotidyl transferase-mediated nick end labeling (TUNEL). It is a method that can detect the degree of DNA strand break in the process of cell death by fluorescence microscopy. We repeated 2 times in the GUS, DAB, and NBT staining, and chose 1 time result to present in the paper.

4.5. Determination of ROS-Related Parameters and Lignin Content

At 10 weeks of age, fresh leaves were randomly selected from three rice plants. The H_2O_2 content, soluble proteins content and MDA content, as well as the activities of APX, CAT, SOD, and POD were determined by employing the respective assay kit (Nanjing Jiancheng Bioengineering Institute, Nanjing, China), following the manufacturer's instructions. Three biological replicates were performed for all measurements. To determine total lignin content, fresh leaves from each sample were dried at 105 °C for 15 min and at 80 °C for 3 days, and then 0.01 g of powder was added to the next test with a kit from Suzhou Mengxi Bio-Pharmaceutical Technology Co., Ltd. (Suzhou, China). Fresh leaves of 10-week-old rice plants were transferred into a biological company (Suzhou Mengxi Bio-Pharmaceutical Technology Co., Ltd., Suzhou, China) using liquid chromatography for the determination of S-lignin. We repeated 3 times in ROS detection and chose 1 time result to present in the paper.

4.6. Disease Evaluation

To detect the resistance of the *spl42* to BB, the species of BB were inoculated with the leaf clipping method. Fourteen days after inoculation, the damage length of the inoculated leaves was measured with a transparent plastic ruler. The average of 6 replicates was used for analysis. The BB subspecies were 16 *Xoo*. They were selected from common subspecies in China and representative subspecies in the Philippines including PXO99, GD1385, NX42, Zhe173, OS-225, PXO71, PXO341, HBl7, PXO86, PX079, PXO112, PXO145, PXO280, PXO339, JS97-2, and LN57 (Table S2).

4.7. Genetic Analysis and Mapping of OsSPL42

In this study, we constructed the F_2 population, from which about 500 plants were selected for genetic analysis and 152 plants for mapping. The DNA was extracted from the leaves with lesion mimic using the cetyltrimethylammonium bromide (CTAB) method [60]. We used about 200 pairs of SSR markers linked to the target genes for initial localization and selected 7 pairs of InDel markers for fine localization. Finally, the candidate genes were determined by comparing the base differences in the target interval between the wild type and the mutant by whole genome sequencing.

4.8. RNA Extraction and qRT-PCR

The NucleoZOL kit was used for the total RNA extraction. For specific steps, refer to the manufacturer's instructions (MACHEREY-NAGEL, Düren, Germany). Then, the RNA reverse transcription and qRT-PCR were performed by a kit (Promega kit, Madison, WI, USA). The real-time fluorescent quantitative PCR was performed using a kit ((Thermo Fisher Scientific, Waltham, MA, USA). For specific operation methods, refer to the instructions. The results of qRT-PCR were calculated according to the formula $2^{-\Delta CT}$, $2^{-(CT\ sample - CT\ Actin)}$, with three biological replicates. We repeated 2 times in qRT-PCR experiment and chose 1 time result to present in the paper.

4.9. Vector Construction

The infusion method was used in all vector constructions, and the primers used were shown in Table S3. The CDS full-length of *OsSPL42* including 1107 bp was cloned into pAN580, and the GFP vector PAN580-OsSPL42 was cloned into rice protoplasts. The fluorescence signal of GFP was detected by a confocal laser scanning microscope (Carl Zeiss, Inc., Jena, Germany). The GUS transient expression vector was constructed by cloning the promoter region of the *OsSPL42* gene with about 2.6 kb onto pCAMBIA1381Z vector. Then, we introduced the recombinant vector into the callus of the embryogenic call induced from the NPB mature seeds. Transgenic plants were stained using the GUS staining kit (COOLABER SCIENCE & TECHNOLOGY Co., Ltd., Beijing, China). In the complementary validation experiment, PrimerSTAR Max DNA Polymerase was used to expand the full length of *OsSPL42* and part of the promoter and region terminator sequence. The full length of *OsSPL42* was 3168 bp, and the upstream promoter region was 2710 bp. The downstream terminator region is 1518 bp, and the total sequence is about 7 kb. It was then cloned into the pCAMBIA1300 vector, constructed into the PCAMBIA1300-OsSPL42 recombinant vector, and transferred into the mutant *spl42* callus by the *Agrobacterium tumefaciens*-mediated method [61]. We amplified the full length of the CDS sequence of the NPB into the pHY-UBIZ vector to form the pHY-UBI-OsSPL42 recombinant and transferred the recombinant plasmid into the callus of the NPB using the *Agrobacterium tumefaciens*-mediated method. The CRISPR/Cas9 knockout vector of *OsSPL42* was constructed using the NPB as the receptor material and U3 as the vector. The method was mainly based on Liu Yaoguang's laboratory method system of South China Agricultural University [62].

4.10. Sequence Analysis and Off-Target Verification of CRISPR/Cas9 Knockout Targets

It is necessary to carry out the sequence analysis and off-target verification of the knockout target of transgenic plants. The primers used for off-target verification are shown in Table S4. A small amount of the leaf genomic DNA of transgenic plants was extracted, and primers containing 200–300 bp fragments, including knockout targets, were designed. Then, the sequence was amplified and sequenced to verify the success of knockout, and the knockout site of each transgenic plant was analyzed. The target sequence can be found on the CRISPR-P website (http://cbi.hzau.edu.cn/cgi-bin/CRISPR, accessed on 19 May 2021) when searching for homologous sequences in other locations of the target sequence in rice. Primers containing 200–300 bp fragments, including a homologous sequence of the knockout target, were designed, and the sequence was amplified in transgenic plants to verify whether the knockout target was off target.

4.11. Phylogenetic Analysis of OsSPL42

The basic information of *OsSPL42* was found in the Rice Database (https://www.ricedata.cn/, accessed on 2 April 2021). We found the domain of *OsSPL42* gene prediction on the SMART website (http://smart.embl-heidelberg.de/, accessed on 18 November 2022). We found the homologous proteins of *OsSPL42* in other species using the NCBI database (http://blast.ncbi.nlm.nih.gov/Blast.cgi, accessed on 18 November 2022). We selected three dicotyledonous plants, *Arabidopsis thaliana*, *Nicotiana tabacum*, and *Glycine max*, and four monocotyledonous plants, *Zizania palustris*, *Zea mays*, *Sorghum bicolor*, and *Triticum aestivum*. These homologous protein sequences were analyzed and compared using DNAMAN v7 (http://www.lynnon.com/ accessed on 18 November 2022). Finally, we used MEGA v6 software (http://www.megasoftware.net/ accessed on 18 November 2022) to construct a phylogenetic tree based on the neighbor-joining method.

Supplementary Materials: The following supporting information can be downloaded at: https://www.mdpi.com/article/10.3390/ijms252211871/s1.

Author Contributions: Investigation, P.L.; investigation, H.S.; methodology, J.-L.W.; project administration, X.X.; funding acquisition, J.G.; writing—original draft, P.L.; writing—review and editing, X.Z. All authors have read and agreed to the published version of the manuscript.

Funding: This work was supported by the Zhejiang Provincial Natural Science Foundation of China under Grant No. LZ24C130004 and No. LQ24C130008 and The National Key Research and Development Program of China (2022YFF1003301).

Institutional Review Board Statement: Not applicable.

Informed Consent Statement: Not applicable.

Data Availability Statement: Data are contained within the article or Supplementary Materials.

Conflicts of Interest: The authors declare no conflicts of interest.

References

1. Yamanouchi, U.; Yano, M.; Lin, H.X.; Ashikari, M.; Yamada, K. A rice spotted leaf gene, *Spl7*, encodes a heat stress transcription factor protein. *Proc. Natl. Acad. Sci USA* **2002**, *99*, 7530–7535. [CrossRef] [PubMed]
2. Li, X.; Zhu, J.D.; Hu, F.Y.; Ge, S.; Ye, M.Z.; Xiang, H.; Zhang, G.J.; Zheng, X.M.; Zhang, H.Y.; Zhang, S.L.; et al. Single–base resolution maps of cultivated and wild rice methylomes and regulatory roles of DNA methylation in plant gene expression. *BMC Genom.* **2012**, *13*, 300. [CrossRef] [PubMed]
3. Jung, K.H.; Cao, P.; Seo, Y.S.; Dardick, C.; Ronald, P.C. The rice kinase phylogenomics database: A guide for systematic analysis of the rice kinase super–family. *Trends Plant Sci.* **2010**, *15*, 595–599. [CrossRef]
4. Wang, C.; Liu, W.J.; Liao, X.W.; Xu, X.; Yang, S.H.; Zhang, X.B.; Zhou, H.; Zhuang, C.X.; Gong, J.Y.; Wu, J.L. The identification and gene mapping of spotted leaf mutant *spl43* in rice. *Int. J. Mol. Sci.* **2024**, *25*, 6637. [CrossRef]

5. Guo, C.H.; Jung, K.H.; Roberts, S.K.; Mcainsh, M.R.; Hetherington, A.M.; Park, Y.I.; Suh, K.H.; An, G.; Nam, H.G. Mitochondria provide the main source of cytosolic ATP for activation of outward–rectifying k+ channels in mesophyll protoplast of chlorophyll–deficient mutant rice (*OsCHLH*) seedlings. *J. Bio Chem.* **2004**, *279*, 6874–6882.
6. Yang, Y.L.; Xu, J.; Huang, L.C.; Leng, Y.J.; Dai, L.P.; Rao, Y.C.; Chen, L.; Wang, Y.Q.; Tu, Z.J.; Hu, J.; et al. *PGL*, encoding chlorophyllide a oxygenase 1, impacts leaf senescence and indirectly affects grain yield and quality in rice. *J. Exp. Bot.* **2016**, *67*, 1297–1310. [CrossRef]
7. Guo, C.H.; Satoh, K.; Kikuchi, S.; Kim, S.C.; Ko, S.M.; Kang, H.G.; Jeon, J.S.; Kim, C.S.; Park, Y.I. Mitochondrial activity in illuminated leaves of chlorophyll–deficient mutant rice (*OsCHLH*) seedlings. *Plant Biotechnol. Rep.* **2010**, *4*, 281–291.
8. Li, Z.Y.; Mo, W.P.; Jia, L.Q.; Xu, Y.C.; Tang, W.J.; Yang, W.Q.; Guo, Y.L.; Lin, R.C. Rice fluorescent1 is involved in the regulation of chlorophyll. *Plant Cell Physiol.* **2019**, *60*, 2307–2318. [CrossRef]
9. Zhang, Y.; Su, J.B.; Duan, S.; Ao, Y.; Dai, J.R.; Liu, J.; Wang, P.; Li, Y.G.; Liu, B.; Feng, D.R.; et al. A highly efficient rice green tissue protoplast system for transient gene expression and studying light/chloroplast–related processes. *Plant Methods* **2011**, *13*, 30. [CrossRef]
10. Nori, K.; Kazumaru, M.; Nonomura, K.I.; Yukiko, Y.; Yukihiro, I. Rice mutants and genes related to organ development, morphogenesis and physiological traits. *Plan Cell Physiol.* **2005**, *46*, 48–62.
11. Choudhury, F.K.; Rivero, R.M.; Blumwald, E.; Mittler, R. Reactive oxygen species, abiotic stress and stress combination. *Plant J.* **2017**, *90*, 856–867. [CrossRef] [PubMed]
12. Gill, S.S.; Tuteja, N. Reactive oxygen species and antioxidant machinery in abiotic stress tolerance in crop plants. *Plant Physiol. Biochem.* **2010**, *48*, 909–930. [CrossRef] [PubMed]
13. Cao, J.; Jiang, F.; Sodmergen; Cui, K. Time–course of programmed cell death during leaf senescence in *Eucommia ulmoides*. *J. Plant Res.* **2003**, *116*, 7–12. [CrossRef] [PubMed]
14. Bouchez, O.; Huard, C.; Lorrain, S.; Roby, D.; Balagué, C. Ethylene is one of the key elements for cell death and defense response control in the *Arabidopsis* lesion mimic mutant *vad1*. *Plant Physiol.* **2007**, *145*, 465–477. [CrossRef]
15. Zhu, X.B.; Ze, M.; Chern, M.; Chen, X.W.; Wang, J. Deciphering rice lesion mimic mutants to understand molecular network governing plant immunity and growth. *Rice Sci.* **2020**, *27*, 278–288.
16. Keisa, A.; Kanberga–Silina, K.; Nakurte, I.; Kunga, L.; Rostoks, N. Differential disease resistance response in the barley necrotic mutant *nec1*. *BMC Plant Biol.* **2011**, *11*, 66. [CrossRef]
17. Amin, G.M.A.; Kong, K.K.; Sharmin, R.A.; Kong, J.J.; Bhat, J.A.; Zhao, T.J. Characterization and rapid gene–mapping of leaf lesion mimic phenotype of *spl–1* mutant in soybean (*Glycine max* (L.) Merr.). *Int. J. Mol. Sci.* **2019**, *20*, 2193. [CrossRef]
18. Abad, M.S.; Hakimi, S.M.; Kaniewski, W.K.; Rommens, C.M.T.; Shulaev, V.; Lam, E.; Shah, D.M. Characterization of acquired resistance in lesion–mimic transgenic potato expressing bacterio–opsin. *Mol. Plant Microbe Interact.* **1997**, *10*, 635–645. [CrossRef]
19. Kumar, V.; Parkhi, V.; Joshi, S.G.; Christensen, S.; Jayaprakasha, G.K.; Patil, B.S.; Kolomiets, M.V.; Rathore, K.S. A Novel, conditional, lesion mimic phenotype in cotton cotyledons due to the expression of an endochitinase gene from *Trichoderma virens*. *Plant Sci.* **2012**, *183*, 86–95. [CrossRef]
20. Greenberg, J.T.; Guo, A.; Klessig, D.F.; Ausubel, F.M. Programmed cell death in plants: A pathogen–triggered response activated coordinately with multiple defense functions. *Cells* **1994**, *77*, 551–563. [CrossRef]
21. Heath, M.C. Nonhost resistance and nonspecific plant defenses. *Curr. Opin. Plant Biol.* **2000**, *3*, 315–319. [CrossRef] [PubMed]
22. Huang, Q.N.; Yang, Y.; Shi, Y.F.; Chen, J.; Wu, J.L. Spotted–leaf mutants of rice (*Oryza sativa*). *Rice Sci.* **2010**, *17*, 247–256. [CrossRef]
23. Huang, Q.N.; Shi, Y.F.; Yang, Y.; Feng, B.H.; Wei, Y.L.; Chen, J.; Baraoidan, M.; Leung, H.; Wu, J.L. Characterization and genetic analysis of a light– and temperature–sensitive spotted–leaf mutant in rice. *J. Integr. Plant Biol.* **2011**, *53*, 671–681. [CrossRef] [PubMed]
24. Chen, Z.; Chen, T.; Sathe, A.P.; He, Y.; Zhang, X.B.; Wu, J.L. Identification of a novel semi–dominant spotted–leaf mutant with enhanced resistance to *Xanthomonas oryzae* pv. *oryzae* in rice. *Int. J. Mol. Sci.* **2018**, *19*, 3766. [CrossRef]
25. Sathe, A.P.; Su, X.; Chen, Z.; Chen, T.; Wu, J.L. Identification and characterization of a spotted–leaf mutant *spl40* with enhanced bacterial blight resistance in rice. *Rice* **2019**, *12*, 68. [CrossRef]
26. Shang, H.H.; Li, P.P.; Zhang, X.B.; Xu, X.; Gong, J.Y.; Yang, S.H.; He, Y.; Wu, J.L. The gain–of–function mutation, *OsSpl26*, positively regulates plant immunity in rice. *Int. J. Mol. Sci.* **2022**, *23*, 14168. [CrossRef]
27. Tan, J.Y.; Zhang, X.B.; Shang, H.H.; Li, P.P.; Wang, Z.H.; Liao, X.W.; Xu, X.; Yang, S.H.; Gong, J.Y.; Wu, J.L. *ORYZA SATIVA SPOTTED–LEAF 41(OsSPL41)* negatively regulates plant immunity in rice. *Rice Sci.* **2023**, *30*, 426–440.
28. Hirano, K.; Aya, K.; Kondo, M.; Okuno, A.; Morinaka, Y.; Matsuoka, M. *OsCAD2* is the major cad gene responsible for monolignol biosynthesis in rice culm. *Plant Cell Rep.* **2012**, *31*, 91–101. [CrossRef]
29. Bansal, R.; Rana, N.; Singh, A.; Dhiman, P.; Mandlik, R.; Sonah, H.; Deshmukh, R.; Sharma, T.R. Evolutionary Understanding of Metacaspase Genes in Cultivated and Wild Oryza Species and Its Role in Disease Resistance Mechanism in Rice. *Genes* **2020**, *11*, 1412. [CrossRef]
30. Walbot, V.; Hoisington, D.A.; Neuffer, M.G. *Disease Lesion Mimic Mutations*; Springer: Boston, MA, USA, 1983.

31. Wu, C.; Bordeos, A.; Madamba, M.R.S.; Baraoidan, M.; Ramos, M.; Wang, G.L.; Leach, J.E.; Leung, H. Rice lesion mimic mutants with enhanced resistance to diseases. *Mol. Genet. Genom.* **2008**, *279*, 605–619. [CrossRef]
32. Ma, J.; Wang, Y.F.; Ma, X.D.; Meng, L.Z.; Jing, R.N.; Wang, F.; Wang, S.; Cheng, Z.J.; Zhang, X.; Jiang, L.; et al. Disruption of gene *SPL35*, encoding a novel CUE domain–containing protein, leads to cell death and enhanced disease response in rice. *Plant Biotechnol. J.* **2019**, *17*, 1679–1693. [CrossRef]
33. Zhang, Y.; Liu, Q.N.; Zhang, Y.X.; Chen, Y.Y.; Yu, N.; Cao, Y.R.; Zhan, X.D.; Cheng, S.H.; Cao, L.Y. *LMM24* encodes receptor–like cytoplasmic kinase 109, which regulates cell death and defense responses in rice. *Int. J. Mol. Sci.* **2019**, *20*, 3243. [CrossRef] [PubMed]
34. Chen, T.; Chen, Z.; Sathe, A.P.; Zhang, Z.H.; Li, L.; Shang, H.H.; Tang, S.; Zhang, X.; Wu, J.L. Characterization of a novel gain-of-function spotted-leaf mutant with enhanced disease resistance in rice. *Rice Sci.* **2019**, *26*, 372–383.
35. Kampire, M.G.; Sanglou, R.K.; Wang, H.; Kazeem, B.B.; Wu, J.L.; Zhang, X.B. A novel allele encoding 7–hydroxymethyl chlorophyll a reductase confers bacterial blight resistance in rice. *Multidiscip. Digit. Publ. Inst.* **2021**, *22*, 7585. [CrossRef]
36. Wang, J.; Ye, B.Q.; Yin, J.J.; Yuan, C.; Zhou, X.G.; Li, W.T.; He, M.; Wang, J.C.; Chen, W.L.; Qin, P.; et al. Characterization and fine mapping of a light dependent leaf lesion mimic mutant 1 in rice. *Plant Physiol. Biochem.* **2015**, *97*, 44–51. [CrossRef] [PubMed]
37. Xiao, G.Q.; Zhang, H.W.; Lu, X.Y.; Huang, R.F. Characterization and mapping of a novel light–dependent lesion mimic mutant lmm6 in rice (*Oryza sativa* L.). *J. Integr. Agric.* **2015**, *14*, 1687–1696. [CrossRef]
38. Feng, B.H.; Yang, Y.; Shi, Y.F.; Shen, H.C.; Wang, H.M.; Huang, Q.N.; Xu, X.; Lü, X.G.; Wu, J.L. Characterization and genetic analysis of a novel rice spotted–leaf mutant *HM47* with broad–spectrum resistance to *Xanthomonas oryzae* pv. *oryzae*. *J. Integr. Plant Biol.* **2013**, *55*, 473–483. [CrossRef]
39. Zurbriggen, M.D.; Carrillo, N.; Hajirezaei, M.R. ROS signaling in the hypersensitive response. *Plant Signal. Behav.* **2010**, *5*, 393–396. [CrossRef]
40. Jacks, T.J.; Davidonis, G.H. Superoxide, hydrogen peroxide, and the respiratory burst of fungally infected plant cells. *Mol. Cell. Biochem.* **1979**, *158*, 77–79. [CrossRef]
41. Dauwe, R.; Morreel, K.; Goeminne, G.; Gielen, B.; Rohde, A.; Van, B.J.; Ralph, J.; Boudet, A.M.; Kopka, J.; Rochange, S.; et al. Molecular phenotyping of lignin–modified tobacco reveals associated changes in cell–wall metabolism, primary metabolism, stress metabolism and photorespiration. *Plant J.* **2007**, *52*, 263–285. [CrossRef]
42. Chun, H.J.; Baek, D.; Cho, H.M.; Lee, S.H.; Jin, B.J.; Yun, D.J.; Hong, Y.S.; Kim, M.C. Lignin biosynthesis genes play critical roles in the adaptation of *Arabidopsis* plants to high–salt stress. *Plant Signal. Behav.* **2019**, *14*, 1625697. [CrossRef] [PubMed]
43. Zhao, D.Q.; Luan, Y.T.; Shi, W.B.; Zhang, X.Y.; Meng, J.S.; Tao, J. A Paeonia ostii caffeoyl–CoA *O*–methyltransferase confers drought stress tolerance by promoting lignin synthesis and ROS scavenging. *Plant Sci.* **2021**, *303*, 110765. [CrossRef] [PubMed]
44. Zhang, X.; Chen, B.W.; Wang, L.S.; Ali, S.H.; Guo, Y.L.; Liu, J.X.; Wang, J.; Xie, L.N.; Zhang, Q.Z. Genome–wide identification and characterization of caffeic acid *O*–methyltransferase gene family in soybean. *Plants* **2021**, *10*, 2816. [CrossRef] [PubMed]
45. Lam, P.Y.; Tobimastsu, Y.; Matsumoto, N.; Suzuki, S.; Lan, W.; Takeda, Y.; Yamamura, M.; Sakamoto, M.; Ralph, J.; Clive, L.; et al. *OsCAldOMT1* is a bifunctional *O*–methyltransferase involved in the biosynthesis of tricin–lignins in rice cell walls. *Sci. Rep.* **2019**, *9*, 11597. [CrossRef] [PubMed]
46. Liang, S.M.; Xu, S.B.; Qu, D.; Yang, L.M.; Wang, J.G.; Liu, H.L.; Xin, W.; Zou, D.T.; Zheng, H.L. Identification and functional analysis of the caffeic acid *O*–methyltransferase (*COMT*) gene family in rice (*Oryza sativa* L.). *Int. J. Mol. Sci.* **2022**, *23*, 8491. [CrossRef]
47. Lin, F.Q.; Yamano, G.; Hasegawa, M.F.; Anzai, H.; Kawasaki, S.; Kodama, O. Cloning and functional analysis of caffeic acid 3-*O*-methyltransferase from rice (*Oryza sativa*). *J. Pestic. Sci.* **2006**, *31*, 47–53. [CrossRef]
48. Taichi, K.; Norie, H.; Mai, M.; Masaomi, Y.; Takefumi, H.; Shiro, S.; Masahiro, S.; Toshiaki, U. Characterization of 5–hydroxyconiferaldehyde *O*–methyltransferase in *Oryza sativa*. *Plant Tissue Cult. Lett.* **2013**, *30*, 157–167.
49. Huangfu, L.X.; Chen, R.J.; Lu, Y.; Zhang, E.Y.; Miao, J.; Zuo, Z.H.; Zhao, Y.; Zhu, M.Y.; Zhang, Z.H.; Li, P.C.; et al. *OsCOMT*, encoding a caffeic acid O–methyltransferase in melatonin biosynthesis, increases rice grain yield through dual regulation of leaf senescence and vascular development. *Plant Biotechnol. J.* **2022**, *20*, 1122–1139. [CrossRef]
50. Kaurilind, E.; Xu, E.; Brosché, M. A genetic framework for H_2O_2 induced cell death in *Arabidopsis thaliana*. *BMC Genom.* **2015**, *16*, 837. [CrossRef]
51. Qiao, Y.L.; Jiang, W.Z.; Lee, J.; Park, B.; Choi, M.S.; Piao, R.H.; Woo, M.O.; Roh, J.H.; Han, L.Z.; Paek, N.C.; et al. *SPL28* encodes a clathrin–associated adaptor protein complex 1, medium subunit micro 1 (AP1M1) and is responsible for spotted leaf and early senescence in rice (*Oryza sativa*). *New Phytol.* **2010**, *185*, 258–274. [CrossRef]
52. Wang, L.L.; Chen, Y.Y.; Guo, L.; Zhang, H.W.; Zhuang, J.Y. Dissection of *qTGW1.2* to three QTLs for grain weight and grain size in rice (*Oryza sativa* L.). *Euphytica* **2015**, *202*, 119–127. [CrossRef]
53. Jones, J.D.G.; Dangl, J.L. The plant immune system. *Nature* **2006**, *444*, 323–329. [CrossRef] [PubMed]
54. Huot, B.; Yao, J.; Montgomery, B.L.; He, S.Y. Growth–defense tradeoffs in plants: A balancing act to optimize fitness. *Mol. Plant* **2014**, *7*, 1267. [CrossRef]
55. Walters, D.R.; Heil, M. Costs and trade–offs associated with induced resistance. *Physiol. Mol. Plant Pathol.* **2007**, *71*, 3–17. [CrossRef]

56. Arnon, D.I. Copper enzymes in isolated chloroplasts. Polyphenoloxidase in *Beta vulgaris*. *Plant Physiol.* **1949**, *24*, 1–15. [CrossRef] [PubMed]
57. Wellburn, A.R. The spectral determination of chlorophyll a and b, as well as total carotenoids, using various solvents with spectrophotometers of different resolution. *Plant Physiol.* **1994**, *144*, 307–313. [CrossRef]
58. Yin, Z.C.; Chen, J.; Zeng, L.R.; Goh, M.; Leung, H.; Khush, G.S.; Wang, G.L. Characterizing rice lesion mimic mutants and identifying a mutant with broad–spectrum resistance to rice blast and bacterial blight. *Mol. Plant Microbe Interact.* **2000**, *13*, 869–876. [CrossRef]
59. Thordal–Christensen, H.; Zhang, Z.G.; Wei, Y.D.; Collinge, D.B. Subcellular localization of H_2O_2 in plants.H_2O_2 accumulation in papillae and hypersensitive response during the barley–powdery mildew interaction. *Plant J.* **1997**, *11*, 1187–1194. [CrossRef]
60. Rogers, S.O.; Bendich, A.J. Extraction of DNA from milligram amounts of fresh, herbarium and mummified plant tissues. *Plant Mol. Biol.* **1985**, *5*, 69–76. [CrossRef]
61. Hiei, Y.; Ohta, S.; Komari, T.; Kumashiro, T. Efficient transformation of rice (*Oryza sativa* L.) mediated by *Agrobacterium* and sequence analysis of the boundaries of the T–DNA. *Plant J.* **1994**, *6*, 271–282. [CrossRef]
62. Ma, X.L.; Zhang, Q.Y.; Zhu, Q.L.; Liu, W.; Chen, Y.; Qiu, R.; Wang, B.; Yang, Z.F.; Li, H.Y.; Lin, Y.R.; et al. A robust CRISPR/Cas9 system for convenient, high–efficiency multiplex genome editing in monocot and dicot plants. *Mol. Plant* **2015**, *8*, 1274–1284. [CrossRef] [PubMed]

International Journal of
Molecular Sciences

MDPI

Article

Study on the Function of *SlWRKY80* in Tomato Defense against *Meloidogyne incognita*

Yinxia Chen [1], Zhize Wang [1], Weidan Nie [1], Tingjie Zhao [1], Yule Dang [1], Chenghao Feng [1], Lili Liu [1], Chaonan Wang [1] and Chong Du [1,2,*]

[1] College of Horticulture, Xinjiang Agricultural University, Urumqi 830052, China; cyx61222@163.com (Y.C.); werick201026@163.com (Z.W.); 13779334686@163.com (W.N.); 13369647866@163.com (T.Z.); 15029523014@163.com (Y.D.); fch20220515@163.com (C.F.); liulili4477@163.com (L.L.); wcn0107@126.com (C.W.)

[2] Postdoctoral Station of Horticulture, Xinjiang Agricultural University, Urumqi 830052, China

* Correspondence: godv2018@163.com; Tel.: +86-0991-8762357

Abstract: WRKY transcription factors (TFs) can participate in plant biological stress responses and play important roles. *SlWRKY80* was found to be differentially expressed in the *Mi-1*- and *Mi-3*-resistant tomato lines by RNA-seq and may serve as a key node for disease resistance regulation. This study used RNAi to determine whether *SlWRKY80* silencing could influence the sensitivity of 'M82' (*mi-1/mi-1*)-susceptible lines to *M. incognita*. Further overexpression of this gene revealed a significant increase in tomato disease resistance, ranging from highly susceptible to susceptible, combined with the identification of growth (plant height, stem diameter, and leaf area) and physiological (soluble sugars and proteins; root activity) indicators, clarifying the role of *SlWRKY80* as a positive regulatory factor in tomato defense against *M. incognita*. Based on this phenomenon, a preliminary exploration of its metabolic signals revealed that *SlWRKY80* stimulates different degrees of signaling, such as salicylic acid (SA), jasmonic acid (JA), and ethylene (ETH), and may synergistically regulate reactive oxygen species (ROS) accumulation and scavenging enzyme activity, hindering the formation of feeding sites and ultimately leading to the reduction of root gall growth. To our knowledge, *SlWRKY80* has an extremely high utilization value for improving tomato resistance to root-knot nematodes and breeding.

Keywords: *SlWRKY80*; *Meloidogyne incognita*; tomato; regulation; disease resistance

1. Introduction

Plant parasitic nematodes (PPNs) account for approximately 10% of the nematode population, and the global economic loss caused by PPNs can reach USD 157 billion per year, making them one of the most important disease agents and pests in agricultural production [1]. Root-knot nematodes (RKNs) rank first among the PPNs in terms of damage caused to agriculture [2]. The common *Meloidogyne* spp. include *M. incognita*, *M. arenaria*, *M. hapla*, and *M. javanica*, among which *M. incognita* is a particularly destructive dominant pathogenic nematode [3] and is widely distributed among a wide variety of hosts [4]. Tomato, a major cash crop worldwide, is the typical host of RKNs. During nematode infection, J2, which represents the infectious state [5], invades the host from the root tip, migrates to the vascular bundle through the intercellular space, searches for suitable feeding cells, absorbs water and nutrients from giant cells through their mouth needle, and eventually induces the occurrence of large root galls [6].

The main genes involved in resistance to RKNs in tomatoes are those of the *Mi* gene family, whose members originated from *S. peruvianum* [7]. The incompatibility of distant hybridization makes it difficult to construct populations and clones. As a result, other members of this family, except *Mi-1*, have not been fully explored [8]. Although *Mi-1* can mediate resistance and produce a hypersensitive response (HR), resistance is regulated by a variety of biological factors, including important WRKY TFs [9]. The N-terminus of the DNA-binding domain of WRKYs contains a conserved heptapeptide sequence, "WRKYGQK", resulting in WRKYs participating in complex biological pathways by specifically binding to W-box elements within the promoter region of the targets [10]. Based on these findings, WRKYs can be modified by phosphorylation of MAPK cascade activation signals to participate in pathways associated with important disease resistance signals, including SA, JA, and ROS [11].

Previous studies have shown that endogenous SA is synthesized by the isochlorate (IC) and phenylalanine ammonia lyase (PAL) pathways [12]. For example, in *Arabidopsis*, WRKY28 directly binds to the IC synthase 1 (*ICS1*) gene promoter to activate expression and mediate SA synthesis [13]. A newly identified WRKY, BnaWSR1ca, binds directly to *ICS1* and promotes SA production, thereby controlling cell death [14]. In addition, WRKYs can directly or indirectly activate *nonexpressers of the pathogenesis-related gene* (*NPR*), tgaccg-binding factor (*TGA*) and pathogenesis-associated protein 1 (*PR1*), which are involved in SA-induced systemic acquired resistance (SAR) [15]. For instance, the interaction between CmWRKY15-1 and CmNPR1 in *Chrysanthemum morifolium* may activate the expression of the downstream *PR1/2/10* gene, thereby enhancing its resistance to *Puccinia horiana* [16]. WRKY53 can interact with TGA3 and rely on *NPR1* to activate the *CmYLCV* promoter and mediate SA signal production to improve disease resistance [17]. In contrast, *GhWRKY70* regulates the accumulation of SA signals by upregulating the expression of the *PR1* and *NPR1* genes and is involved as a negative regulator of the defense response against *Verticillium dahliae* in cotton [18]. WRKYs can also bind to the ZIM domain of jasmonic acid (JAZ) to regulate the accumulation of JA signals and mediate the resistance process [19]. In the defense response to *Phytophthora sojae*, GmWRKY40 can interact with the JAZ2/4/5/6/7/8 proteins and mediate the production of induced systemic resistance (ISR) by regulating the accumulation of H_2O_2 and the activity of the JA signaling cascade [20]. PnWRKY9 activates the defense of *Panax notoginseng* against *Fusarium solani* by binding to the promoter of the JA reactive defense gene *PnDEFL1* [21]. In contrast, OsWRKY72 in rice can directly bind the promoter of the JA biosynthesis gene *AOS1* and negatively regulate JA synthesis and resistance to *Xanthomonas oryzae* [22]. ROS often function as marker signals in the SA/JA cascade, and programmed cell death (PCD) mediated by ROS is the main factor involved in the defense of many plants against pathogens via HR [23,24].

In *S. lycopersicum*, researchers also explored the WRKYs' involvement in defense against RKNs. WRKY72 TFs (WRKY72, WRKY73, and WRKY74) all actively regulate the level of resistance to RKNs and potato aphids through *Mi-1*-mediated effector-triggered immunity (ETI) processes. Among them, the direct homolog *AtWRKY72* utilizes SA-independent defense mechanisms [25]. Previous studies have found that overexpression of *SlWRKY45* can decrease the expression levels of marker genes (*PR-1* and *Pin2*) in JA and SA signals. Further studies have found that *SlWRKY45* can specifically bind to the promoter of the JA synthesis gene *SlAOC* and inhibit its expression. In addition, the susceptibility of tomato plants to RKNs was enhanced [26,27]. Another study also found that *SlWRKY3* and *SlWRKY35* were significantly induced at the tomato root feeding sites after RKN infection, and *SlWRKY3* responded positively to the SA signal. Overexpression and knockout of *SlWRKY3* indicated that it positively regulated resistance to *M. javanica* [28].

In summary, WRKYs, as a very important family of TFs, play important roles in the process of disease resistance and in regulating resistance levels together with SA, JA, and ROS signaling. The above studies have shown that WRKYs are very necessary to regulate the resistance of tomatoes to RKNs. Based on the RNA-seq results for different resistance materials ('Motelle' (*Mi-1/Mi-1*) and 'LA3858' (*Mi-3/Mi-3*)) from the previous period, *SlWRKY80* was found to be differentially expressed according to both sets of sequencing results ($FC_{(\text{'Motelle'})} = 4.25$, SRA: SRP502510, $FC_{(\text{'LA3858'})} = 3.82$, SRA: SRP355506) (Figure S1). Combined with the research of its highly homologous gene *SlWRKY70* [29], the results of a qPCR study showed that the expression level of *SlWRKY70* in the root was significantly upregulated after RKN infection, and exogenous SA could induce the accumulation of this gene. RNAi further indicated that *SlWRKY70* positively regulates the resistance level in the process of *Mi-1*-mediated resistance. Therefore, *SlWRKY80* may be involved in disease resistance to RKNs as a key regulatory node. In this study, *SlWRKY80* was selected as the research target. Through functional identification and statistical analysis of the growth and physiological indices and accumulation levels of various resistance signals, the possible disease resistance regulatory pathway of *SlWRKY80* was preliminarily identified, laying a solid foundation for further exploration of the molecular mechanism of this valuable gene.

2. Results

2.1. Silencing of SlWRKY80 Increased the Sensitivity of Tomato Plants to RKNs

The virus-induced gene silencing (VIGS) strategy was used to preliminarily verify the regulatory function of *SlWRKY80*. In a previous study, *SlWRKY80* was silenced in resistant tomato plants 'Motelle', and *SlWRKY80* was shown to act as a positive regulator to significantly reduce *Mi-1*-mediated resistance [30]. In this study, silencing *SlWRKY80* in susceptible tomato plants 'M82' again verified the directionality of its regulatory feedback. Therefore, 'M82'-silenced individuals with a greater than 50% decrease in *SlWRKY80* expression were obtained and inoculated with *M. incognita* to detect disease resistance (Figure 1A). The results showed that, compared with those in plants not loaded with *pTRV2* and CK ('M82'), the number of galls on the roots of 'M82' individuals after silencing increased, although the disease index (DI) and resistance level (RL) did not change significantly. However, overall disease susceptibility was enhanced by *SlWRKY80* silencing (Figure 1B).

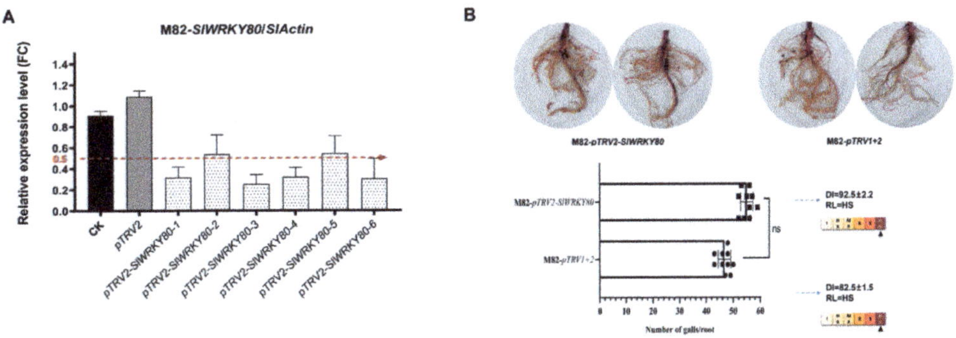

Figure 1. The function of *SlWRKY80* in the defense against RKNs was preliminarily verified by RNAi. (**A**) Silencing efficiency test (partial individuals), using 'M82' as controls (CK), was performed for RT-qPCR detection of *SlWRKY80* in both empty and silenced strains. The red dotted line represents a decrease in relative expression to 50%; (**B**) 'M82' were inoculated with J2s after silencing *SlWRKY80*. Statistics on gall numbers: the results are presented as the means ± SEM (n = 10, $p < 0.05$), through one-way ANOVA statistics. DI calculation and RL of the silenced and unloaded *pTRV2* strains were identified.

2.2. Identification of Disease Resistance in the Overexpression Lines

SlWRKY80 was genetically transformed into the susceptible tomato variety 'M82' as a receptor material, and 17 lines were obtained; of these, eight lines were successfully transformed by PCR (Figure 2A). The positive plants were identified by RT-qPCR, and the plants with relatively high *SlWRKY80* expression were selected (for which the differential expression ratio was more than 4, OE-2, OE-9, OE-15, and OE-17) for T_1 generation reproduction for the resistance identification test (Figure 2B,C).

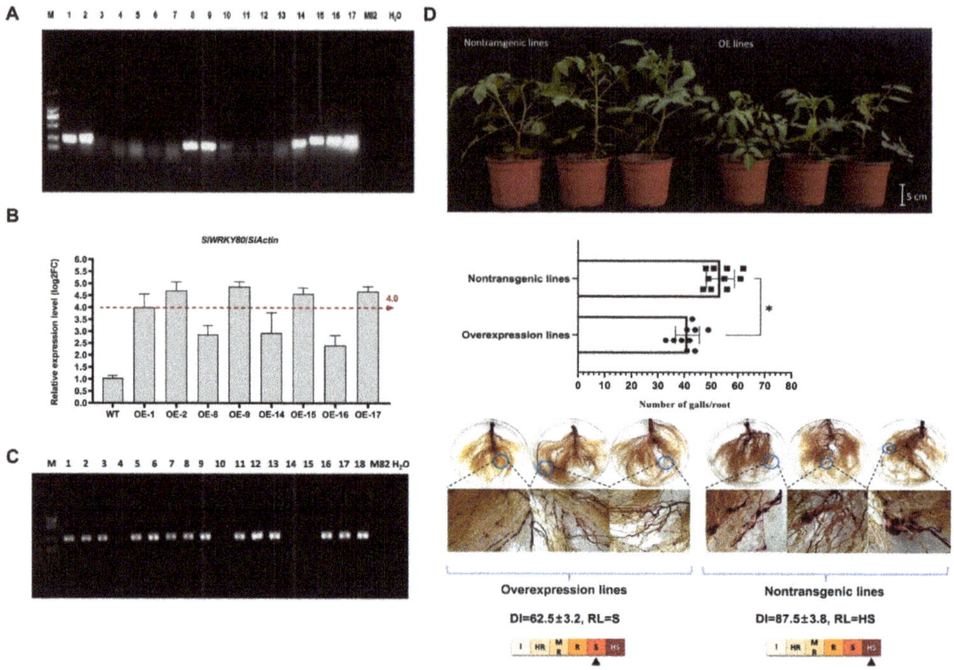

Figure 2. Identification of disease resistance in 'M82' after overexpression of *SlWRKY80*. (**A**) Positive identification of 1–17 transgenic lines. Marker: 2000 bp, the target band size is 249 bp. Among them, 1, 2, 8, 9, and 14–17 were positive lines. (**B**) The expression of *SlWRKY80* in positive plants was detected by RT-qPCR. The red dotted line indicates that the difference is more than 4 times expressed; (**C**) OE-2, OE-9, OE-15, and OE-17 transgenic plants were screened for T_1 generation propagation. Marker: 2000 bp, the target band size is 249 bp. (**D**) After overexpressing *SlWRKY80*, individuals were inoculated with 1000 J2s. After 40 d, plant morphology and root gall morphology were observed, DI was counted, and RL was identified. Root gall statistics are presented as the means ± SEM (n = 10, * $p < 0.05$), through one-way ANOVA statistics.

The test plants were inoculated with *M. incognita* at 1000 J2s per plant, after which the roots were collected and the DI was calculated (Figure 2D). The results showed that the number of root galls in overexpression plants was significantly lower than those in nontransgenic plants, and a small number of small galls with short diameters between 1 and 2 mm were distributed in the root systems of those. The diseased area accounted for 20% to 35% of the total root system. The most nontransgenic plant roots showed dense and large galls with a short diameter of approximately 5 mm. The affected area accounted for 40% to 50% of the total root system. The root-knot level of most of the overexpression plants reached 1–3, and the RL was S (DI = 62.5). The root-knot level of most of the nontransgenic plants reached 3–4, and the RL was HS (DI = 87.5). In summary, overexpression of *SlWRKY80* is beneficial for reducing the sensitivity of tomato plants to *M. incognita*.

2.3. Determination of the Growth Indices of the Overexpression Lines after Inoculation

The plant height, stem diameter, and leaf area of *SlWRKY80*-overexpressing T_1 plants and nontransgenic plants were measured after inoculation with *M. incognita*. The results showed that the plant height of overexpression lines increased slowly after inoculation, while that of nontransgenic lines increased rapidly, being significantly higher than that of overexpression tomatoes at each measurement, and 1.88 times higher than that of overexpression lines at 28 days (Figure 3A). The stem diameter of the overexpression and nontransgenic lines increased slowly, and the stem diameter of the overexpression lines was always greater than that of the nontransgenic lines. There was a significant difference between the two groups at 8 d, 16 d, and 24 d after inoculation, but the stem diameters of the two treatment groups gradually became similar overall (Figure 3B). A similar trend was observed for leaf area, as there was no significant difference at each treatment stage (Figure 3C). Obviously, after *SlWRKY80* overexpression, the overall disease resistance level of the plants was improved. In particular, the plants did not show an excessive growth trend, and the relatively healthy growth state was maintained.

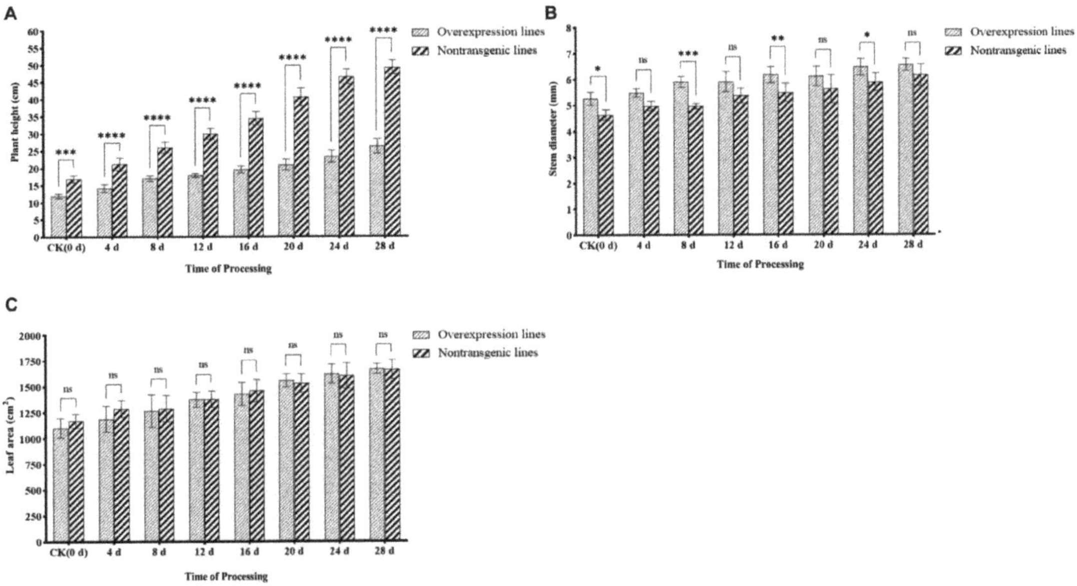

Figure 3. Determination of growth index of overexpression lines after inoculation. The horizontal coordinate represents the processing time of 0–28 d, and the vertical coordinate represents (**A**) plant height. The difference multiples of nontransgenic lines and overexpression lines before and after inoculation at 4 d, 8 d, 12 d, 16 d, 20 d, 24 d, and 28 d were, respectively, 1.42, 1.49, 1.53, 1.67, 1.76, 1.94, 1.99, and 1.87 times; (**B**) stem diameter; (**C**) leaf area. The results are presented as the means ± SEM (n = 3, * $p < 0.05$, ** $p < 0.01$, *** $p < 0.001$ and **** $p < 0.0001$), through one-way ANOVA statistics.

2.4. Determination of the Physiological Indices of the Overexpression Lines after Inoculation

Soluble sugar, protein, and root activity can reflect the growth state of the plant. After overexpression and nontransgenic tomatoes were inoculated, the soluble protein content exhibited a continuous trend of first increasing and then decreasing (Figure 4A). In the first 20 days after inoculation, there was no significant difference in the content of soluble protein in the overexpression lines, while in the nontransgenic lines, there was still no significant change at all stages. The overall trend of soluble protein content in overexpression lines was always higher than that in nontransgenic lines. In addition, at 4 d and 16 d, there was a significant difference between the two treatments. At 28 d, the protein content of the overexpression plants was 3.27 Ug/g FW, 2.44 times that of the nontransgenic lines.

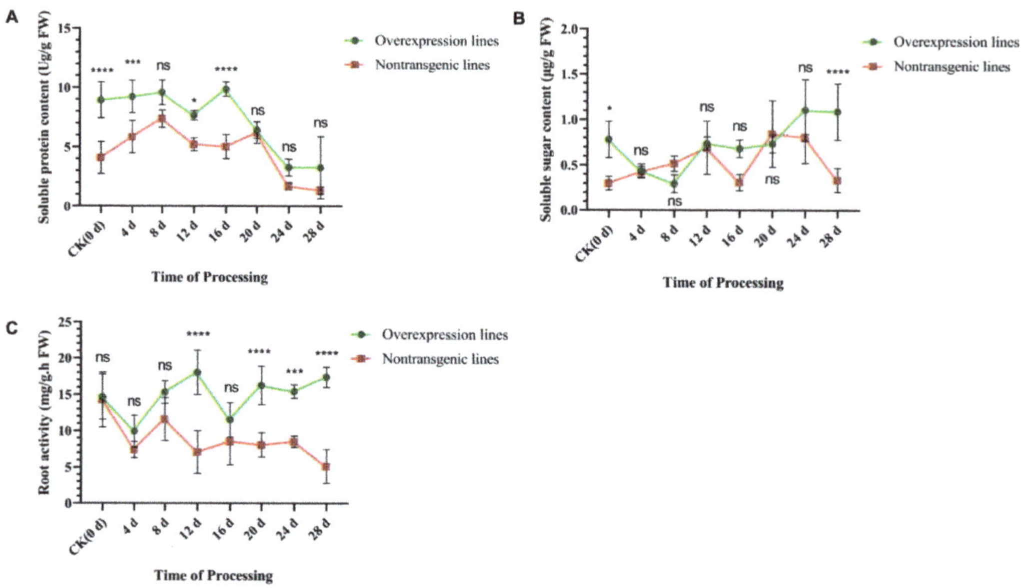

Figure 4. Determination of physiological index of overexpression lines after inoculation. The horizontal coordinate represents the processing time 0–28 d, and the vertical coordinate represents (**A**) soluble protein; (**B**) soluble sugar; (**C**) root activity. The lines of different colors represent significant differences in comparison with CK (0 d) at different stages after inoculation in the two treated lines, respectively. The results are presented as the means ± SEM (n = 3, * $p < 0.05$, *** $p < 0.001$ and **** $p < 0.0001$), through one-way ANOVA statistics.

The soluble sugar content of the overexpression lines generally decreased first and then increased (Figure 4B), with a minimum value of 0.3 μg/g FW occurring at 8 d after inoculation, after which it reached the highest level at 24 d (1.16 μg/g FW), which was 1.43 times greater than the same stage of the nontransgenic plants. At 28 d, the concentration was reduced to 1.09 μg/g FW, but compared with the nontransgenic plants, the content showed a significant difference, being 3.21 times higher. The soluble sugar content of the nontransgenic lines showed an overall increasing trend, reaching the highest value at 20 d (0.84 μg/g FW), then decreasing, and finally dropping to 0.34 μg/g FW at 28 d.

Root activity can directly reflect the degree of infestation and destruction from RKNs. After inoculation (Figure 4C), the activity in the roots of the overexpression lines was greater than that in the roots of the nontransgenic lines at all stages and reached the highest content of 18.05 mg/g/h FW at 12 d, which was 2.55 times greater than in the nontransgenic plants. At 28 d, the root activity of the overexpression plants was 17.39 mg/g FW, which was 3.38 times greater than that of the nontransgenic plants. However, the activity of the roots of the nontransgenic plants tended to decrease and was only 5.14 mg/g FW at 28 d. In conclusion, after overexpression of *SlWRKY80*, the physiological indices maintained higher levels than those of the WT, especially regarding better root activity, which reflects the disease resistance ability of plants, while maintaining better nutrient absorption and plant growth state.

2.5. Measurement of Endogenous Hormone Accumulation in SlWRKY80-Overexpressing Plants after Inoculation

By identifying the accumulation levels of endogenous SA, JA, ETH, and abscisic acid (ABA) in the overexpression and nontransgenic lines at 0–48 h after inoculation, it was found that the accumulation levels of the four hormones varied slightly in the nontransgenic lines (Figure 5). After 12 h in the overexpression lines, endogenous SA,

JA, and ETH biosynthesis was rapidly induced, and at 24 h, the levels of both SA and ETH were significantly greater than those in the nontransgenic lines, with increases of 50.2% and 24.6%, respectively. SA was found to accumulate, reaching 89.87 ng/g at 48 h in the overexpression plants, which was extremely significantly greater than in the nontransgenic plants. The ETH signal rapidly peaked (0.44 pmol/g) at 36 h, and at 48 h, it decreased to 0.42 pmol/g, which was still extremely significantly greater than in the nontransgenic plants. JA levels peaked at 36 h (21.87 pmol/g) in the overexpression plants and were approximately twice as high as in the nontransgenic plants. Afterward, the accumulation level decreased, and at 48 h, it decreased to 18.08 pmol/g but was still extremely significantly greater than in nontransgenic plants. ABA signaling was induced and continuously increased 24 h after the inoculation of the overexpression lines, whose expression levels were extremely significantly greater than those in the nontransgenic lines at 36 h and reached 765.96 ng/g at 48 h. The above results demonstrated that *SlWRKY80* might mediate important signals, such as SA and JA, to induce defense responses against RKN invasion.

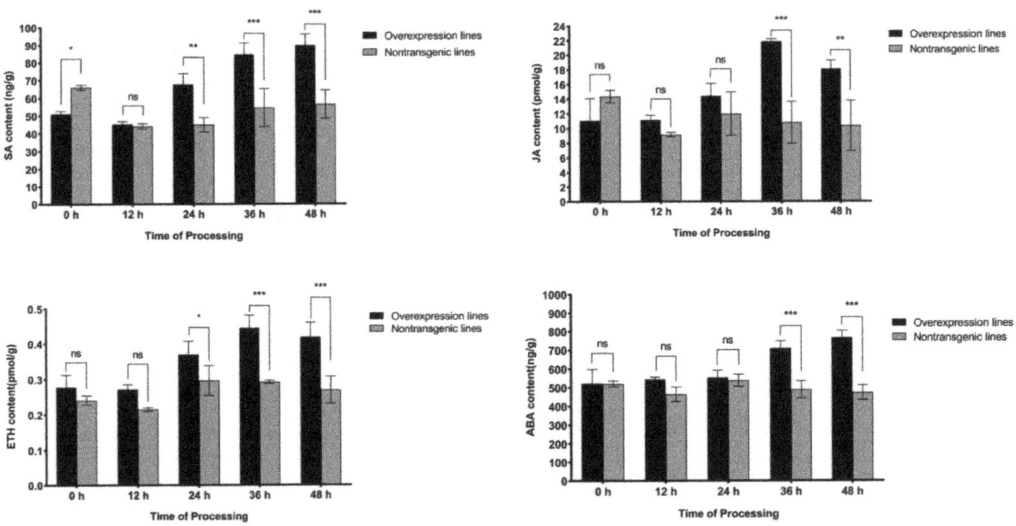

Figure 5. Statistics of SA, JA, ETH, and ABA accumulation in overexpression lines. The horizontal coordinate represents the processing time of 0–48 h, and the vertical coordinate represents the content of SA, JA, ETH, and ABA. The results are presented as the means ± SEM (n = 3, * $p < 0.05$, ** $p < 0.01$ and *** $p < 0.001$), through one-way ANOVA statistics.

2.6. Determination of ROS System Levels in SlWRKY80-Overexpressing Plants after Inoculation

ROS is a marker signal indicating the degree of stress in plants, so the accumulation trend of ROS content is particularly important for the evaluation of disease resistance. Research has shown that the overall metabolism of ROS in overexpressing plants after inoculation with *M. incognita* was induced, with a significantly greater level at 24 h than in nontransgenic plants, and the level of ROS eventually reached 1495.62 ng/g at 48 h. H_2O_2 is the main form of ROS that plays a role in plants. In this study, at 12 h, the H_2O_2 content in the overexpression lines rapidly increased, and at this point, there was a significant difference compared to the nontransgenic plants. After 36 h, the accumulation of H_2O_2 tended to stabilize, reaching 63.61 μmol/g at 48 h, but was still extremely significantly greater than in nontransgenic plants (Figure 6).

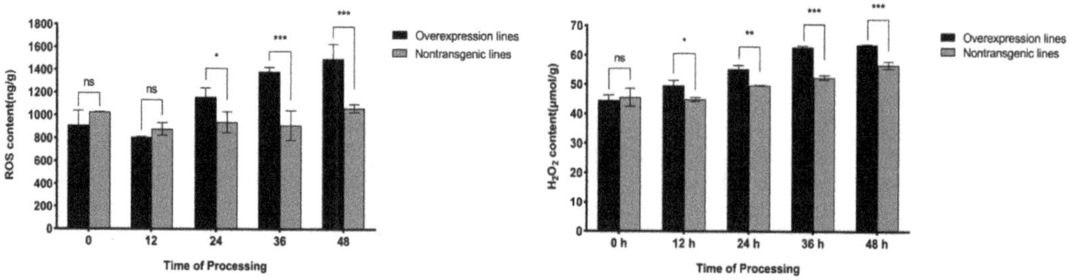

Figure 6. Statistics of the ROS system in overexpression lines. The horizontal coordinate represents the processing time of 0–48 h, and the vertical coordinate represents the content of total ROS and H_2O_2. The results are presented as the means ± SEM (n = 3, * $p < 0.05$, ** $p < 0.01$ and *** $p < 0.001$), through one-way ANOVA statistics.

ROS-scavenging enzymes play an important role in regulating the levels of ROS (H_2O_2). The determination of SOD, CAT, and POD (Figure 7) activities revealed that POD activity in the overexpression lines was rapidly induced at 24 h after inoculation, followed by a continuous increase in activity, reaching 3.14 U/g at 48 h, which was approximately 1.8 times greater than in the nontransgenic plants. In contrast, the activities of SOD and CAT showed a downward trend after inoculation; both presented extremely significantly lower activity than that of nontransgenic lines at 36 and 48 h, and at the 48 h stage, the activity of the two enzymes decreased by 40.3% and 43.2%, respectively, compared to the nontransgenic plants.

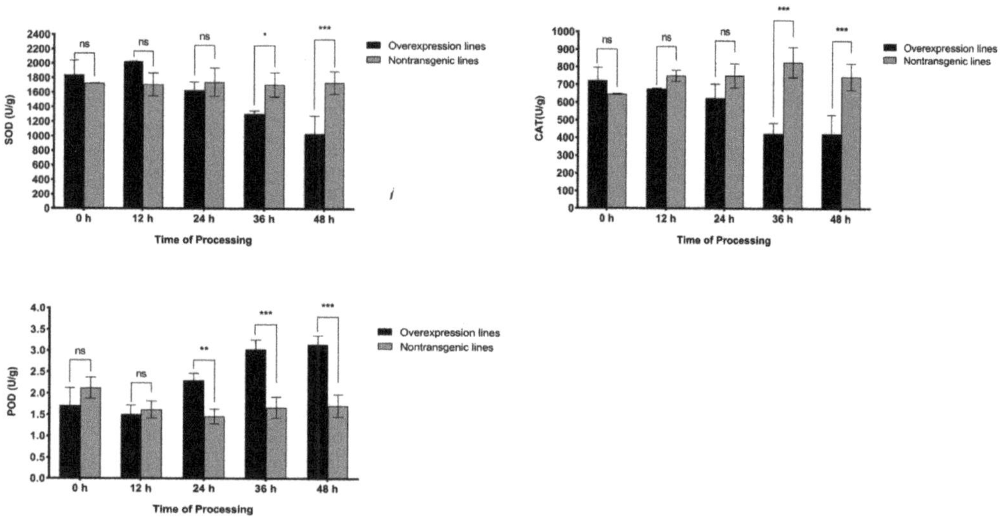

Figure 7. Statistics of the ROS system in overexpression lines. The horizontal coordinate represents the processing time of 0–48 h, and the vertical coordinate represents the activity of POD, SOD, and CAT. The results are presented as the means ± SEM (n = 3, * $p < 0.05$, ** $p < 0.01$ and *** $p < 0.001$), through one-way ANOVA statistics.

3. Discussion

WRKY TFs are widely involved in plant biotic stress responses. This study was based on RNA-seq data from different disease-resistant tomato plants (*Mi-1* and *Mi-3*), and *SlWRKY80*, which was differentially expressed, was used as the research object. Furthermore, RNAi and overexpression strategies were utilized to determine the function of *SlWRKY80* in tomato plants. By measuring the degree of change in disease resistance, it was clarified that *SlWRKY80* can serve as a positive regulatory factor in tomato defense against RKNs, which is clearer in terms of growth and physiological indicators. The dynamic change of soluble sugars and proteins in the *SlWRKY80*-overexpressing lines was overall greater than in the nontransgenic lines, which proved that the overall health of the *SlWRKY80*-overexpressing plants was relatively better [31]. Root activity can directly reflect the degree of root injury. The root activity of the overexpression lines remained high and relatively stable throughout the whole inoculation period, while the root activity of the nontransgenic lines was not only low but also exhibited a downward trend. A decrease in root activity over time made it difficult for the plants to absorb water and nutrients, eventually leading to a continuous abnormal increase in the plant height of the nontransgenic lines, and severe excessive growth [32].

During the disease resistance process, defense-related hormones such as SA and JA and other signals are often activated, leading to the production of HR in plants [33]. SA participates in SAR in plants [34], while JA and ETH are involved mainly in the ISR process of combating necrotrophic pathogen infection [35]. Previous studies have shown that overexpression of *CmWRKY15-1* in *chrysanthemum* can activate CmNPR1 and induce downstream disease resistance-related genes, which enhance resistance to *P. horiana* through the SAR system [16]. Overexpression of *NbWRKY40* increased the transcription levels of an SA biosynthesis gene (*ICS1*) and SAR marker genes (*PR1b* and *PR2*) and inhibited tobacco infection with TMV [36]. OsWRKY67 can activate the expression of *PR1a* and *PR10* by binding to their promoters, increasing the accumulation of SA and thereby enhancing plant resistance to leaf blast and bacterial blight disease [37]. In *Arabidopsis*, *WRKY11* triggers ISR by activating the JA signaling pathway, thereby increasing plant resistance to *Pst* DC3000 [38]. Overexpression of *GhWRKY70* in *Arabidopsis* can increase the expression of the JA synthesis-related gene *AtAOS1*, ultimately improving plant resistance to *Verticillium dahliae* [39].

In the early stage of RKN infection, the accumulation of plant disease resistance signals plays an important regulatory role in the defense against RKNs [40]. In the present study, after 12 h of infection with *M. incognita*, SA, JA, and ETH were induced in the SlWRKY80-overexpressing lines, increasing tomato resistance to varying degrees, possibly via the SAR and ISR pathways. In particular, SA and JA presented high levels of expression at 36 h and 48 h, respectively, showing almost twice the accumulation compared to the nontransgenic lines. However, when SA was consistently upregulated, JA and ETH accumulation levels tended to increase after 36 h. There was no significant difference in the accumulation of JA between the two treatments at 0–24 h after inoculation, and a significant difference did not appear until 36 h, which is consistent with the results obtained with ABA. In addition, during the early stage of inoculation, the increase in ABA concentration was not significant, and ABA induction was relatively delayed. Previous studies have shown that ABA indirectly participates in plant immune regulation by activating the expression of genes such as *C3H* and promoting the biosynthesis of SA [41]. These findings further suggest that ABA is not the main signal stimulating disease resistance. However, the specific molecular mechanism underlying the findings of this study needs further investigation. In summary, *SlWRKY80* can mediate the defense of tomato plants against RKN invasion through various resistance processes by regulating the accumulation of signaling pathways such as SA, JA, and ETH.

Plants can respond to pathogen attacks by generating oxidative stress through the continuous accumulation of ROS, which play a core role in plant immune responses. H_2O_2 is the most stable ROS and often serves as an intercellular and intracellular signal to trigger downstream stress and disease resistance responses [42]. In this study, the accumulation of total ROS and H_2O_2 in the overexpression lines inoculated with *M. incognita* showed an upward trend (0–48 h), and both showed significant differences compared to those in the nontransgenic lines after 12 h and 24 h, respectively. Taken together, these findings indicate that ROS (H_2O_2) are rapidly induced in roots and that high-level ROS production hinders the survival of J2s in this environment [43]. In addition, ROS can indirectly induce PCD at the site of infection by combining the above hormone signals to ultimately cause HR at the early stages of infection [44], reducing root gall production and improving plant disease resistance. This resistance mechanism can force local healthy cells to die, preventing J2s from establishing effective feeding sites and obtaining nutrients, thus preventing them from continuing to grow and reproduce [45]. Notably, excessive accumulation of ROS can cause severe cell damage and increase the sensitivity of plants to pathogens [46]. In this study, during the 0–48 h stage after the inoculation of the overexpression lines, SOD and CAT activities exhibited a decreasing trend, while POD activity exhibited the opposite trend. Together, these three factors regulate the accumulation of ROS (H_2O_2), ultimately leading to a steady-state level of ROS (H_2O_2) during the 36–48 h period, preventing ROS (H_2O_2)-related damage in tomato plants. Overall, *SlWRKY80* has a positive regulatory function in improving tomato resistance to RKNs and has extremely high potential for use in subsequent disease resistance breeding.

4. Materials and Methods

4.1. Plant Germplasm and Growth Conditions

The disease-resistant material 'Motelle' (*Mi-1/Mi-1*) was used for cDNA acquisition in the RNAi experiment, and the susceptible material 'M82' (*mi-1/mi-1*) was used for the construction of overexpression lines. The above materials were placed in a composite substrate consisting of peat/perlite/vermiculite at a ratio of 3:1:1, with an ambient temperature of 25 °C and a relative humidity of approximately 60%. The plants were housed under a 16 h/8 h light/dark cycle. After the plants reached the four-leaf and one-heart stage, molecular experiments and nematode inoculation tests were conducted.

4.2. Nematode Assay and Identification of Disease Resistance

Diseased tomato roots that were infected with *M. incognita* and contained many egg masses were selected, and the egg masses were removed and placed in a culture dish. Sterile water was added to cover the egg masses, after which the dish was incubated at 28 °C. Then, the J2-stage nematodes that had hatched were collected. The above J2s were inoculated during the four-leaf and one-heart stage of tomato plants, with an inoculation amount of 1000 J2s per individual. After 40 days of inoculation, the nematodes in the tomato root tissue were stained and observed using the acid fuchsin staining method, after which the number of root galls was calculated. Based on the number of root galls and the range of distribution, the root-knot levels were divided into a total of 6 levels, from 0 to 5 (0: root system with no galls; 1: 1–20% roots with galls; 2: 21–40% roots with galls; 3: 41–60% roots with galls; 4: 61–80% roots with galls; 5: 81–100% roots with galls). The disease index (DI) was calculated according to the following formula, and the resistance level (RL) was evaluated. DI = [$\sum$ (number of plants at each level) × (corresponding number of levels)/(total number of surveyed plants × highest level numerical value) × 100]. The evaluation criteria for resistance were as follows: immunity (I): DI = 0, high resistance (HR): $0 < DI \leq 20$, moderate resistance (MR): $20 < DI \leq 40$, resistance (R): $40 < DI \leq 60$, susceptible (S): $60 < DI \leq 80$, and highly susceptible (HS): $80 < DI$.

4.3. Construction of RNAi and Overexpression Vectors

The CDS of *SlWRKY80* was obtained from the SGN database (https://solgenomics.net (accessed on 12 March 2022)) (Table S1), total RNA was extracted from 'Motelle' tomato leaves using a reagent kit (Vazyme, Nanjing, China), and reverse-transcribed cDNA was obtained using a reagent kit (Vazyme, Nanjing, China). Using *Bam*H I as the cleavage site, specific upstream and downstream primers with a 20 bp recombinant homologous arm were designed, and the high-fidelity enzyme Phanta Max (Vazyme, Nanjing, China) was used for PCR amplification of the CDS region.

The *pTRV2* interference vector and *pCAMBIA-1300-mCherry* overexpression vector were linearized by single-enzyme digestion using *Bam*H I, after which an Infusion cloning kit (Vazyme, Nanjing, China) was used to clone *SlWRKY80*, completing the construction of the recombinant vector (Figure S2). Subsequently, the recombinant vector was transformed into *E. coli* DH5α, which was subsequently cultured in solid LB media (containing 50 μg/mL kanamycin) for 12 h in the dark. Monoclonal colonies were selected and cultured in a shaker at 37 °C and 220 rpm for 13 h. Finally, PCR identification and sequencing were performed.

4.4. RNAi and Overexpression Line Construction

The recombinant *pTRV2-SlWRKY80*, unloaded *pTRV2*, and *pTRV2-PDS* vectors were subsequently transformed into *Agrobacterium* GV3101, which was subsequently injected into the leaves of 'M82' tomato plants at the four-leaf and one-heart stage, with 60 plants treated with each. After infection, the plants were first stored at 18–21 °C. The tomatoes were cultured in darkness with a humidity greater than 80% for 72 h and then cultured under a photoperiod of 16 h/8 h at 23~25 °C and 60% humidity. After the *pTRV2-PDS* plants exhibited bleaching (after approximately 2 wks), the expression level of *SlWRKY80* was measured, and tomato plants with a silencing efficiency greater than 50% were selected for follow-up tests (10 individuals were selected for each material).

Clean and full 'M82' tomato seeds were selected and placed in a tissue culture bottle containing MS medium for dark culture at 25 °C for 2–3 d. After approximately 75% of the seeds were germinated, they were cultured at 23 °C with a photoperiod of 16 h/8 h. Approximately 7–9 d after normal culture, the true leaves, cotyledon, and hypocotyl were cut into 0.5 cm × 0.5 cm leaf blocks on a sterile superclean workbench and placed on preculture medium (MS + 2.0 mg/L ZT + 0.2 mg/L IAA) for low-light culture for 24–48 h. The explants were then infected with *Agrobacterium* GV3101 containing the plasmid *pCAMBIA1300-SlWRKY80-EGFP*, placed in coculture medium (MS + 2.0 mg/L ZT + 0.2 mg/L IAA), and cocultured under low light for 24–48 h. The cocultured explants were transferred successively into screening medium (MS + 2.0 mg/L ZT + 0.2 mg/L IAA + 100 mg/L Kan + 500 mg/L carbenicillin (Cb)) and bud elongation medium (MS + 1.0 mg/L ZT + 0.05 mg/L IAA + 100 mg/L Kan + 500 mg/L Cb) for continued culture, and subculture was performed once at 15–20 d. Finally, when the advected buds grew to approximately 1 cm, healthy regenerated young buds were selected, and the base calli were transferred to the root medium for culture (1/2 ms + 0.1 mg/L IAA + 50 mg/L Kan + 300 mg/L Cb) to form a complete plant.

4.5. RT-qPCR Assay

When the tomato materials to be tested grew to the four-leaf and one-heart stage, the plant leaves were selected and the total RNA and cDNA were obtained by the previous kit. The cDNA was diluted 10-fold and used as a template for RT-qPCR detection using Top Green qPCR Super Mix (Vazyme, Nanjing, China). Using the *SlActin* gene as an internal reference gene [47], primers were designed to determine the expression of the target gene at the transcriptional level (Table S1). The cycling conditions were 5 min at 95 °C, followed by 40 cycles of 5 s at 95 °C and 10 s at 60 °C. Relative expression values were calculated using the $2^{-\Delta\Delta CT}$ method [48]. Three biological replications, each with three technical repetitions, were performed for all reactions.

4.6. Determination of Plant Hormones and the ROS System

The transgenic plants and the control plants were inoculated with *M. incognita*. Leaves at different inoculation time points were selected for the determination of SA, JA, ETH, and ABA contents, while roots were selected for the determination of H_2O_2 level. The HPLC strategy was used for the above determination [49]. The contents of ROS and the activities of ROS-scavenging enzymes (SOD, POD, and CAT) in the roots were determined via ELISA kits (Tiangen, G0104, G0106, G0107, Beijing, China).

4.7. Measurement of Growth and Physiological Indices

The transgenic plants and the control plants were inoculated with *M. incognita*, after which the growth indices, including plant height, stem diameter, and leaf area, were measured. The leaves were selected for the determination of related physiological indices, including soluble protein, soluble sugar, and root activity, which were determined by Liu's and Ma's method [31,50]. The measurement period was once every 4 days for a total of 28 days, with three biological replicates for each treatment.

5. Conclusions

In this study, we selected *SlWRKY80* with highly differential expression from the RNA-seq data of different disease-resistant materials (*Mi-1* and *Mi-3*) as the research target. RNAi and overexpression strategies validated *SlWRKY80* as a positive regulator involved in tomato defense against *M. incognita*. Soluble protein and sugar and root activity reflected the improvement of overall plant disease resistance after overexpression of *SlWRKY80*, which was also reflected in the growth indices. In terms of disease resistance signals, *SlWRKY80* is likely to mediate disease resistance responses through the accumulation of SA, JA, and ETH signals, which results in the reduction of the invasion efficiency of RKNs by relying on the continuous accumulation of ROS.

Supplementary Materials: The following supporting information can be downloaded at: https://www.mdpi.com/article/10.3390/ijms25168892/s1.

Author Contributions: Conceptualization, C.D. and C.W.; methodology, L.L.; software, T.Z.; validation, Z.W., Y.C. and Y.D.; formal analysis, C.F. and W.N.; investigation, Y.C.; resources, L.L.; writing—original draft preparation, Y.C.; writing—review and editing, C.D.; visualization, C.D.; funding acquisition, C.D. All authors have read and agreed to the published version of the manuscript.

Funding: This research was funded by the National Natural Science Foundation of China, grant number 32302652, the Natural Science Foundation of Xinjiang Uygur Autonomous Region, grant number 2022D01B95, Sub-Project of Key Research and Development Plan of Autonomous Region, grant number 2022B02032-2.

Institutional Review Board Statement: Not applicable.

Informed Consent Statement: Not applicable.

Data Availability Statement: The original contributions presented in the study are included in the article/Supplementary Materials, further inquiries can be directed to the corresponding author.

Acknowledgments: Thanks to Shaohui Wang from Beijing University of Agriculture for providing sufficient pathogenic roots for this study.

Conflicts of Interest: The authors declare no conflicts of interest.

References

1. Coyne, D.L.; Cortada, L.; Dalzell, J.J.; Claudius-Cole, A.O.; Haukeland, S.; Luambano, N.; Talwana, H. Plant-Parasitic Nematodes and Food Security in Sub-Saharan Africa. *Annu. Rev. Phytopathol.* **2018**, *56*, 381–403. [CrossRef] [PubMed]
2. Jones, J.T.; Haegeman, A.; Danchin, E.G.; Gaur, H.S.; Helder, J.; Jones, M.G.; Kikuchi, T.; Manzanilla-López, R.; Palomares-Rius, J.E.; Wesemael, W.M. Top 10 plant-parasitic nematodes in molecular plant pathology. *Mol. Plant Pathol.* **2013**, *14*, 946–961. [CrossRef] [PubMed]
3. Jagdale, S.; Rao, U.; Giri, A.P. Effectors of Root-Knot Nematodes: An Arsenal for Successful Parasitism. *Front. Plant Sci.* **2021**, *12*, 800030. [CrossRef] [PubMed]
4. Dong, L.; Li, X.; Huang, C.; Lu, Q.; Li, B.; Yao, Y.; Liu, T.; Zuo, Y. Reduced Meloidogyne incognita infection of tomato in the presence of castor and the involvement of fatty acids. *Sci. Hortic.* **2018**, *237*, 169–175. [CrossRef]
5. Caillaud, M.-C.; Dubreuil, G.; Quentin, M.L.; Perfus-Barbeoch, L.; Lecomte, P.; Engler, J.D.A.; Abad, P.; Rosso, M.N.L.; Favery, B. Root-knot nematodes manipulate plant cell functions during a compatible interaction. *J. Plant Physiol.* **2008**, *165*, 104–113. [CrossRef] [PubMed]
6. Ibrahim, H.M.M.; Ahmad, E.M.; Martínez-Medina, A.; Aly, M.A.M. Effective approaches to study the plant-root knot nematode interaction. *Plant Physiol. Biochem.* **2019**, *141*, 332–342. [CrossRef] [PubMed]

7. Wang, Y.; Yang, W.; Zhang, W.; Han, Q.; Feng, M.; Shen, H. Mapping of a Heat-Stable Gene for Resistance to Southern Root-Knot Nematode in Solanum lycopersicum. *Plant Mol. Biol. Report.* **2013**, *31*, 352–362. [CrossRef]
8. Du, C.; Jiang, J.; Zhang, H.; Zhao, T.; Yang, H.; Zhang, D.; Zhao, Z.; Xu, X.; Li, J. Transcriptomic profiling of Solanum peruvianum LA3858 revealed a Mi-3-mediated hypersensitive response to Meloidogyne incognita. *BMC Genom.* **2020**, *21*, 250. [CrossRef]
9. Wani, S.H.; Anand, S.; Singh, B.; Bohra, A.; Joshi, R. WRKY transcription factors and plant defense responses: Latest discoveries and future prospects. *Plant Cell Rep.* **2021**, *40*, 1071–1085. [CrossRef]
10. Javed, T.; Gao, S.J. WRKY transcription factors in plant defense. *Trends Genet.* **2023**, *39*, 787–801. [CrossRef]
11. Pritha, K.; Jyothilakshmi, V. Role of WRKY transcription factors in plant defense against lepidopteran insect herbivores: An overview. *J. Plant Biochem. Biotechnol.* **2021**, *30*, 698–707.
12. Anket, S.; Kaur, K.S.; Kanika, K.; Muthusamy, R.; Vinod, K.; Renu, B.; Marian, B.; Milan, S.; Marco, L.; Bingsong, Z. Salicylic Acid: A Phenolic Molecule with Multiple Roles in Salt-Stressed Plants. *J. Plant Growth Regul.* **2023**, *42*, 4581–4605.
13. van Verk, M.C.; Bol, J.F.; Linthorst, H.J. WRKY transcription factors involved in activation of SA biosynthesis genes. *BMC Plant Biol.* **2011**, *11*, 89. [CrossRef] [PubMed]
14. Cui, X.; Zhao, P.; Liang, W.; Cheng, Q.; Mu, B.; Niu, F.; Yan, J.; Liu, C.; Xie, H.; Kav, N.N.V.; et al. A Rapeseed WRKY Transcription Factor Phosphorylated by CPK Modulates Cell Death and Leaf Senescence by Regulating the Expression of ROS and SA-Synthesis-Related Genes. *J. Agric. Food Chem.* **2020**, *68*, 7348–7359. [CrossRef] [PubMed]
15. Peng, Y.; Yang, J.; Li, X.; Zhang, Y. Salicylic Acid: Biosynthesis and Signaling. *Annu. Rev. Plant Biol.* **2021**, *72*, 761–791. [CrossRef] [PubMed]
16. Gao, G.; Jin, R.; Liu, D.; Zhang, X.; Sun, X.; Zhu, P.; Mao, H. CmWRKY15-1 Promotes Resistance to Chrysanthemum White Rust by Regulating CmNPR1 Expression. *Front. Plant Sci.* **2022**, *13*, 865607. [CrossRef] [PubMed]
17. Sarkar, S.; Das, A.; Khandagale, P.; Maiti, I.B.; Chattopadhyay, S.; Dey, N. Interaction of Arabidopsis TGA3 and WRKY53 transcription factors on Cestrum yellow leaf curling virus (CmYLCV) promoter mediates salicylic acid-dependent gene expression in planta. *Planta* **2018**, *247*, 181–199. [CrossRef] [PubMed]
18. Xiong, X.; Sun, S.; Li, Y.; Zhang, X.; Sun, J. The cotton WRKY transcription factor GhWRKY70 negatively regulates the defense response against Verticillium dahliae. *Crop J.* **2019**, *7*, 123–132. [CrossRef]
19. Gilroy, E.; Breen, S. Interplay between phytohormone signalling pathways in plant defence—Other than salicylic acid and jasmonic acid. *Essays Biochem.* **2022**, *66*, 657–671.
20. Cui, X.; Yan, Q.; Gan, S.; Xue, D.; Wang, H.; Xing, H.; Zhao, J.; Guo, N. GmWRKY40, a member of the WRKY transcription factor genes identified from *Glycine max* L., enhanced the resistance to Phytophthora sojae. *BMC Plant Biol.* **2019**, *19*, 598. [CrossRef]
21. Zheng, L.; Qiu, B.; Su, L.; Wang, H.; Cui, X.; Ge, F.; Liu, D. Panax notoginseng WRKY Transcription Factor 9 Is a Positive Regulator in Responding to Root Rot Pathogen Fusarium solani. *Front. Plant Sci.* **2022**, *13*, 930644. [CrossRef]
22. Hou, Y.; Wang, Y.; Tang, L.; Tong, X.; Wang, L.; Liu, L.; Huang, S.; Zhang, J. SAPK10-Mediated Phosphorylation on WRKY72 Releases Its Suppression on Jasmonic Acid Biosynthesis and Bacterial Blight Resistance. *iScience* **2019**, *16*, 499–510. [CrossRef]
23. Qi, J.; Wang, J.; Gong, Z.; Zhou, J.M. Apoplastic ROS signaling in plant immunity. *Curr. Opin. Plant Biol.* **2017**, *38*, 92–100. [CrossRef]
24. Ravi, B.; Foyer, C.H.; Pandey, G.K. The integration of reactive oxygen species (ROS) and calcium signalling in abiotic stress responses. *Plant Cell Environ.* **2023**, *46*, 1985–2006. [CrossRef]
25. Bhattarai, K.K.; Atamian, H.S.; Kaloshian, I.; Eulgem, T. WRKY72-type transcription factors contribute to basal immunity in tomato and Arabidopsis as well as gene-for-gene resistance mediated by the tomato R gene Mi-1. *Plant J.* **2010**, *63*, 229–240. [CrossRef]
26. Chinnapandi, B.; Bucki, P.; Braun Miyara, S. SlWRKY45, nematode-responsive tomato WRKY gene, enhances susceptibility to the root knot nematode; M. javanica infection. *Plant Signal. Behav.* **2017**, *12*, e1356530. [CrossRef]
27. Huang, H.; Zhao, W.; Qiao, H.; Li, C.; Sun, L.; Yang, R.; Ma, X.; Ma, J.; Song, S.; Wang, S. SlWRKY45 interacts with jasmonate-ZIM domain proteins to negatively regulate defense against the root-knot nematode Meloidogyne incognita in tomato. *Hortic. Res.* **2022**, *9*, uhac197. [CrossRef]
28. Chinnapandi, B.; Bucki, P.; Fitoussi, N.; Kolomiets, M.; Borrego, E.; Braun Miyara, S. Tomato SlWRKY3 acts as a positive regulator for resistance against the root-knot nematode Meloidogyne javanica by activating lipids and hormone-mediated defense-signaling pathways. *Plant Signal. Behav.* **2019**, *14*, 1601951. [CrossRef]
29. Atamian, H.S.; Eulgem, T.; Kaloshian, I. SlWRKY70 is required for Mi-1-mediated resistance to aphids and nematodes in tomato. *Planta* **2012**, *235*, 299–309. [CrossRef] [PubMed]
30. Nie, W.; Liu, L.; Chen, Y.; Luo, M.; Feng, C.; Wang, C.; Yang, Z.; Du, C. Identification of the Regulatory Role of SlWRKYs in Tomato Defense against Meloidogyne incognita. *Plants* **2023**, *12*, 2416. [CrossRef] [PubMed]
31. Ma, Q.J.; Sun, M.H.; Lu, J.; Liu, Y.J.; Hu, D.G.; Hao, Y.J. Transcription Factor AREB2 Is Involved in Soluble Sugar Accumulation by Activating Sugar Transporter and Amylase Genes. *Plant Physiol.* **2017**, *174*, 2348–2362. [CrossRef]
32. Xu, Y.; Xu, R.; Li, S.; Ran, S.; Wang, J.; Zhou, Y.; Gao, H.; Zhong, F. The mechanism of melatonin promotion on cucumber seedling growth at different nitrogen levels. *Plant Physiol. Biochem.* **2024**, *206*, 108263. [CrossRef] [PubMed]
33. Shu, P.; Zhang, S.; Li, Y.; Wang, X.; Yao, L.; Sheng, J.; Shen, L. Over-expression of SlWRKY46 in tomato plants increases susceptibility to Botrytis cinerea by modulating ROS homeostasis and SA and JA signaling pathways. *Plant Physiol. Biochem.* **2021**, *166*, 1–9. [CrossRef]

34. Hartmann, M.; Zeier, J. N-hydroxypipecolic acid and salicylic acid: A metabolic duo for systemic acquired resistance. *Curr. Opin. Plant Biol.* **2019**, *50*, 44–57. [CrossRef]
35. Andersen, E.J.; Ali, S.; Byamukama, E.; Yen, Y.; Nepal, M.P. Disease Resistance Mechanisms in Plants. *Genes* **2018**, *9*, 339. [CrossRef]
36. Jiang, Y.; Zheng, W.; Li, J.; Liu, P.; Zhong, K.; Jin, P.; Xu, M.; Yang, J.; Chen, J. NbWRKY40 Positively Regulates the Response of Nicotiana benthamiana to Tomato Mosaic Virus via Salicylic Acid Signaling. *Front. Plant Sci.* **2020**, *11*, 603518. [CrossRef]
37. Liu, Q.; Li, X.; Yan, S.; Yu, T.; Yang, J.; Dong, J.; Zhang, S.; Zhao, J.; Yang, T.; Mao, X.; et al. OsWRKY67 positively regulates blast and bacteria blight resistance by direct activation of PR genes in rice. *BMC Plant Biol.* **2018**, *18*, 257. [CrossRef]
38. Jiang, C.H.; Huang, Z.Y.; Xie, P.; Gu, C.; Li, K.; Wang, D.C.; Yu, Y.Y.; Fan, Z.H.; Wang, C.J.; Wang, Y.P.; et al. Transcription factors WRKY70 and WRKY11 served as regulators in rhizobacterium Bacillus cereus AR156-induced systemic resistance to Pseudomonas syringae pv. tomato DC3000 in Arabidopsis. *J. Exp. Bot.* **2016**, *67*, 157–174. [CrossRef]
39. Zhang, S.; Dong, L.; Zhang, X.; Fu, X.; Zhao, L.; Wu, L.; Wang, X.; Liu, J. The transcription factor GhWRKY70 from gossypium hirsutum enhances resistance to verticillium wilt via the jasmonic acid pathway. *Bmc Plant Biol.* **2023**, *23*, 141. [CrossRef]
40. Liu, S.; Kandoth, P.K.; Warren, S.D.; Yeckel, G.; Heinz, R.; Alden, J.; Yang, C.; Jamai, A.; El-Mellouki, T.; Juvale, P.S.; et al. A soybean cyst nematode resistance gene points to a new mechanism of plant resistance to pathogens. *Nature* **2012**, *492*, 256–260. [CrossRef] [PubMed]
41. Rekhter, D.; Lüdke, D.; Ding, Y.; Feussner, K.; Zienkiewicz, K.; Lipka, V.; Wiermer, M.; Zhang, Y.; Feussner, I. Isochorismate-derived biosynthesis of the plant stress hormone salicylic acid. *Science* **2019**, *365*, 498–502. [CrossRef]
42. Camejo, D.; Guzmán-Cedeño, Á.; Moreno, A. Reactive oxygen species, essential molecules, during plant-pathogen interactions. *Plant Physiol. Biochem.* **2016**, *103*, 10–23. [CrossRef]
43. Gilroy, S.; Białasek, M.; Suzuki, N.; Górecka, M.; Devireddy, A.R.; Karpiński, S.; Mittler, R. ROS, Calcium, and Electric Signals: Key Mediators of Rapid Systemic Signaling in Plants. *Plant Physiol.* **2016**, *171*, 1606–1615. [CrossRef] [PubMed]
44. Rodrigues, J.M.; Coutinho, F.S.; Dos Santos, D.S.; Vital, C.E.; Ramos, J.; Reis, P.B.; Oliveira, M.G.A.; Mehta, A.; Fontes, E.P.B.; Ramos, H.J.O. BiP-overexpressing soybean plants display accelerated hypersensitivity response (HR) affecting the SA-dependent sphingolipid and flavonoid pathways. *Phytochemistry* **2021**, *185*, 112704. [CrossRef]
45. Lv, J.; Zhang, Y.; Sun, M.; Chen, J.; Ge, Y.; Li, J. 1-Methylcyclopropene (1-MCP) treatment differentially mediated expression of vacuolar processing enzyme (VPE) genes and delayed programmed cell death (PCD) during ripening and senescence of apple fruit. *Sci. Hortic.* **2023**, *307*, 111489. [CrossRef]
46. Wei, C.; Song, L.; Qin, L. Heterologous expression of HsPstS gene reducing arsenic accumulation and improving As-tolerance in transgenic tobaccos by enhancing CAT activity. *J Plant Physiol.* **2023**, *282*, 153940. [CrossRef]
47. Liu, M.; Diretto, G.; Pirrello, J.; Roustan, J.P.; Li, Z.; Giuliano, G.; Regad, F.; Bouzayen, M. The chimeric repressor version of an Ethylene Response Factor (ERF) family member, Sl-ERF.B3, shows contrasting effects on tomato fruit ripening. *New Phytol.* **2014**, *203*, 206–218. [CrossRef]
48. Livak, K.J.; Schmittgen, T.D. Analysis of relative gene expression data using real-time quantitative PCR and the 2(-Delta Delta C(T)) Method. *Methods* **2013**, *25*, 402–408. [CrossRef] [PubMed]
49. Eitle, M.W.; Griesser, M.; Vankova, R.; Dobrev, P.; Aberer, S.; Forneck, A. Grape phylloxera (*D. vitifoliae*) manipulates SA/JA concentrations and signalling pathways in root galls of *Vitis* spp. *Plant Physiol. Biochem.* **2019**, *144*, 85–91. [CrossRef]
50. Liu, Y.; Zhang, J. Lanthanum Promotes Bahiagrass (*Paspalum notatum*) Roots Growth by Improving Root Activity, Photosynthesis and Respiration. *Plants* **2022**, *11*, 382. [CrossRef] [PubMed]

International Journal of
Molecular Sciences

MDPI

Article

Genome-Wide Detection of SPX Family and Profiling of *CoSPX-MFS3* in Regulating Low-Phosphate Stress in Tea-Oil *Camellia*

Juanjuan Chen [1,2], Xiaojiao Han [1], Linxiu Liu [1], Bingbing Yang [1], Renying Zhuo [1,*] and Xiaohua Yao [1,*]

[1] Key Laboratory of Tree Breeding of Zhejiang Province, Research Institute of Subtropical Forestry, Chinese Academy of Forestry, Hangzhou 311400, China; chenjuan@caf.ac.cn (J.C.); hanxj@caf.ac.cn (X.H.); lxliu9968@163.com (L.L.); bingbingyang1224@163.com (B.Y.)

[2] Forestry Faculty, Nanjing Forestry University, Nanjing 210037, China

[*] Correspondence: zhuory@caf.ac.cn (R.Z.); yaoxh@caf.ac.cn (X.Y.); Tel.: +86-0571-6331-1860 (R.Z. & X.Y.)

Abstract: *Camellia oleifera* a member of the family Theaceae, is a phosphorus (P) tolerator native to southern China. The SPX gene family critically regulates plant growth and development and maintains phosphate (Pi) homeostasis. However, the involvement of *SPX* genes in Pi signaling in Tea-Oil *Camellia* remains unknown. In this work, 20 *SPX* genes were identified and categorized into four subgroups. Conserved domains, motifs, gene structure, chromosomal location and gene duplication events were also investigated in the SPX gene family. Defense and stress responsiveness cis-elements were identified in the *SPX* gene promoters, which participated in low-Pi stress responses. Based on transcriptome data and qRT-PCR results, nine *CoSPX* genes had similar expression patterns and eight genes (except *CoPHO1H3*) were up-regulated at 30 days after exposure to low-Pi stress. *CoSPX-MFS3* was selected as a key candidate gene by WGCNA analysis. CoSPX-MFS3 was a tonoplast protein. Overexpression of *CoSPX-MFS3* in *Arabidopsis* promoted the accumulation of total P content and decreased the anthocyanin content. Overexpression of *CoSPX-MFS3* could enhance low-Pi tolerance by increased biomass and organic acid contents in transgenic *Arabidopsis* lines. Furthermore, the expression patterns of seven phosphate starvation genes were higher in transgenic *Arabidopsis* than those in the wild type. These results highlight novel physiological roles of the *SPX* family genes in *C. oleifera* under low-Pi stress, and lays the foundation for a deeper knowledge of the response mechanism of *C. oleifera* to low-Pi stress.

Keywords: SPX gene family; Tea-Oil *Camellia*; Pi stress; expression profiles; *CoSPX-MFS3*

1. Introduction

Phosphorus (P) is among the necessary nutritional elements for the development and metabolism of all living organisms and plays an essential role in all life activities. As the second most important nutrient element, P is not only a constituent of important compounds (e.g., nucleic acids and proteins) in plants, but also regulates nearly all energy metabolism processes such as photosynthesis, respiration, energy conversion and enzymatic reactions [1–3]. Soil P is recognized as an integral part in Al, Fe or Ca salt. P can be taken up by plants in the form of orthophosphate (Pi, H_2PO^{4-}/HPO_4^{2-}) [4]. The concentration of available Pi is less than 10 μM in most soils [5]. Under different Pi environmental stresses, plants have evolved a sequence of adaptative mechanisms in morphological, physiological and molecular functions [6]. These mechanisms in plants are all controlled by complex and exquisite molecular regulatory networks, including an increasing number of transcription factors and Pi starvation responsive genes, microRNAs and other functional genes in the plant Pi signaling regulatory pathways.

The SPX domain (Pfam PF03105) was coined based on the first three letters of Suppressor of Yeast gpa1 (SYG1), Phosphatase 81 (PHO81) and Xenotropic and Polytropic

Retrovirus receptor1 (XPR1) [7,8]. The protein-containing SPX domain is vital for the regulation of Pi signaling and Pi homeostasis. In yeast cells, *Pho84* is a high-affinity inorganic Pi sensor to maintain intracellular inorganic Pi uptake and transport [9]. As a low-affinity transporter, *Pho91* is in charge of the Pi allocation from vacuoles to cytosol [10]. The SPX protein domain is typically restricted to the N-termini of SPX gene family members. SPX domain proteins are assigned to four categories, namely, SPX, SPX-RING (SPX-Really Interesting New Gene), SPX-MFS (SPX-Major Facility Superfamily) and SPX-EXS (SPX-yeast ERD1, human XPR1 and yeast SYG1 protein subfamilies).

The SPX subfamily members contain only the SPX domain. In *Arabidopsis*, there are four SPX members, namely, *AtSPX1–4*. *AtSPX1* was distinctly upregulated by Pi starvation, and *AtSPX1* overexpression also increased the expression of phosphate starvation response (PSR) genes [11]. In potato (*Solanum tuberosum*), *StSPX2/3/5* are transcriptionally up-regulated under low-Pi stress [12]. The expression patterns of *ZmSPX4.1* and *ZmSPX4.2* were significantly different in maize (*Zea mays*) sensitive and insensitive to low-Pi condition [13].

The SPX-RING subfamily regulates Pi homeostasis under N limitation condition, and is commonly recognized as nitrogen limiting adaptation (*NLA*). The *NLA* gene can mediate the response of plants to nitrogen limitation, while *nla* mutant hyperaccumulates Pi, whose phenotype is similar to the *pho* mutant [14]. Meanwhile, loss of function of *NLA* enhances Pi accumulation by increasing several *PHT1s* at the protein level instead of the transcript level. *NLA* can direct the ubiquitination of *PHT1* in plasma membrane, by triggering clathrin-dependent endocytosis as well as transportation from endosome to vacuole [15].

The SPX-MFS subfamily contains the MFS domain. The MFS subfamily proteins contain 12–14 transmembrane domains and transport a broad range of substrates, including nucleotides, amino acids, ions and peptides [7]. In plants, the SPX-MFS family has been demonstrated to function as a Pi transporter [16–18]. *Arabidopsis* has three putative genes in this family (*At1g63010*, *At4g11810* and *At4g22990*). The SPX-MFS protein was also designated as a member of the phosphate transporter 5 family (*PHT5*) according to the systematic nomenclature of the *PHT1–PHT4* phosphate transporter protein [19,20]. In *Arabidopsis*, high expression of *PHT5* led to Pi accumulation, arrested growth and attenuated Pi influx into vacuoles [16].

The SPX-EXS subfamily consists of a hydrophobic region, namely the EXS domain (PF03124), and the PHO1 gene family members containing SPX and EXS domains are only found in eukaryotes [21]. There are 11 *Arabidopsis* (*AtPHO1*, *AtPHO1;H1–H10*) and 3 *Oryza sativa* (*OsPHO1;1/2/3*) members that are responsible for Pi homeostasis by transferring Pi from the root xylem to the shoot [22,23]. In *Arabidopsis*, *AtPHO1* overexpression led to a 2- to 3-fold rise in shoot Pi levels and significantly inhibited shoot development [24]. *OsPHO1;2* also regulated Pi transportation from rice plant roots to stems [25].

Tea-oil camellia (*Camellia oleifera*) is regarded as one of the four largest woody oil trees worldwide, along with coconut, olive and oil palm [26]. Tea-oil is obtained from the seeds, which contain various secondary metabolites and unsaturated fatty acids. It has been called "Oriental olive oil" due to its antioxidant activity and high oil content [27,28]. *C. oleifera* has been shown to have a high comprehensive utilization value. The fruit and seed shell of *C. oleifera* comprise a large amount of lignin, hemicellulose and cellulose, which have a wide range of industrial uses, and can be used to produce furfural, xylitol and activated carbon [29]. Tea-seed meal is rich in protein, polysaccharide and saponin and is used as raw material for light industry and chemical industry products [30]. *Camellia lanceoleosa*, a diploid wild species in the *Camellia* Sect. *Oleifera*, which is closely related to polyploid *C. oleifera* Abel, has gained increasing attention from scientific communities in recent years [31]. Although *SPX* family members have been diffusely studied in some crops and model plants, including rice [32], *Arabidopsis* [33], maize [13], wheat [34], rapeseed (*Brassica napus*) [35] and potato [36], a comprehensive analysis of *SPX* family members has not been performed in woody plants. Therefore, we initially explored the mechanism of *C. oleifera* response to reduced Pi stress in woody plants. Herein, all putative *SPX* genes in

C. lanceoleosa genome were identified, and their structural characteristics and conserved motifs were analyzed. Moreover, the expression profiles of homologous *CoSPX* genes in the hexaploidy *C. oleifera* "Changlin166" roots under low-Pi stress were determined. The findings provide a reference basis for perspective functional experiments on the SPX gene family and a new theoretical framework for identifying novel genes, exploring genetic resources and improving Pi tolerance in tea-oil *Camellia*.

2. Results

2.1. Detection of *C. lanceoleosa SPX* Genes

The 20 SPX protein sequences of *Arabidopsis* were downloaded and utilized as query sequences. To determine putative SPX protein in *C. lanceoleosa*, the SPX (PF03105) domain was confirmed by Pfam, CDD and SMART (http://smart.embl-heidelberg.de/, accessed on 22 April 2023). There were 20 *SPX* genes (*ClSPXs*) identified in the genome of *C. lanceoleosa*, which were named based on the corresponding *Arabidopsis* homologs. Similar to wheat (*Triticum aestivum*), tomato (*Solanum lycopersicum*) and rice (*Oryza sativa*), the SPX subfamily contained the largest clade with 12 members, followed by 3 SPX-MFS, 3 SPX-EXS and 2 SPX-RING members (Table 1). The detailed information of all 20 SPX members, such as the number of gene ID, location, exons, protein properties, molecular weight, predicted isoelectric point values and estimated subcellular location, are presented in Supplemental Table S1. Further analysis uncovered that the ClSPX proteins consisted of 237–699 amino acids, carrying a molecular weight of 19.6–78.1 kDa. EXPASY analysis indicated that the SPX protein sequences had different isoelectric point (pI) values (range = 4.89–9.27). Subcellular location analysis showed that ClSPX proteins were localized in cell membrane, tonoplast, nucleus and chloroplast. This characteristic implies that different subfamilies may exhibit varying biology functions.

Table 1. Number of *SPX* genes in seven plant species.

Species	SPX	SPX-EXS	SPX-MFS	SPX-RING	Total	Reference
Arabidopsis	4	11	3	2	20	[11]
Wheat	15	12	12	7	46	[34]
Maize	7	15	9	2	33	[13]
Brassica napus	11	43	8	7	69	[35]
Solanum lycopersicum	7	6	4	2	19	[37]
Rice	6	3	4	2	15	[38]
C. lanceoleosa	12	3	3	2	20	This study

2.2. Phylogenetic Analysis of *ClSPXs*

To illustrate the evolutionary relationship among SPX homologs in different plants, a phylogenetic tree was constructed comprising 20 ClSPXs in *C. lanceoleosa*, 20 AtSPXs in *Arabidopsis* and 24 CsSPXs in *C. sinensis*. The SPXs were clustered into four subclasses, namely, groups SPX, SPX-RING, SPX-MFS and SPX-EXS (Figure 1). The SPX group contained the most members, whereas the SPX-RING group contained the least, only two members.

2.3. *ClSPX* Protein Domain, Motif Composition and Gene Structure

The 20 ClSPX proteins were classified into four subfamilies according to the supplemental domains in protein structure, including SPX, SPX-RING, SPX-MFS and SPX-EXS (Figure 2B). To further examine the characteristics of *SPX* genes in *C. lanceoleosa*, 25 conserved motifs were identified by MEME analysis (Figure 2C). The specific information about the 25 putative motifs was listed in Supplementary Table S2. It was found that only the SPX subfamily contained motif 2. Motif 1 was specific to almost all ClSPXs (except SPX-EXS subfamilies), while motif 7 was only unique to the SPX-RING and SPX-MFS subfamilies (Figure 2C). Next, the structure of *ClSPX* genes was analyzed based on their exons and introns. Most *ClSPXs* exhibited similar exon–intron organization. ClSPXs with only the SPX domain have three exons, while ClSPX-RING, ClSPX-MFS2b/2c and ClSPX-MFS2a contain five, ten and nine exons, respectively (Figure 2D).

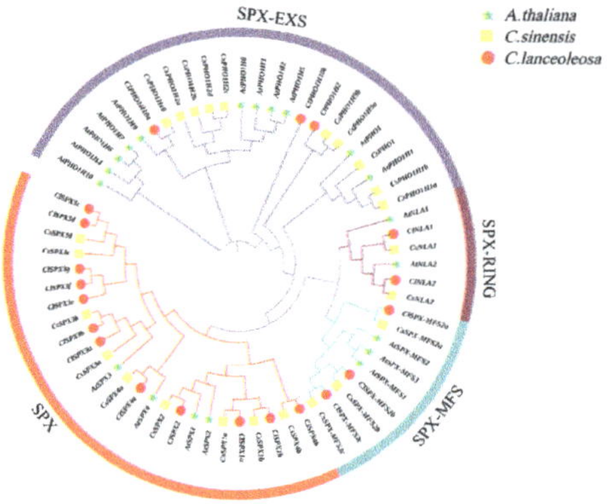

Figure 1. Phylogenetic relationships among the *SPX* genes from *C. lanceoleosa* (*Cl*), *C. sinensis* (*Cs*) and *Arabidopsis* (*At*). The phylogenetic tree of the three species was constructed according to the maximum-likelihood method with 1000 bootstrap replicates. Green stars represent *At*; yellow squares are *Cs*; red circles indicate *Cl*.

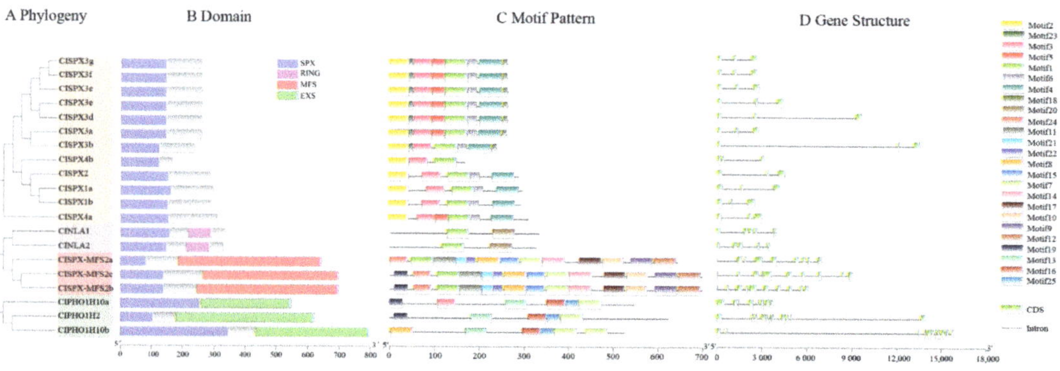

Figure 2. Phylogenetic relationships, domain, motif compositions and gene structures of *ClSPX* genes in *C. lanceoleosa*. (**A**) Phylogenetic analysis. (**B**) Domain analysis. (**C**) All conserved motifs in the ClSPX proteins were identified using the MEME program. Different motifs are highlighted with different colored boxes (numbered 1–25). (**D**) Gene structure. Exons are indicated by green, whereas gray lines represent introns.

2.4. Assessment of Cis-Acting Elements in ClSPX Promoters

The promoter regions of the upstream 2000 bp of 20 *ClSPXs* were assessed with PlantCARE. All cis-regulatory elements, such as TC-rich repeat, MBS, ARE, W-box, CAT-box, GCN4-motif, GARE-motif, ABRE, P-box, ERE and TATC-box elements, were assigned to three various categories: hormones, development and stress (Figure 3A). In addition, we identified 11 cis-regulatory elements, including defense and stress, MeJA stress response, auxin-responsive elements and others (Figure 3B), suggesting that *SPX* genes play vital roles in regulating plant abiotic stress [39,40].

2.5. Chromosomal Location and Synteny-Based Evaluation of ClSPX Genes

The physical locations of *SPX* genes were localized to the chromosomes (Chr) of *C. lanceoleosa* with TBtools [41]. The *ClSPX* genes were mapped to eight chromosomes, but were unevenly distributed. Non-uniform *SPX* gene distribution was observed on the chromosomes (Figure S1). Three *SPX* genes were found on chromosomes 2, 11 and 13, while six *SPX* genes were on chromosome 3. However, collinear genes were not present on the same chromosome. In fact, most genes possessed two similar genes on different chromosomes (e.g., *ClSPX-MFS2b-ClMFS-MFS2c/ClMFS-MFS2a* and *ClSPX4b-ClSPX2/ClSPX1b*), which formed two collinear gene pairs (Figure 4A). We found 7 segmental duplication gene pairs on six chromosomes. The Ka/Ks ratios of gene pairs were all <0.5, demonstrating that these genes may experience intense selection during the evolution process (Supplementary Table S3). Thus, assessing the evolutionary relationship of SPX family members may help to determine the functions of *ClSPX* genes. Syntenic maps of *C. sinensis* and *Arabidopsis* were constructed with *C. lanceoleosa* (Figure 4B). The results showed that 9 *ClSPX* genes exhibited syntenic relationships with 8 and 11 genes in *C. sinensis* and *Arabidopsis*, respectively (Supplementary Table S4). Moreover, 23 orthologous gene pairs were observed between *C. lanceoleosa* and *Arabidopsis*, and 21 orthologous gene pairs between *C. lanceoleosa* and *C. sinensis*. Six SPX collinear gene pairs were detected among *C. lanceoleosa*, *Arabidopsis* and *C. sinensis*. Most sites in these genes experienced intense purifying selection (Supplementary Table S4).

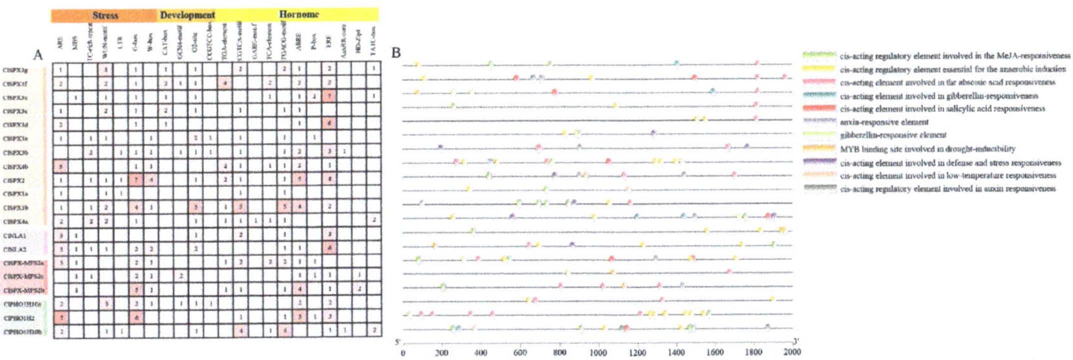

Figure 3. Analysis of cis-acting elements in the *ClSPX* promoter region. (**A**) The number of each cis-acting elements in the promoter region (2 kb upstream of the translation start site) of *ClSPX* genes. The depth of the color represents the number of cis-acting elements (**B**) Distribution of related cis-acting elements in *ClSPX* promoters.

2.6. The Expression Profiles of CoSPX Genes in C. oleifera Tissues

As previously mentioned, *C. lanceoleosa* is closely related to polyploid *C. oleifera* Abel. At present, the main cultivar of *C. oleifera* is a polyploid plant. To further understand the *CoSPX* gene functions in polyploid *C. oleifera*, the expression profiles of all putative genes were analyzed in various tissues (Figure S2). The results demonstrated that most *CoSPX* genes have higher expression level in roots and stems. The same subfamily members of *CoPHO1H3*/*CoPHO1H5*, *CoSPX-MFS1*/*CoSPX-MFS2* and *CoNLA1*/*CoNLA2* had similar expression patterns.

2.7. The Expression Distribution of CoSPXs under Pi-Deficiency

The expression patterns of *CoSPX* genes were determined according to our previously reported RNA-seq data. Using the Fragments Per Kilobase of transcript sequence per Millions of base pairs sequenced (FPKM) values of our previous RNA-seq data, we found that almost all *SPX* genes could respond to low-Pi stress (Figure 5A). The results of transcriptome data and qRT–PCR (quantitative real-time polymerase chain reaction) showed that nine *CoSPX* genes had the same expression tendency, and eight genes (except *CoPHO1H3*) were upregulated at 30 days after exposure to low-Pi stress (Figure 5B).

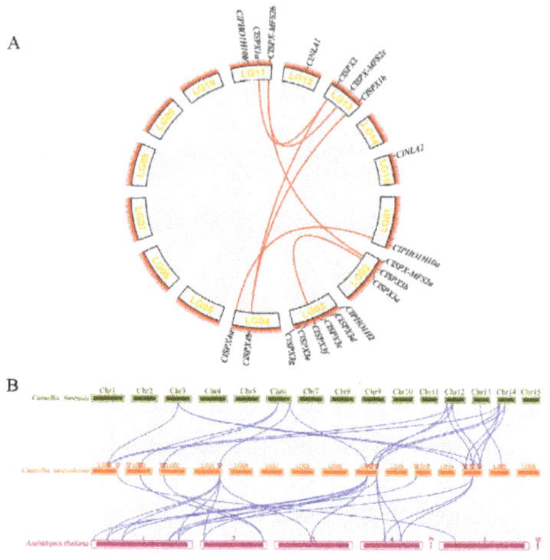

Figure 4. Synteny relationships. (**A**) Genome location and synteny of *SPX* genes in *C. lanceoleosa*. Gray lines indicate syntenic blocks in the *C. lanceoleosa* genome, while the red lines between chromosomes indicate segmentally duplicated gene pairs. (**B**) Synteny between *ClSPX* genes and genes in other species (*Arabidopsis* and *C. sinensis*). Number 1–5 means 5 chromosomes in *Arabidopsis*. Number LG01-LG15 means 15 chromosomes in *C. lanceoleosa*. Gray lines in the background represent collinear blocks in *C. lanceoleosa* and the other species, while blue lines indicate syntenic *SPX* gene pairs.

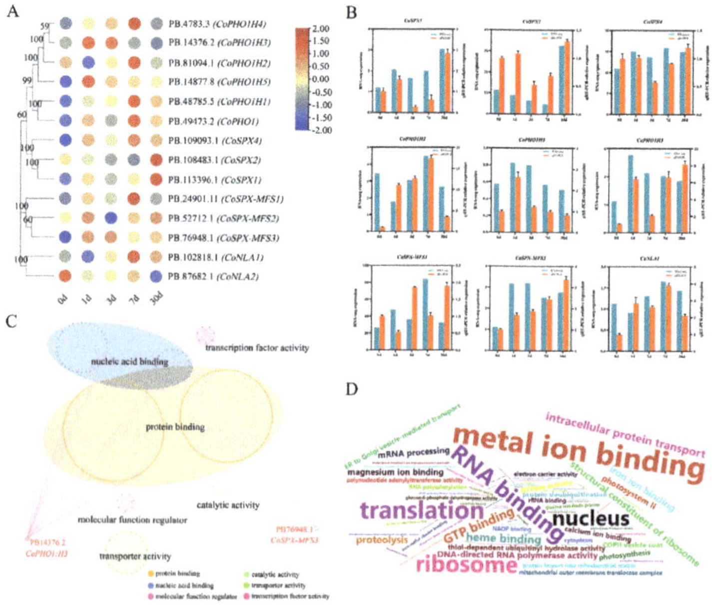

Figure 5. Expression profiles of *CoSPX* genes in plant roots under Pi stress conditions. (**A**) Gene expression data at 0, 1, 3, 7 and 30 days after the 5 μM Pi treatment were retrieved from an RNA-seq database. Expression levels are indicated by a gradient from low (blue) to high (red). (**B**) qRT-PCR validation and RNA-seq data of nine *CoSPX* genes. (**C**) *CoSPX-MFS3* gene co-expression regulatory network. The node genes are divided on the basis of the following six gene ontology (GO) terms, which are represented by different colors: protein binding, nucleic acid binding, molecular transducer activity, catalytic activity, transporter activity and transcription factor activity. (**D**) The frequency of GO terms.

2.8. Co-Expression Network of *CoSPX* Genes

To explore the roles of *CoSPX* genes in regulating Pi response gene expression, a co-expression modulatory axis was constructed according to the previous transcriptome analysis under low-Pi stress, among which two genes were identified as hub genes (Figure 5C). The hub gene-related nodes were involved in nucleic acid binding, protein binding, molecular function regulation, as well as transcription factor, catalytic, and transporter activities (Figure 5C). The frequency of GO terms was visualized using a word cloud (Figure 5D).

2.9. Sequence Alignment of *ClSPX-MFS2b* Homologous Genes

Based on our transcriptome data, a homologous gene *CoSPX-MFS3* in *C. oleifera* was selected as a key hub gene. Then, through multiple sequence alignment, we found the four closest species of CoSPX-MFS3 proteins with SPX and MFS domain (Figure 6A). By analyzing the SPX-MFS amino acid sequences of the 19 species, we found that CoSPX-MFS3 was the most closely related to ClSPX-MFS2b (Figure 6B). Thus, the functions of *CoSPX-MFS3* were analyzed by transcriptome combined with *C. lanceoleosa* diploid genome.

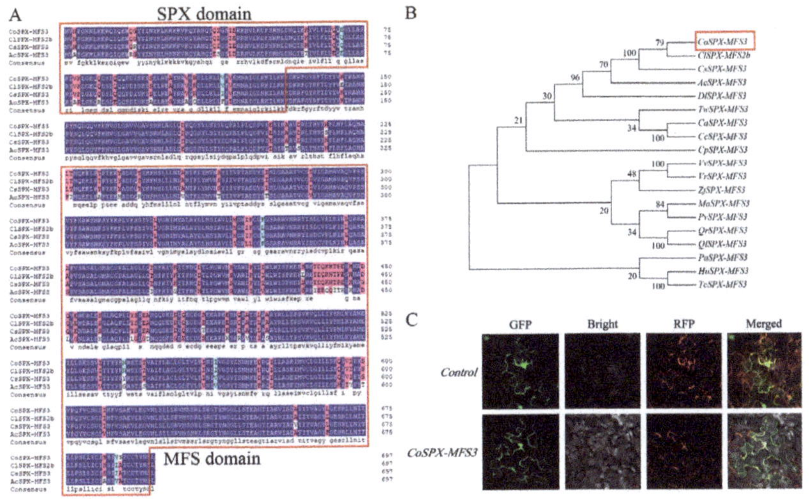

Figure 6. Identification of CoSPX-MFS3. (**A**) The amino acid sequence alignment from CoSPX-MFS3 and its orthologs. Red box means SPX domain and MFS domain, respectively (**B**) The phylogenetic relationships among the *SPX-MFS3* gene among 19 species. Red box means CoSPX-MFS3 (**C**) The subcellular localization of CoSPX-MFS3.

2.10. Localization of *CoSPX-MFS3* in the Tonoplast

Subcellular localization indicated that CoSPX-MFS3-EGFP fusion proteins were gathered in the tonoplast (Figure 6C), which is in line with the prediction from Cell-PLoc.

2.11. Overexpression of *CoSPX-MFS3* Enhances Pi Tolerance in *Arabidopsis*

To evaluate *CoSPX-MFS3* action during reduced Pi stress, *Arabidopsis* plants with *CoSPX-MFS3* overexpression (OE) were established. After qRT-PCR analysis of *CoSPX-MFS3* expression, the homozygous T3 lines were generated for further analysis (Figure S3). To examine potential *CoSPX-MFS3*-mediated regulation of Pi signaling and reduced Pi tolerance, we examined the phenotype of *Arabidopsis* plants overexpressing *CoSPX-MFS3* under Pi-deficiency (LP, 5 μM). The results showed a strong tolerance of *Arabidopsis* transgenic line with longer roots (Figure 7A,C). The accumulation of anthocyanin content is a marker of low-Pi stress in plants [42–44]. The degree of damage in leaves at 21 days after exposure to low-Pi stress was strongly elevated in wild-type (WT) versus *CoSPX-MFS3*

OE lines (Figure 7B). The total anthocyanin concentration of transgenic plants was considerably diminished, compared to WT plants (Figure 7F). The root biomass was clearly distinct between WT and transgenic lines, regardless of low or normal Pi (NP, 1 mm) stress treatment (Figure 7D,E). The total P content in *CoSPX-MFS3* OE lines was increased (Figure 7G,H). Moreover, the malic and citric acid contents, as well as acid phosphatase activity were significantly higher in roots and leaves of the OE line, relative to WT plants (Figure 8A–F). All results demonstrated that *CoSPX-MFS3* could enhance Pi tolerance by increased biomass and organic acid contents. In addition, seven Pi-responsive homologous genes were selected for expression level verifications in transgenic *Arabidopsis*. The qRT-PCR results showed that all of them were up-regulated in transgenic *Arabidopsis* compared to WT (Figure 9). In summary, we conclude that *CoSPX-MFS3* may play an important role in mediating Pi tolerance in transgenic *Arabidopsis*.

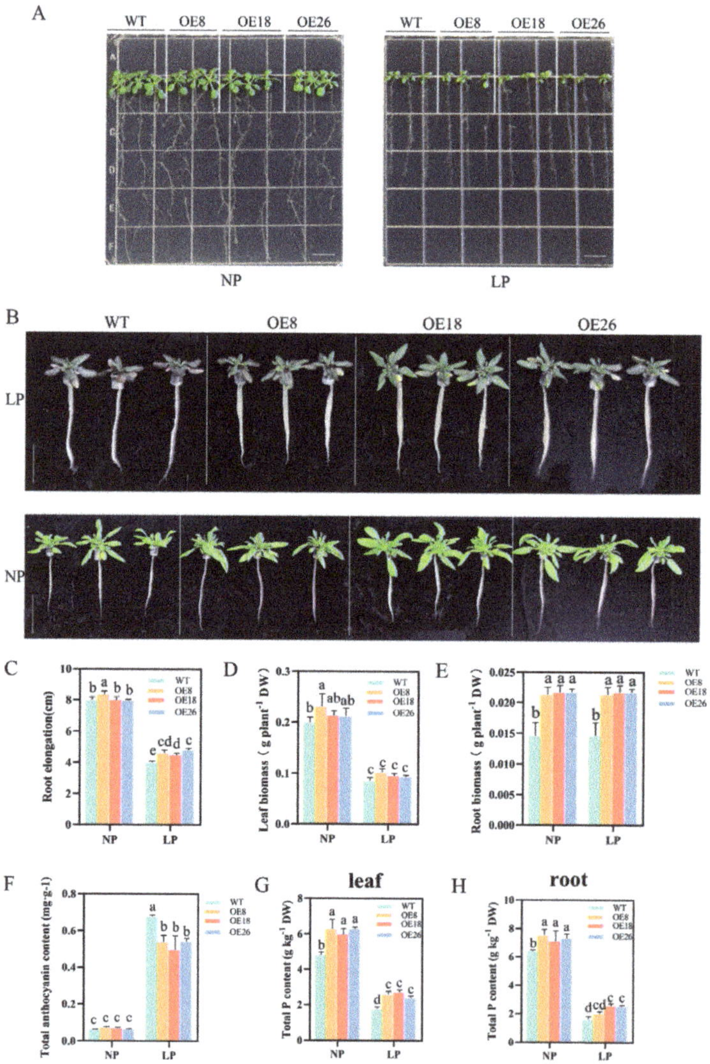

Figure 7. *CoSPX-MFS3*-overexpressing *Arabidopsis* plants exhibited strong tolerance to low-Pi stress. (**A**) Phenotypic analysis of *CoSPX-MFS3* overexpressing *Arabidopsis* under LP conditions. Seeds were

germinated on 1/2MS agar medium for 5 d, then seedlings were transferred to NP (1 mM) and LP (5 μM) medium for 7 d. (**B**) Phenotypes of *CoSPX-MFS3* overexpressing lines and wild-type (WT) plants under NP and LP conditions. Bar = 5 cm. (**C**) Root elongation of seedlings Bar = 1.5 cm. (**D**,**E**) The biomass of roots and leaves. (**F**) The anthocyanin content of the WT and transgenic lines. (**G**,**H**) The total P content of leaves and roots. Bars represent the mean ± standard deviation (SD) of at least three independent biological replicates. Different letters above the bars represent significant differences at $p < 0.05$.

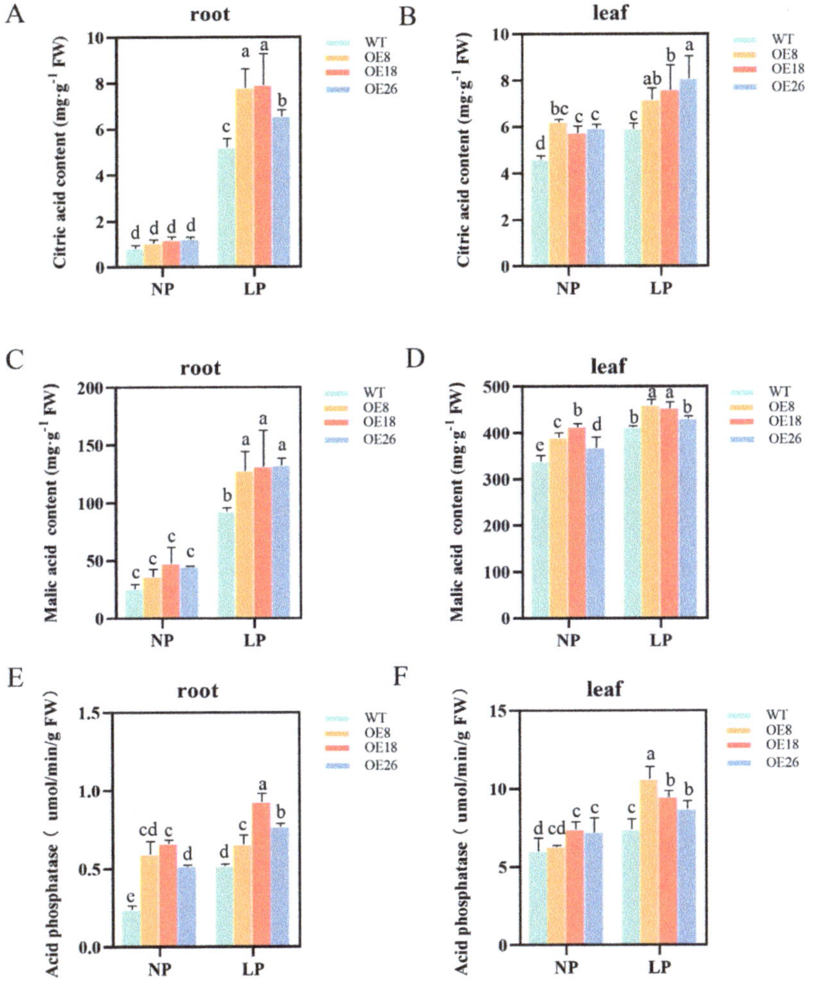

Figure 8. The organic acid content of *CoSPX-MFS3*-overexpressing *Arabidopsis* plants. (**A**,**B**) The citric acid content of the WT and transgenic lines. (**C**,**D**) The malic acid content. (**E**,**F**) The acid phosphatase content. Bars represent the mean ± standard deviation (SD) of at least three independent biological replicates. Different letters above the bars represent significant differences at $p < 0.05$.

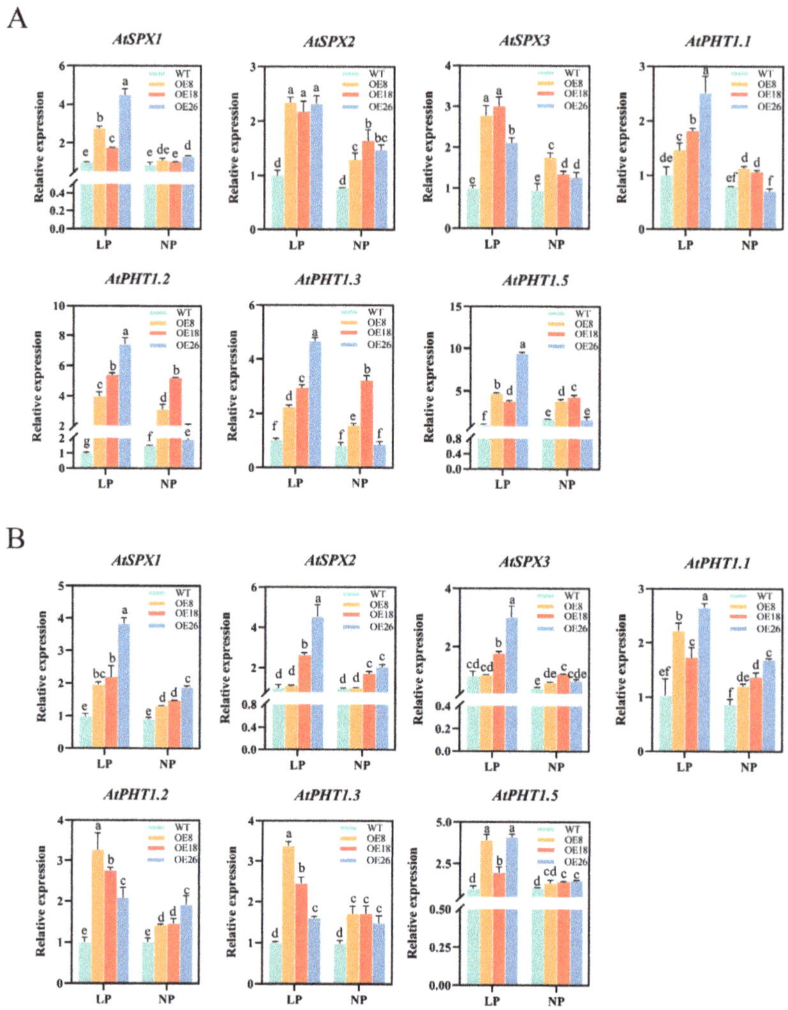

Figure 9. qRT-PCR analysis of the transcript expression of the homologous gene of WT and transgenic plants under two Pi levels in (**A**) root and (**B**) leaves. Different letters above the bars represent significant differences at $p < 0.05$.

3. Discussion

In response to low-Pi tolerance, plants form a series of biochemical and physiological adaptive mechanisms, which in turn enhance the absorption and utilization of soil Pi. Adaptive changes in plants under low-Pi stress are finely regulated by Pi signaling networks, in which *SPX* can regulate Pi-signaling networks in plants. Tea-oil *Camellia* mainly grows in acidic soil with very low P content in the southern area of China. Thus, assessing the biological functions of SPX proteins can elucidate the mechanisms underlying the adaption of *C. oleifera* to low-Pi pressure.

In our study, the total number of *SPX* genes in *C. lanceoleosa* was relatively similar to that in *Arabidopsis* (20) [11] and *Solanum lycopersicum* (19) [37], but less compared to wheat (46) [34], maize (33) [13] and *Brassica napus* (69) [35]. These differences in family members of different species may be due to environmental changes during evolution. The relationship between orthologs and paralogs shows that the diversity among the members of this gene family occurred before and after the divergences of monocots and dicots [45]. On the other hand, mutations in regulatory regions such as the promoter region and coding DNA regions probably caused the duplicated genes to have different expression patterns. [46,47]. Gene duplication is a crucial mechanism that can acquire new genes and create genetic novelty in organisms [48]. Eight putative *ClSPXs* were mapped to the corresponding chromosome according to the localization data of the *C. lanceoleosa* genome. The findings demonstrated that *ClSPXs* were non-uniformly distributed on the chromosomes of *C. lanceoleosa*. Meanwhile, it was found that segmental duplication occurred in *C. lanceoleosa* for expanding all *SPX* members during the process of evolution. The Ka/Ks rate is of great significance in reconstructing phylogeny and evolutionary selection [49]. Ka/Ks rates of <1, 1 and >1 represent purifying selection, neutral selection and positive selection, respectively [50]. The mean Ka/Ks value (0.177) of *C. lanceoleosa* demonstrated a strong purifying selection of *SPX* genes. In the same way, the Ka/Ks values of syntenic gene pairs between *C. lanceoleosa* and *Arabidopsis* or *C. sinensis* were less than 1. The Ka/Ks results imply that these paralogous gene pairs have experienced intense purifying selection during the evolutionary process.

The SPX conserved domain (SPX domain) plays a crucial role in modulating plant nutrient stress. In *Arabidopsis*, *AtSPX1* interacts with a unique four-stranded coiled-coil domain in *AtPHR1* [51]. In rice, SPX proteins (phosphate sensor) bind to OsPHR2 in high-phosphate condition, thus inhibiting OsPHR2 binding to P1BS elements in promoters of *OsRAM1*, *OsPT11*, *OsWRI5A* and *OsAMT3;1*, thus decreasing mycorrhizal-related gene expression and inhibiting mycorrhizal colonization [52]. In maize, *ZmSPX* genes (except *ZmSPX3*) were markedly induced by low-Pi stress [13]. In wheat, five *TaSPX* genes were up-regulated after Pi starvation treatment [34]. In this study, nine *CoSPX* genes had the same expression tendency, and eight genes (except *CoPHO1H3*) were up-regulated at 30 days after exposure to low-Pi stress, which was similar to other plants with low-Pi treatment. In general, SPX domain proteins as a low-Pi sensor. Two hub genes (*CoPHO1H3* and *CoSPX-MFS3*) were identified by WGCNA analysis. Meanwhile, co-expression analysis revealed that edge genes were involved in protein binding, nucleic acid binding, molecular function regulation, transporter activity, catalytic activity and transcription factor activity. These results demonstrate that SPX domain proteins serve an essential function in low-Pi stress perception and response.

In rice, *OsSPX-MFS3* was the first reported as a vacuolar Pi efflux transporter that modulates Pi homeostasis [18] and *OsSPX-MFS1* as a major Pi transporter that regulates leaf Pi homeostasis [38]. In *Brassica napus*, *BnA09PHT5;1b* and *BnCnPHT5;1b* were identified as two vacuolar Pi influx transporters [53]. The previous study revealed that SPX-MFS proteins primarily modulate Pi transport and homeostasis. Herein, we conducted a preliminary assessment of *CoSPX-MFS3* by heterologous transformed *Arabidopsis*. Overexpression of *CoSPX-MFS3* in *Arabidopsis* mainly enhanced low-Pi tolerance by increased biomass, organic acid content and total Pi content. Citric acid is an organic compound found in the roots of various plant species in response to Pi stress [54]. In white lupin, the citric acid content was increased at the later stage of Pi deficiency [55]. Although our results are different from rice and *Brassica napus*, this may be due to the long-term evolution of woody and herbaceous plants. Furthermore, overexpression of *CoSPX-MFS3* in *Arabidopsis* increased the transcription levels of a series of Pi starvation genes, including *PTs*, *SPX1*, *SPX2* and *SPX3*. This result is consistent with the previous findings that *AtSPX1* overexpression in *Arabidopsis* could increase the mRNA levels of *AtPAP2*, *AtACP5* and *AtRNS1* [11]. In summary, *CoSPX-MFS3* is a positive tolerator that responds to low-Pi stress. However, the detailed molecular mechanism of *CoSPX-MFS3* regulation in Pi acquisition and translocation need to be further identified.

4. Materials and Methods

4.1. Assessment of the SPX Family Genes in C. lanceoleosa

The *C. lanceoleosa* genome sequence and protein sequence information files were sourced from the NCBI database (https://www.ncbi.nlm.nih.gov/; accession number: PRJNA780224, accessed on 22 April 2023). To identify all SPX family members in *C. lanceoleosa*, the hidden Markov model (HMM) profiles of SPX domain (PF03105) were obtained using the Pfam database (http://pfam.xfam.org/, accessed on 22 April 2023). The 20 known SPX domain sequences from *Arabidopsis* were retrieved using the BLASTP search in TAIR (https://www.arabidopsis.org/, accessed on 22 April 2023) according to previous methods. After manually removing the redundant protein sequences of *C. lanceoleosa*, the candidate protein sequence containing a complete SPX domain (PF03105) was regarded as the final SPX protein sequence according to the NCBI Conserved Domain Database (CDD; https://www.ncbi.nlm.nih.gov/cdd/, accessed on 22 April 2023). These *C. lanceoleosa SPX* genes were renamed *ClSPXs*. The characteristics and subcellular localization of ClSPXs protein sequences were assessed using the ExPASY (http://Web.ExPASY.Org/protparam/, accessed on 22 April 2023) and Cell-PLoc 2.0 (http://www.csbio.sjtu.edu.cn/bioinf/Cell-PLoc/, accessed on 22 April 2023) tools, respectively.

4.2. Construction of Phylogenetic Trees

Full-length protein sequences from *Arabidopsis* and *Camellia sinensis* were acquired from the NCBI protein database to determine their evolution. The definitized amino acid sequences of ClSPXs were subjected to multiple sequence alignment via ClustalW (https://www.genome.jp/tools-bin/clustalw, accessed on 22 April 2023). The maximum-likelihood criteria in MEGA 7.0 were used to establish phylogenetic trees, with Poisson corrections and 1000 bootstrap replicates. The recognized *ClSPX* genes were assigned to distinct categories based on the *AtSPXs* stratification. Interactive Tree of Life (iTOL.7) software (https://itol.embl.de/, accessed on 22 April 2023) was employed for visualization and modification of the phylogenetic tress.

4.3. Assessment of ClSPX Gene Domain, Structure and Motif Sequences

The intron/exon map structure was drawn using the tools of the Gene Structure Display Server v2.0 (http://gsds.gao-lab.org/, accessed on 22 April 2023) platform. The conserved motifs in ClSPX proteins were determined using the MEME program v5.0.5 (http://meme-suite.org/tools/meme, accessed on 22 April 2023), with a maximum of 25 motifs and optimal motif width range of 6–50 amino acid residues.

4.4. Determination of Cis-Acting Elements in ClSPX Promoter

The 2000-bp sequences upstream of the *ClSPX* translation initiation codon were analyzed through the PlantCARE database in order to determine the cis-regulatory elements.

4.5. Chromosomal Location and Synteny-Based Evaluation of ClSPX Genes

The chromosomal location data of *ClSPXs* were extracted from the *C. lanceoleosa* genome annotation file. This information was also used to construct chromosomal mapping by TBtools [41]. Gene duplication assessment was conducted via the One-Step MCScanX function in TBtools, and the result was visualized by the CGview tool [41].

4.6. CoSPXs Expression Profiles

To determine the expression profiles of *CoSPX* at different stages (0, 1, 3, 7 and 30 days), the FPKM values of *CoSPX* genes at five treatment stages of *C. oleifera* were obtained from our previous RNA-seq data [56]. The expression profiles of *CoSPXs* at different stress stages were analyzed based on their FPKM values, and a heatmap was produced using the R heatmap function (gplots). Z-score normalization was used to normalize the expression values. Total RNA isolation was employed by the RNA kit (Aidlab Biotechnologies Co., Ltd., Beijing, China) following the manufacturer's instructions. To conduct reverse transcription,

PrimeScriptTM RT Master Mix (TaKaRa, Dalian, China) was used. Then, quantitative real-time PCR (qRT-PCR) analysis was conducted on the 7300 Real-Time PCR System (Applied Biosystems, Shanghai, CA, USA) with the help of the TB Green® Premix Ex Taq™ II (Tli RNaseH Plus, Takara). The "Genes" module in SPDE software (Version 2.0)was used to design gene-specific primers [57]. The *GAPDH* gene served as the endogenous control for *CoSPXs* [58]. Actin gene was chosen as an internal reference for AtSPXs. Relative gene expression was computed via the $2^{-\Delta\Delta Ct}$ formula [59]. The employed primers are detailed in Supplemental Table S5.

4.7. Co-Expression Regulatory Network of CoSPX Genes

Among all co-expressed genes, *CoSPX* genes exhibited the largest interconnection. Thus, they were established as hub genes. Edge genes were annotated by GO terms, and their frequency was visualized in a cloud word (https://www.bioladder.cn/web/#/chart/20, accessed on 22 April 2023). Finally, the co-expression regulatory network was performed in Cytoscape 3.8.2.

4.8. Multiple Sequence Alignment, Subcellular Localization of CoSPX-MFS3

The SPX-MFS3 amino acid sequences of 19 different species were downloaded from NCBI for evolutionary analysis, and a multiple sequence alignment was constructed. The CoSPX-MFS3 coding sequences were fused with the mGFP-encoding sequences in the pMDC43 expression vector with the CloneExpress II One Step Cloning Kit (Vazyme, Nanjing, China). For transient expression analysis, the recombinant plasmids were incorporated into *Agrobacterium tumefaciens* EHA105 cells, prior to transfer to *Nicotiana benthamiana* leaves. The empty vectors were employed as controls. After three days, the transient expression of GFP-fusion proteins was determined using an LSM900 confocal microscope imaging system (Zeiss, Shanghai, China). mCherry-labeled vacuole markers were used to visualize the tonoplast.

4.9. Ectopic CoSPX-MFS3 Expression in Arabidopsis and Pi Exposure

The full-length CDS of *CoSPX-MFS3* was amplified from *C. oleifera* and inserted into the *CaMV 35S* promoter-regulated binary vector pEXT06 employing the ClonExpress II One Step Cloning Kit (Vazyme). Subsequently, *A. tumefaciens* GV3101-regulated transformation of *Arabidopsis* was employed. The transgenic *Arabidopsis* seeds were harvested from individual plants, and positive lines were identified on 1/2 MS medium using hygromycin until homozygous transgenic *Arabidopsis* lines were obtained. Using 75% ethanol, the homozygous seeds of *Arabidopsis* underwent surface sterilization three times, prior to a 24-h incubation at 4 °C. After incubation, the seeds were placed on petri dishes containing Pi-sufficient (normal Pi [NP], 1 mm Pi) medium for 5 days, prior to transfer to Pi-deficient (low Pi [LP], 5 μM Pi) medium for 7 days with a 16:8-h light/dark cycle at 22 °C. Wild-type (WT) plants and T3 homozygous transgenic lines were cultivated in 1/2-strength Hoagland solution (Coolaber, China) for one week. Subsequently, the plants were placed in 1/2-strength phosphorus-free solution and grown for three weeks. The fresh roots and leaves were immediately frozen in liquid nitrogen and kept at −80 °C. The roots and leaves were harvested for the determination of biomass.

4.10. Measurement of Total Pi Content

For determination total P content, the method described in a previous study [60] were strictly followed.

4.11. Determination of Anthocyanin and Organic acid Content

Fresh roots and leaves (0.1 g) of *Arabidopsis* were crushed in liquid nitrogen and incubated with 1% HCl-methanol for 24 h. The optical density values of anthocyanin were recorded at 530 nm (Tang et al., 2022). Commercial kits (Geruisi-bio, Suzhou, China) were employed for the detection of malic acid (Lot. G0862W48), citric acid (Lot. G0864F) and acid phosphatase (Lot.G0903W) contents. Briefly, *Arabidopsis* samples (0.1 g) were extracted using the kit's buffer, followed by centrifugation (10,000 rpm, 15 min, 4 °C) [41]. The assays were performed in triplicate.

5. Conclusions

In summary, 20 *SPX* genes were identified in the *C. lanceoleosa* genome, and their conserved domain, motif distribution, gene duplication, cis-acting elements and chromosomal distribution were analyzed. Segmental duplication event was the key factor affecting the evolutionary process of the SPX gene family in *C. lanceoleosa* based on collinearity analysis. Additionally, the expression distribution of *SPX* genes in low-Pi stress conditions were analyzed using transcriptome data and qRT-PCR experiment. A hub gene *CoSPX-MFS3* induced by low-Pi stress could enhance low-Pi tolerance in transgenic *Arabidopsis*. Altogether, these findings present a reference basis for the enhanced understanding of the physiological roles of *SPX* genes in *C. oleifera*.

Supplementary Materials: The following supporting information can be downloaded at: https://www.mdpi.com/article/10.3390/ijms241411552/s1.

Author Contributions: J.C., X.Y. and R.Z. designed and conducted the experiments. J.C., B.Y. and L.L. performed the experiments. J.C. anlayzed the data and wrote the manuscript. J.C. and X.H. contributed to discussion in the writing process. J.C., X.H., X.Y. and R.Z. revised the manuscript. All authors have read and agreed to the published version of the manuscript.

Funding: This study was supported by the National Key R&D Program of China (2019YFD1001602).

Institutional Review Board Statement: Not applicable.

Informed Consent Statement: Not applicable.

Data Availability Statement: The *C. lanceoleosa* genome sequence and protein sequence information files were sourced from the NCBI database (accession number: PRJNA780224).

Conflicts of Interest: The authors declare no conflict of interest.

Abbreviations

P	Phosphours
Pi	Phosphate
LP	Low Pi
NP	Normal Pi
PSR	Phosphate starvation respsonse
SYG1	Suppressor of Yeast gpa1
XPR1	Xenotropic and Polytropic Retrovirus receptor1
NLA	Nitrogen Limiting Adaptation
ABRE	Abscisic acid response element
PHT5	Phosphate transporter 5 family
FPKM	Fragments Per Kilobase of transcript sequence per Millions base pairs
NJ	Neighbor-joining
HMM	Hidden Markov Model
qRT-PCR	Quantitative real-time polymerase chain reaction

References

1. Lambers, H. Phosphorus acquisition and utilization in plants. *Annu. Rev. Plant Biol.* **2022**, *73*, 11–126. [CrossRef]
2. Malhotra, H.; Sharma, S.; Pandey, R. Phosphorus nutrition: Plant growth in response to deficiency and excess. In *Plant Nutrients and Abiotic Stress Tolerance*; Springer: Berlin/Heidelberg, Germany, 2018; pp. 171–190.
3. Plaxton, W.C.; Tran, H.T. Metabolic adaptations of phosphate-starved plants. *Plant Physiol.* **2011**, *156*, 1006–1015. [CrossRef] [PubMed]
4. Holford, I. Soil phosphorus: Its measurement, and its uptake by plants. *Soil Res.* **1997**, *35*, 227–240. [CrossRef]
5. Wang, Z.; Kuo, H.-F.; Chiou, T.-J. Intracellular phosphate sensing and regulation of phosphate transport systems in plants. *Plant Physiol.* **2021**, *187*, 2043–2055. [CrossRef]

6. Raghothama, K.; Karthikeyan, A. Phosphate acquisition. *Plant Soil* **2005**, *274*, 37–49. [CrossRef]
7. Secco, D.; Wang, C.; Arpat, B.A.; Wang, Z.; Poirier, Y.; Tyerman, S.D.; Wu, P.; Shou, H.; Whelan, J. The emerging importance of the SPX domain-containing proteins in phosphate homeostasis. *New Phytol.* **2012**, *193*, 842–851. [CrossRef]
8. Secco, D.; Wang, C.; Shou, H.; Whelan, J. Phosphate homeostasis in the yeast *Saccharomyces cerevisiae*, the key role of the SPX domain-containing proteins. *FEBS Lett.* **2012**, *586*, 289–295. [CrossRef]
9. Wykoff, D.D.; O'Shea, E.K. Phosphate transport and sensing in *Saccharomyces cerevisiae*. *Genetics* **2001**, *159*, 1491–1499. [CrossRef] [PubMed]
10. Hürlimann, H.C.; Stadler-Waibel, M.; Werner, T.P.; Freimoser, F.M. *Pho91* is a vacuolar phosphate transporter that regulates phosphate and polyphosphate metabolism in *Saccharomyces cerevisiae*. *Mol. Biol. Cell* **2007**, *18*, 4438–4445. [CrossRef] [PubMed]
11. Duan, K.; Yi, K.; Dang, L.; Huang, H.; Wu, W.; Wu, P. Characterization of a sub-family of *Arabidopsis* genes with the SPX domain reveals their diverse functions in plant tolerance to phosphorus starvation. *Plant J.* **2008**, *54*, 965–975. [CrossRef] [PubMed]
12. Qi, W. Characterization of SPX Exclusive Family Members in Plant Pi Sensing and Regulation. Ph.D. Thesis, University of Leeds, Leeds, UK, 2016.
13. Xiao, J.; Xie, X.; Li, C.; Xing, G.; Cheng, K.; Li, H.; Liu, N.; Tan, J.; Zheng, W. Identification of SPX family genes in the maize genome and their expression under different phosphate regimes. *Plant Physiol. Biochem.* **2021**, *168*, 211–220. [CrossRef] [PubMed]
14. Park, B.S.; Seo, J.S.; Chua, N.-H. NITROGEN LIMITATION ADAPTATION recruits PHOSPHATE2 to target the phosphate transporter PT2 for degradation during the regulation of *Arabidopsis* phosphate homeostasis. *Plant Cell* **2014**, *26*, 454–464. [CrossRef] [PubMed]
15. Lin, W.-Y.; Huang, T.-K.; Chiou, T.-J. NITROGEN LIMITATION ADAPTATION, a target of microRNA827, mediates degradation of plasma membrane–localized phosphate transporters to maintain phosphate homeostasis in *Arabidopsis*. *Plant Cell* **2013**, *25*, 4061–4074. [CrossRef] [PubMed]
16. Liu, J.; Yang, L.; Luan, M.; Wang, Y.; Zhang, C.; Zhang, B.; Shi, J.; Zhao, F.-G.; Lan, W.; Luan, S. A vacuolar phosphate transporter essential for phosphate homeostasis in *Arabidopsis*. *Proc. Natl. Acad. Sci. USA* **2015**, *112*, E6571–E6578. [CrossRef]
17. Liu, F.; Xu, Y.; Jiang, H.; Jiang, C.; Du, Y.; Gong, C.; Wang, W.; Zhu, S.; Han, G.; Cheng, B. Systematic identification, evolution and expression analysis of the Zea mays PHT1 gene family reveals several new members involved in root colonization by arbuscular mycorrhizal fungi. *Int. J. Mol. Sci.* **2016**, *17*, 930. [CrossRef]
18. Wang, C.; Yue, W.; Ying, Y.; Wang, S.; Secco, D.; Liu, Y.; Whelan, J.; Tyerman, S.D.; Shou, H. Rice SPX-Major Facility Superfamily3, a vacuolar phosphate efflux transporter, is involved in maintaining phosphate homeostasis in Rice1. *Plant Physiol.* **2015**, *169*, 2822–2831.
19. Poirier, Y.; Bucher, M. Phosphate transport and homeostasis in *Arabidopsis*. *Arab. Book/Am. Soc. Plant Biol.* **2002**, *1*, e0024. [CrossRef]
20. Rausch, C.; Bucher, M. Molecular mechanisms of phosphate transport in plants. *Planta* **2002**, *216*, 23–37. [CrossRef]
21. Wang, Y.; Ribot, C.; Rezzonico, E.; Poirier, Y. Structure and expression profile of the *Arabidopsis* PHO1 gene family indicates a broad role in inorganic phosphate homeostasis. *Plant Physiol.* **2004**, *135*, 400–411. [CrossRef]
22. Secco, D.; Baumann, A.; Poirier, Y. Characterization of the rice PHO1 gene family reveals a key role for *OsPHO1; 2* in phosphate homeostasis and the evolution of a distinct clade in dicotyledons. *Plant Physiol.* **2010**, *152*, 1693–1704. [CrossRef]
23. Wege, S.; Khan, G.A.; Jung, J.-Y.; Vogiatzaki, E.; Pradervand, S.; Aller, I.; Meyer, A.J.; Poirier, Y. The EXS domain of PHO1 participates in the response of shoots to phosphate deficiency via a root-to-shoot signal. *Plant Physiol.* **2016**, *170*, 385–400. [CrossRef]
24. Stefanovic, A.; Arpat, A.B.; Bligny, R.; Gout, E.; Vidoudez, C.; Bensimon, M.; Poirier, Y. Over-expression of *PHO1* in *Arabidopsis* leaves reveals its role in mediating phosphate efflux. *Plant J.* **2011**, *66*, 689–699. [CrossRef]
25. Chaiwong, N.; Prom-u-Thai, C.; Bouain, N.; Lacombe, B.; Rouached, H. Individual versus combinatorial effects of silicon, phosphate, and iron deficiency on the growth of lowland and upland rice varieties. *Int. J. Mol. Sci.* **2018**, *19*, 899. [CrossRef] [PubMed]
26. Ma, J.; Ye, H.; Rui, Y.; Chen, G.; Zhang, N. Fatty acid composition of *Camellia oleifera* oil. *J. Für Verbraucherschutz Und Leb.* **2011**, *6*, 9–12. [CrossRef]
27. Yu, J.; Yan, H.; Wu, Y.; Wang, Y.; Xia, P. Quality Evaluation of the Oil of *Camellia* spp. *Foods* **2022**, *11*, 2221. [CrossRef] [PubMed]
28. Zhang, F.; Zhu, F.; Chen, B.; Su, E.; Chen, Y.; Cao, F. Composition, bioactive substances, extraction technologies and the influences on characteristics of *Camellia oleifera* oil: A review. *Food Res. Int.* **2022**, *156*, 111159. [CrossRef] [PubMed]
29. Quan, W.; Wang, A.; Gao, C.; Li, C. Applications of Chinese *Camellia oleifera* and its by-products: A review. *Front. Chem.* **2022**, *10*, 921246. [CrossRef]
30. Luan, F.; Zeng, J.; Yang, Y.; He, X.; Wang, B.; Gao, Y.; Zeng, N. Recent advances in Camellia oleifera Abel: A review of nutritional constituents, biofunctional properties, and potential industrial applications. *J. Funct. Foods* **2020**, *75*, 104242. [CrossRef]
31. Gong, W.; Xiao, S.; Wang, L.; Liao, Z.; Chang, Y.; Mo, W.; Hu, G.; Li, W.; Zhao, G.; Zhu, H. Chromosome-level genome of *Camellia lanceoleosa* provides a valuable resource for understanding genome evolution and self-incompatibility. *Plant J.* **2022**, *110*, 881–898. [CrossRef]
32. Liu, N.; Shang, W.; Li, C.; Jia, L.; Wang, X.; Xing, G.; Zheng, W. Evolution of the SPX gene family in plants and its role in the response mechanism to phosphorus stress. *Open Biol.* **2018**, *8*, 170231. [CrossRef]

33. Luan, M.; Zhao, F.; Sun, G.; Xu, M.; Fu, A.; Lan, W.; Luan, S. A SPX domain vacuolar transporter links phosphate sensing to homeostasis in Arabidopsis. *Mol. Plant* **2022**, *15*, 1590–1601. [CrossRef]
34. Kumar, A.; Sharma, M.; Gahlaut, V.; Nagaraju, M.; Chaudhary, S.; Kumar, A.; Tyagi, P.; Gajula, M.P.; Singh, K.P. Genome-wide identification, characterization, and expression profiling of SPX gene family in wheat. *Int. J. Biol. Macromol.* **2019**, *140*, 17–32. [CrossRef] [PubMed]
35. Du, H.; Yang, C.; Ding, G.; Shi, L.; Xu, F. Genome-wide identification and characterization of SPX domain-containing members and their responses to phosphate deficiency in Brassica napus. *Front. Plant Sci.* **2017**, *8*, 35. [CrossRef]
36. Liu, B.; Zhao, S.; Wu, X.; Wang, X.; Nan, Y.; Wang, D.; Chen, Q. Identification and characterization of phosphate transporter genes in potato. *J. Biotechnol.* **2017**, *264*, 17–28. [CrossRef] [PubMed]
37. Li, M.; Yu, B. Recent advances in the regulation of plant miRNA biogenesis. *RNA Biol.* **2021**, *18*, 2087–2096. [CrossRef] [PubMed]
38. Wang, C.; Huang, W.; Ying, Y.; Li, S.; Secco, D.; Tyerman, S.; Whelan, J.; Shou, H. Functional characterization of the rice SPX-MFS family reveals a key role of *OsSPX-MFS1* in controlling phosphate homeostasis in leaves. *New Phytol.* **2012**, *196*, 139–148. [CrossRef]
39. Singh, I.; Smita, S.; Mishra, D.C.; Kumar, S.; Singh, B.K.; Rai, A. Abiotic stress responsive miRNA-target network and related markers (SNP, SSR) in Brassica juncea. *Front. Plant Sci.* **2017**, *8*, 1943. [CrossRef]
40. Yang, J.; Zhao, X.; Chen, Y.; Li, G.; Li, X.; Xia, M.; Sun, Z.; Chen, Y.; Li, Y.; Yao, L. Identification, Structural, and Expression Analyses of SPX Genes in Giant Duckweed (Spirodela polyrhiza) Reveals Its Role in Response to Low Phosphorus and Nitrogen Stresses. *Cells* **2022**, *11*, 1167. [CrossRef]
41. Jiang, C.; Gao, X.; Liao, L.; Harberd, N.P.; Fu, X. Phosphate Starvation Root Architecture and Anthocyanin Accumulation Responses Are Modulated by the Gibberellin-DELLA Signaling Pathway in *Arabidopsis*. *Plant Physiol.* **2007**, *145*, 1460–1470. [CrossRef]
42. Wang, X.; Du, G.; Wang, X.; Meng, Y.; Li, Y.; Wu, P.; Yi, K. The Function of LPR1 is controlled by an element in the promoter and is independent of SUMO E3 Ligase SIZ1 in response to low Pi stress in *Arabidopsis thaliana*. *Plant Cell Physiol.* **2010**, *51*, 380–394. [CrossRef]
43. Xu, L.; Jin, L.; Long, L.; Liu, L.; He, X.; Gao, W.; Zhu, L.; Zhang, X. Overexpression of *GbWRKY1* positively regulates the Pi starvation response by alteration of auxin sensitivity in *Arabidopsis*. *Plant Cell Rep.* **2012**, *31*, 2177–2188. [CrossRef] [PubMed]
44. Ahmadizadeh, M.; Rezaee, S.; Heidari, P. Genome-wide characterization and expression analysis of fatty acid desaturase gene family in *Camelina sativa*-ScienceDirect. *Gene Rep.* **2020**, *21*, 100894. [CrossRef]
45. Yaghobi, M.; Heidari, P. Genome-Wide Analysis of Aquaporin Gene Family in *Triticum turgidum* and its expression profile in response to salt stress. *Genes* **2023**, *14*, 202. [CrossRef] [PubMed]
46. Hashemipetroudi, S.H.; Arab, M.; Heidari, P.; Kuhlmann, M. Genome-wide analysis of the laccase (LAC) gene family in Aeluropus littoralis: A focus on identification, evolution and expression patterns in response to abiotic stresses and ABA treatment. *Front. Plant Sci.* **2023**, *14*, 1112354. [CrossRef]
47. Magadum, S.; Banerjee, U.; Murugan, P.; Gangapur, D.; Ravikesavan, R. Gene duplication as a major force in evolution. *J. Genet.* **2013**, *92*, 155–161. [CrossRef]
48. Zhang, Z.; Li, J.; Zhao, X.-Q.; Wang, J.; Wong, G.K.-S.; Yu, J. KaKs_Calculator: Calculating Ka and Ks through model selection and model averaging. *Genom. Proteom. Bioinform.* **2006**, *4*, 259–263. [CrossRef]
49. Li, J.; Zhang, Z.; Vang, S.; Yu, J.; Wong, G.K.-S.; Wang, J. Correlation between Ka/Ks and Ks is related to substitution model and evolutionary lineage. *J. Mol. Evol.* **2009**, *68*, 414–423. [CrossRef] [PubMed]
50. Ried, M.K.; Wild, R.; Zhu, J.; Pipercevic, J.; Sturm, K.; Broger, L.; Harmel, R.K.; Abriata, L.A.; Hothorn, L.A.; Fiedler, D. Inositol pyrophosphates promote the interaction of SPX domains with the coiled-coil motif of PHR transcription factors to regulate plant phosphate homeostasis. *Nat. Commun.* **2021**, *12*, 384. [CrossRef]
51. Shi, J.; Zhao, B.; Zheng, S.; Zhang, X.; Wang, X.; Dong, W.; Xie, Q.; Wang, G.; Xiao, Y.; Chen, F. A phosphate starvation response-centered network regulates mycorrhizal symbiosis. *Cell* **2021**, *184*, 5527–5540.e18. [CrossRef]
52. Han, B.; Wang, C.; Wu, T.; Yan, J.; Jiang, A.; Liu, Y.; Luo, Y.; Cai, H.; Ding, G.; Dong, X. Identification of vacuolar phosphate influx transporters in Brassica napus. *Plant Cell Environ.* **2022**, *45*, 3338–3353. [CrossRef]
53. Neumann, G.; Massonneau, A.; Martinoia, E.; Römheld, V. Physiological adaptations to phosphorus deficiency during proteoid root development in white lupin. *Planta* **1999**, *208*, 373–382. [CrossRef]
54. Neumann, G.; Römheld, V. Root excretion of carboxylic acids and protons in phosphorus-deficient plants. *Plant Soil* **1999**, *211*, 121–130. [CrossRef]
55. Chen, C.; Chen, H.; Zhang, Y.; Thomas, H.R.; Frank, M.H.; He, Y.; Xia, R. TBtools: An integrative toolkit developed for interactive analyses of big biological data. *Mol. Plant* **2020**, *13*, 1194–1202. [CrossRef]
56. Chen, J.; Han, X.; Ye, S.; Liu, L.; Yang, B.; Cao, Y.; Zhuo, R.; Yao, X. Integration of small RNA, degradome, and transcriptome sequencing data illustrates the mechanism of low phosphorus adaptation in Camellia oleifera. *Front. Plant Sci.* **2022**, *13*, 932926. [CrossRef] [PubMed]
57. Xu, D.; Lu, Z.; Jin, K.; Qiu, W.; Qiao, G.; Han, X.; Zhuo, R. SPDE: A multi-functional software for sequence processing and data extraction. *Bioinformatics* **2021**, *37*, 3686–3687. [CrossRef]
58. Zhou, C.; Lin, P.; Yao, X.; Wang, K.; Chang, J.; Han, X. Selection of reference genes for quantitative real-time PCR in six oil-tea *camellia* based on RNA-seq. *Mol. Biol.* **2013**, *47*, 836–851. [CrossRef]

59. Livak, K.J.; Schmittgen, T.D. Analysis of relative gene expression data using real-time quantitative PCR and the $2^{-\Delta\Delta CT}$ method. *Methods* **2001**, *25*, 402–408. [CrossRef] [PubMed]
60. Jia, H.; Ren, H.; Gu, M.; Zhao, J.; Sun, S.; Zhang, X.; Chen, J.; Wu, P.; Xu, G. The phosphate transporter gene OsPht1; 8 is involved in phosphate homeostasis in rice. *Plant Physiol.* **2011**, *156*, 1164–1175. [CrossRef]

International Journal of
Molecular Sciences

MDPI

Communication

ABF1 Positively Regulates Rice Chilling Tolerance via Inducing Trehalose Biosynthesis

Yazhou Shu [1,2,†], Wensheng Zhang [3,†], Liqun Tang [1], Zhiyong Li [1], Xinyong Liu [1], Xixi Liu [1], Wanning Liu [1], Guanghao Li [1], Jiezheng Ying [1], Jie Huang [1], Xiaohong Tong [1], Honghong Hu [2], Jian Zhang [1,*] and Yifeng Wang [1,*]

[1] State Key Laboratory of Rice Biology and Breeding, China National Rice Research Institute, Hangzhou 311400, China; mm123456m@126.com (Y.S.); liquntang2013@126.com (L.T.); lzhy1418@163.com (Z.L.); liuxinyong1234@gmail.com (X.L.); 18338690086@163.com (X.L.); dearliuwanning@126.com (W.L.); guanghao_12@163.com (G.L.); yingjiezheng@caas.cn (J.Y.); huangjie@caas.cn (J.H.); tongxiaohong@caas.cn (X.T.)

[2] College of Life Science and Technology, Huazhong Agricultural University, Wuhan 430070, China; huhh@mail.hzau.edu.cn

[3] School of Life Sciences, Hubei University, Wuhan 430062, China; zz964419115@163.com

* Correspondence: zhangjian@caas.cn (J.Z.); wangyifeng@caas.cn (Y.W.); Tel./Fax: +86-571-6337-0277 (J.Z.); +86-571-6337-0206 (Y.W.)

† These authors contributed equally to this work.

Abstract: Chilling stress seriously limits grain yield and quality worldwide. However, the genes and the underlying mechanisms that respond to chilling stress remain elusive. This study identified ABF1, a cold-induced transcription factor of the bZIP family. Disruption of *ABF1* impaired chilling tolerance with increased ion leakage and reduced proline contents, while *ABF1* over-expression lines exhibited the opposite tendency, suggesting that ABF1 positively regulated chilling tolerance in rice. Moreover, SnRK2 protein kinase SAPK10 could phosphorylate ABF1, and strengthen the DNA-binding ability of ABF1 to the G-box *cis*-element of the promoter of *TPS2*, a positive regulator of trehalose biosynthesis, consequently elevating the *TPS2* transcription and the endogenous trehalose contents. Meanwhile, applying exogenous trehalose enhanced the chilling tolerance of *abf1* mutant lines. In summary, this study provides a novel pathway 'SAPK10-ABF1-*TPS2*' involved in rice chilling tolerance through regulating trehalose homeostasis.

Keywords: ABF1; trehalose; chilling tolerance; phosphorylation; rice (*Oryza sativa* L.)

1. Introduction

As staple food for more than half of the world's population, rice (*Oryza sativa* L.) is highly sensitive to chilling stress [1]. With an expanding rice cultivation area from the original tropical and sub-tropical regions to high-latitude areas, chilling stress becomes a major factor that damages crop growth and development, restricts geographical distribution, and eventually constrains yield and quality [2]. Therefore, it is of great value to uncover the molecular mechanism in response to chilling stress for breeding chilling-tolerant varieties.

In the past decades, evidence has shown that trehalose enhances plants' growth and development in response to cold stress [3–5]. The synthesis of trehalose mainly includes two steps, the first is the conversion of UDP (uridine diphosphate)-glucose and glucose-6-phosphate to T6P (Trehalose-6-phosphate) and UDP by TPS (Trehalose-6-phosphate synthase), then TPP (Trehalose-6-phosphate phosphatase) dephosphorylates T6P to trehalose, eventually trehalase (TRE) degraded trehalose into two glucose [6,7]. In Arabidopsis thaliana, over-expression of *TPPI*, a member of the TPP family, improves seedling chilling tolerance through accumulating Jasmonic acid and soluble sugar, such as trehalose, while *tppI* mutant lines are sensitive to low temperatures [8]. Overexpression of *TPS1* or *TPP1* increases trehalose content and strengthens seedlings chilling adaptation in rice [3,9]. Similarly, the over-expression of wheat *TPS11* in Arabidopsis shows a higher survival rate in response to chilling stress compared with the wild type [10]. On the other hand, applying exogenous trehalose could enhance chilling tolerance by increasing endogenous trehalose levels, root growth, and water uptake [3,5,11]. The application of trehalose could also improve the efficiency and quantum yield of PS-II of the photosynthetic system, increase NO and H_2O_2 levels to enhance antioxidant activity and suppress membrane lipid peroxidation, whereas reducing the electrolyte leakage to maintain plant growth and harvest in response to cold stress [4].

The basic leucine zipper (bZIP) transcription factors have a bZIP domain composed of two structural features: a DNA-binding base region and a leucine zipper dimerization region [12]. Numerous bZIP transcription factors have been reported to play essential roles in plants' survival under cold environments [13–15]. In rice, the expression of *bZIP52* is strongly induced by low temperatures. Several abiotic adversity-related genes such as *LEA3*, *TPP1*, and *Rab25* are down-regulated in *bZIP52* over-expression (*OxbZIP52*) lines, and *OxbZIP52* shows increased cold sensitivity, suggesting that bZIP52 acts as a negative regulator of cold stress response [16]. bZIP73 regulates rice cold tolerance not only in the seedling period, but also during the heading period. Japonica type $bZIP73^{Jap}$ could interact with bZIP71 to control the balance of abscisic acid (ABA) and reactive oxygen species, thereby improving rice chilling tolerance [13,17]. In wheat, *bZIP60* is induced by ethylene glycol, salt, cold, and ABA, and over-expression of *bZIP60* in Arabidopsis significantly improves the seedling chilling tolerance [18]. bZIP96 is reported to interact with ICE1 (inducer of CBF expression) to positively regulate wheat cold adaptation [19]. In maize, bZIP68, a negative regulator inhibiting seedling chilling tolerance, is phosphorylated by MPK8 to promote protein stability and the DNA binding affinity to the promoter of *DREB 1.7* under chilling stress [14]. In chrysanthemums, Bai et al. (2022) found that bZIP3 could interact with bZIP2 to activate the expression of *POD*, and then promote the peroxidase activity to modulate the balance of reactive oxygen species to improve the cold stress tolerance [15].

The above research shows that both trehalose and bZIP transcription factors are vital for plant chilling tolerance. However, whether a relationship exists between them in response to cold stress is still uncovered. This study identified that ABF1 (bZIP12) acts as a positive regulator in rice chilling tolerance by inducing trehalose biosynthesis through the novel pathway 'SAPK10-ABF1-*TPS2*'. Briefly, *abf1* mutant lines exhibited increased chilling sensitivity by interfering with trehalose biosynthesis. SnRK2 protein kinase SAPK10 could phosphorylate ABF1 and promote the latter's binding ability to *TPS2*, a member of the trehalose-6-phosphate synthase family, ultimately inducing trehalose biosynthesis under chilling stress.

2. Results and Discussion

2.1. ABF1 Acts as a Positive Regulator of Rice Chilling Tolerance

In this study, we first analyzed the expression level of *ABF1* in response to cold treatment at 0, 1, 3, 6 and 12 h. Upon cold treatment, the transcription level of *ABF1* sharply rose to about 3 fold at 1 h, and kept steady until 6 h, then dropped slowly to about 2.5 fold at 12 h, which indicated that ABF1 might participate in rice chilling adaption (Figure S1). To explore the biological role of ABF1, two independent homozygous mutant lines (*abf1-1* and *abf1-2*) and two *ABF1* over-expression lines (*OxABF1-1* and *OxABF1-9*) were generated (Figure S2). The *abf1-1* and *abf1-2* mutant lines harbored a TC deletion and a TCGC deletion in the first exon and shifted the open reading frame of *ABF1*, respectively (Figure S2). As shown in Figure 1, under normal conditions, no significant difference in seedling growth was distinguished between the wild-type and mutant lines (Figure 1). However, the leaves of the mutant lines showed more severe wilting and crimping than the wild-type under cold treatment for 5 days after 7-days recovery, with the survival rates (less than 23%) notably lower than that of the wild type (average 32%), while the survival rates of *ABF1* over-expression lines (average 49%) were significantly higher than the wild type (Figures 1 and 2A,B). Previous reports have shown that chilling stress affects numerous physiological and cellular processes in plants, including inhibiting photosynthesis, regulating calcium signals, and altering the organelle's ultrastructure, especially the chloroplast lipid membrane [20]. Accordingly, plants employ several strategies when recovering from chilling to normal temperatures, such as inducing the expression of photosynthetic enzymes, enhancing the linoleic acid metabolism and inhibiting phenylpropanoid biosynthesis and ribosome synthesis [21]. Among these processes, maintaining the cell membrane integrity and stability is vital for chilling tolerance, and ion leakage reflects the cell membrane's damage degree, while proline declines the electrolyte leakage level to protect membrane stability [22,23]. In accordance with the increased cold sensitivity of *abf1*, the mutant lines' electrolyte leakage was higher than the wild type, accompanied by reduced proline contents (Figure 2B–D). In contrast, *ABF1* over-expression lines exhibited the opposite tendency under cold stress (Figure 2B–D). These results demonstrated that ABF1 positively participated in rice chilling tolerance.

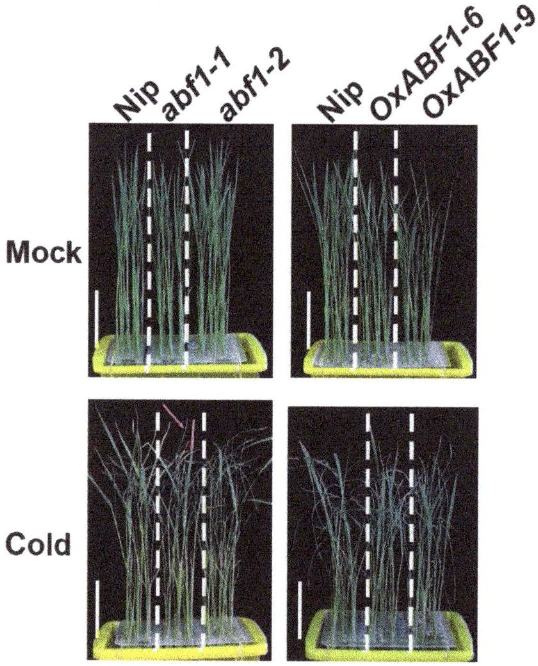

Figure 1. ABF1 positively regulates rice seedling chilling tolerance. *abf1* mutant lines are less tolerant to chilling stress with more severe wilting and crimping leaves than the wild type. *ABF1* overexpression lines showed the opposite tendency. Seven-day-old seedlings were hydroponically cultured under chilling stress (4 °C) for 5 days and photographed. Red arrow indicated the wilting and crimping leaves. Bar = 3 cm.

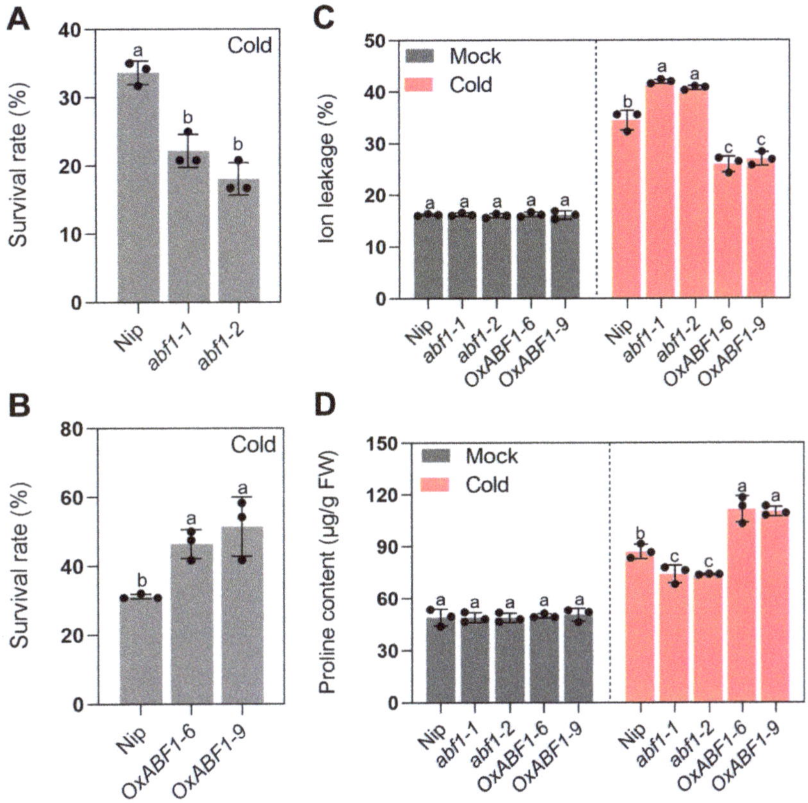

Figure 2. (**A**) Survival rates of wild-type, *abf1* mutant lines under chilling stress (4 °C) for 5 days. Survival rates (%) = the number of seedlings that survived after chilling stress treatment/total number of the seedlings (%). (**B**) Survival rates of wild type, *ABF1* overexpression lines under chilling stress (4 °C) for 5 days. (**C**) Ion leakage of seedling leaves from wild type, *abf1* mutant lines, and *ABF1* overexpression lines under chilling stress (4 °C) for 5 days. (**D**) Proline contents of seedling leaves from wild type, *abf1* mutant lines and *ABF1* overexpression lines under chilling stress (4 °C) for 5 days. Error bars indicate SD with biological triplicates ($n = 3$). Tukey's test with one-way analysis of variance (ANOVA). Different letters indicate statistical differences at $p < 0.05$.

Previous reports have shown that ABF1 is involved in various abiotic stress responses, such as hypoxia, salt, drought, and cold adversity stress [24]. *abf1* mutant lines are hypersensitive to drought and salt stresses, and *ABF1* over-expression lines significantly improve drought tolerance [25,26]. Moreover, ABF1 acts as a negative regulator for flowering transition and is functionally redundant with bZIP40, which may delay heading by directly activating *WRKY104* and indirectly inhibiting *Ehd1* under drought conditions [26]. Our lab recently reported that ABF1 regulates plant height and seed germination by inhibiting gibberellin synthesis in rice [27]. Overexpression of *ABF1* results in a typical gibberellin deficient phenotype, exhibiting semi-dwarfism and retarded seed germination, which could be restored by applying exogenous GA_3 [27]. In this study, we have detected that *ABF1* was induced by chilling stress, which was consistent with the described previously, and further found a novel role of ABF1 involved in elevating rice chilling adaption, indicating ABF1 could play pleiotropic roles in regulating various abiotic stresses tolerance.

2.2. SAPK10 Directly Interacts with and Phosphorylates ABF1

After screening a rice seedling-derived cDNA library using a yeast two-hybrid system, we found that SAPK10, a member of the specific Ser/Thr SnRK2 (Sucrose non-fermenting-1-related protein kinase 2) type kinase family, could interact with ABF1 (Figure 3A). Intriguingly, the transcription level of *SAPK10* was also significantly elevated under cold treatment (Figure S1). A bimolecular fluorescence complementation (BiFC) assay detected no YFP fluorescence signals when YN-ABF1 or YC-SAPK10 was expressed alone, and ABF1 and SAPK10 could interact to generate YFP fluorescence in the nuclei of tobacco leaf cells (Figure 4). In the GST pull-down assay, the purified His-SAPK10 was pulled down with GST-ABF1, but not with GST, and Coimmunoprecipitation (Co-IP) assay exhibited a similar result, ABF1-GFP was immunoprecipitated by SAPK10-FLAG, while GFP was not (Figure 3B,C). Moreover, in vitro kinase assays detected a phosphorylation band of purified GST-ABF1 (p-GST-ABF1) when GST-ABF1 co-expressed with His-SAPK10 in *E. coli*, and the relative strength of p-GST-ABF1 band was reduced about 23% of that of GST-ABF1 band after CIAP treatment (Figure 3D). Noteworthy, the band shift mobility of p-GST-ABF1 is significantly slower than that of GST-ABF1, further supporting the highly specific of the SAPK10-mediated ABF1 phosphorylation (Figure 3D). Additionally, we also detected the auto-phosphorylation band of SAPK10-FLAG, which was consistent with previous work [28].

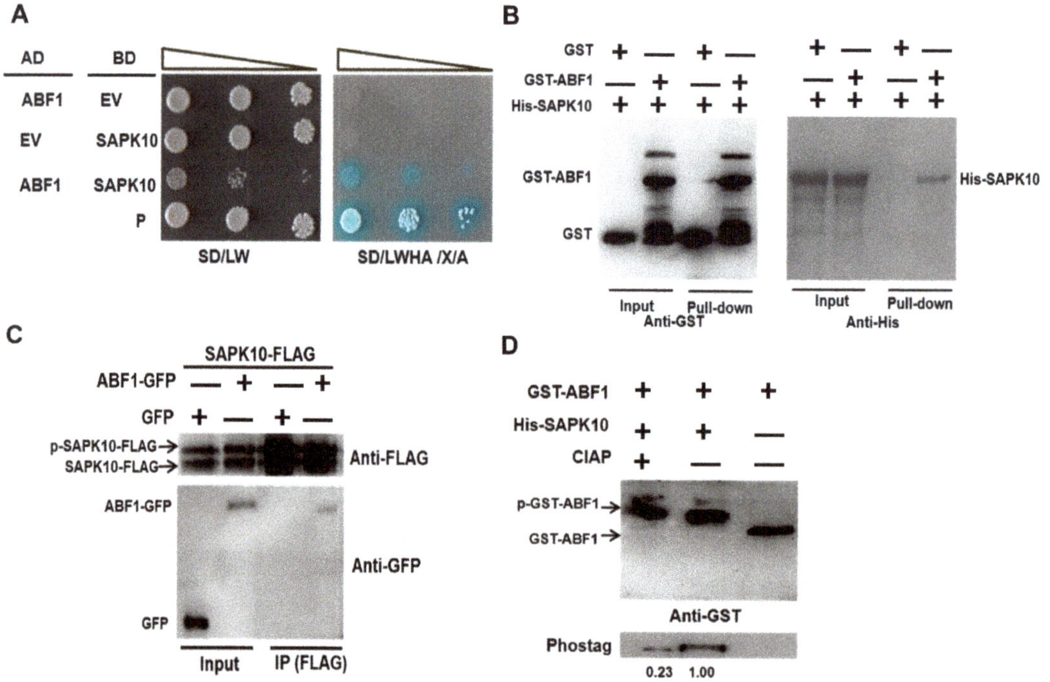

Figure 3. SAPK10 physically binds to and phosphorylates ABF1. (**A**) Yeast two-hybrid assays. Yeast cells cotransformed with SAPK10-AD and ABF1 fused to the GAL4-binding domain (ABF1-BD) were grown on selective media. BD, pGBKT7; AD, pGADT7; EV, empty vector; SD/LW, -Leu-Trp; SD/LWHA, -Leu-Trp-His-Adenine; P, positive control using pGADT7-T + pGBKT7-53; X: X-a-Gal in 0.04 mg mL^{-1}; A: Aureobasidin A in 100 ng mL^{-1}. (**B**) Pull-down assay. Purified His-SAPK10, GST and GST-ABF1 were subjected to pull-down assays and detected with anti-His and anti-GSTantibodies, respectively. (**C**) Coimmunoprecipitation (Co-IP) assay. GFP, ABF1-GFP and SAPK10-FLAG extracted from infiltrated tobacco (Nicotiana benthamiana) leaves were used in a Co-IP assay. Precipitates were

detected with anti-GFP and anti-FLAG antibodies, respectively. p-SAPK10-FLAG, SAPK10-FLAG phosphorylated band. (**D**) Kinase assay of SAPK10 on ABF1. ABF1 was phosphorylated by SAPK10. Equal amounts of the recombinant proteins were detected with an anti-GST antibody (top panel). The phosphorylated proteins were detected with biotinylated Phos-tag zinc BTL111 complex (bottom panel). p-GST-ABF1, GST-ABF1 phosphorylated band. CIAP, calf intestinal phosphatase. The relative intensity of the GST-ABF1 phosphorylated band was set to 1.00.

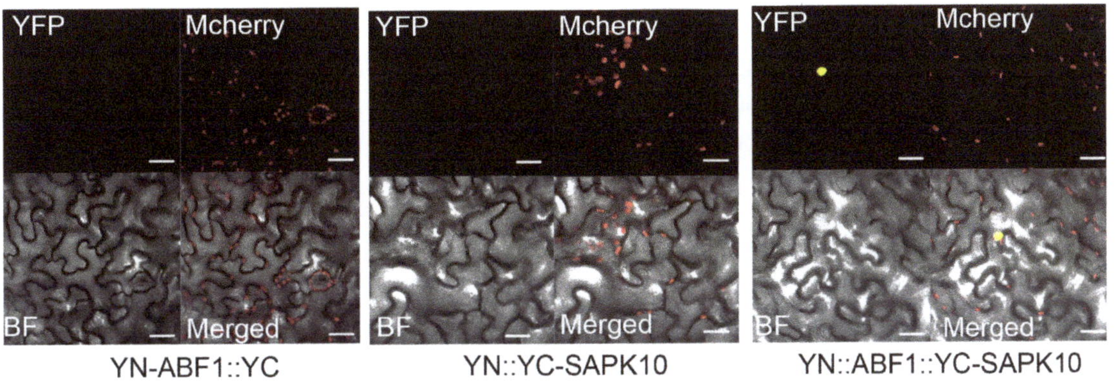

Figure 4. BiFC assay showed the interaction of SAPK10 and ABF1 in tobacco leaf epidermal cells. Red fluorescence represents the auto-fluorescence of chloroplasts, and yellow fluorescence in the nuclei of tobacco leaf cells represents the fluorescence generated by the interaction between ABF1 and SAPK10. Bar = 10 μm.

2.3. Disruption of SAPK10 Impairs Rice Chilling Tolerance

To further clarify the role of SAPK10 in regulating rice chilling tolerance, two independent homozygous mutant lines (*sapk10-2* and *sapk10-3*) and two *SAPK10* overexpression lines (*OxSAPK10-1* and *OxSAPK10-3*) were used for further analysis, respectively (Figure S3). The *sapk10-2* and *sapk10-3* mutant lines harbored an A insertion or an A/G insertion in the first exon and shifted the open reading frame of *SAPK10* (Figure S3). Interestingly, *sapk10* mutant lines displayed a phenotype similar to *abf1* under cold stress. *sapk10* mutant lines also exhibited more severe wilting and crimping than the wild type, with the survival rates (less than 25%) lower than the wild type (Figures 5 and 6A). Although the height of *SAPK10* overexpression lines was shorter than the wild type under normal conditions as previous report [28], the survival rates (average 40%) of *SAPK10* overexpression lines were higher compared with the wild type under cold treatment (Figures 5 and 6B). In accordance with the above observations, the electrolyte leakage of sapk10 increased more rapidly than in the wild type. In contrast, the electrolyte leakage was reduced in *SAPK10* overexpression lines under cold treatment, whereas the proline contents of these transgenic lines showed an opposite tendency (Figure 6C,D). These experiments showed that SAPK10 enhanced rice chilling adaptation.

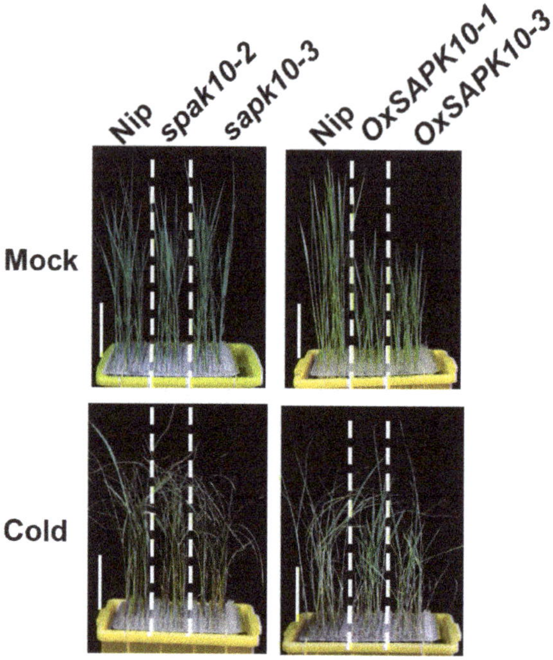

Figure 5. SAPK10 plays a positive role in rice seedling chilling tolerance. *sapk10* mutant lines are less tolerant to chilling stress with more severe wilting and crimping leaves than the wild type. *SAPK10* overexpression lines showed the opposite tendency. Seven-day-old seedlings were hydroponically cultured under chilling stress (4 °C) for 5 days and photographed. Bar = 3 cm.

SnRK2 is a specific Ser/Thr protein kinase that plays essential roles in osmotic stress response and other abiotic stress processes [29]. In rice, there are ten SnRK2 family members, namely SAPK1-SAPK10, of which SAPK6 and SAPK8 act as crucial protein kinases in the low-temperature signal transduction pathway [30,31]. Under cold stress, SAPK6 interacts with and phosphorylates IPA1 at Serine 201 and 213 sites to stabilize the protein stability of IPA1, thereafter enhancing IPA1 accumulation and activating the expression level of *CBF3*, ultimately enhancing rice cold stress adaption [31]. Wang et al. (2021) report that SAPK8 phosphorylates a cyclic nucleotide-gated channel CNGC9, activating its channel activity, and promoting cold-induced calcium influx and cytoplasmic calcium elevation in rice [30]. In this study, we found a novel role of SAPK10 involved in positively regulating rice chilling tolerance, though it has been reported to either phosphorylate bZIP72 or bZIP20 to participate in abscisic acid and jasmonic acid synergistically inhibiting seed germination or enhancing drought and salt tolerance processes [28,32]. Interestingly, SAPK6, SAPK8, SAPK10, and ABF1 are all induced by abscisic acid (ABA) and act as positive regulators in the ABA signal pathway [25,29,33]. Since ABA biosynthesis is strengthened during cold stress and a series of ABA-responsive genes are induced under cold treatment, whereas the mutant lines interfering with ABA biosynthesis exhibit impaired cold adaptation, and application of exogenous ABA enhances chilling tolerance, suggesting that ABA functions as a vital phytohormone positively regulating cold stress response [34,35]. Whether the three protein kinases or other members of the SnRK2 family interact with ABF1 involved in the crosstalk between ABA and chilling tolerance would be an interesting topic to be further investigated.

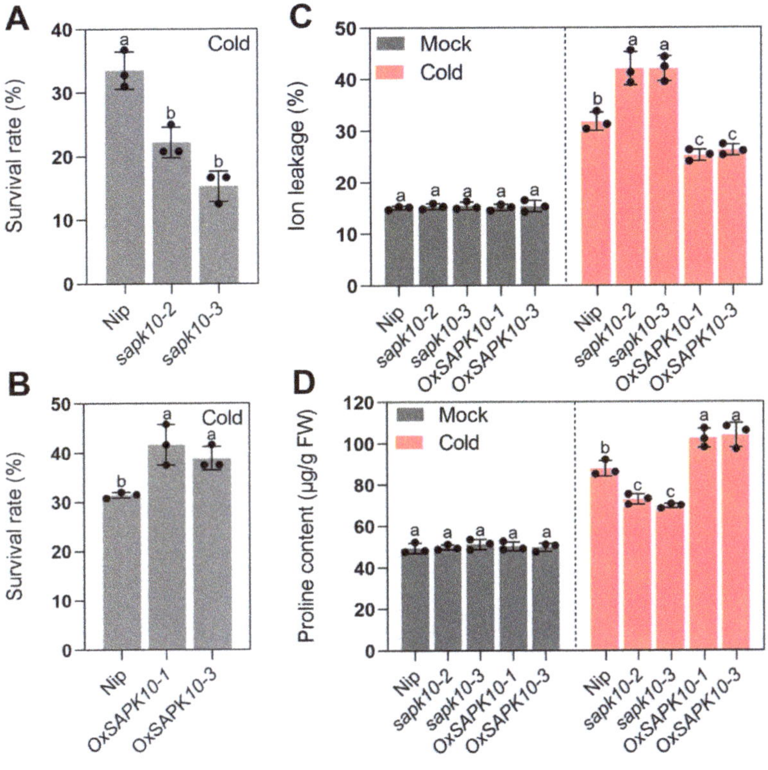

Figure 6. (**A**) Survival rates of wild-type, *sapk10* mutant lines under chilling stress (4 °C) for 5 days. (**B**) Survival rates of wild-type, *SAPK10* overexpression lines under chilling stress (4 °C) for 5 days. (**C**) Ion leakage of seedling leaves from wild type, *sapk10* mutant lines, and *SAPK10* overexpression lines under chilling stress (4 °C) for 5 days. (**D**) Proline contents of seedling leaves from wild type, *sapk10* mutant lines and *SAPK10* overexpression lines under chilling stress (4 °C) for 5 days. Error bars indicate SD with biological triplicates ($n = 3$). Tukey's test with one-way analysis of variance (ANOVA). Different letters indicate statistical differences at $p < 0.05$.

2.4. *abf1* Inhibits Seedling Cold Adaptation by Interfering with Trehalose Homeostasis

Previous reports have shown that trehalose improves chilling tolerance in several species [3,8,36], which triggered us to speculate that ABF1-mediated chilling alleviation may be partially based on trehalose accumulation. We first detect the endogenous trehalose level in *abf1* mutant lines to verify this hypothesis. As anticipated, the *abf1* seedlings had lower trehalose levels than the wild-type under cold treatment, while there was no difference under normal conditions (Figure 7A). Furthermore, applying exogenous trehalose could enhance the chilling tolerance of *abf1*, relieving severe wilting and crimping phenomena (Figure 8). Notably, the relative survival rates of *abf1* were significantly higher than the wild type when exogenous trehalose was added (Figure 7B,C). Accordingly, the endogenous trehalose contents in *ABF1* over-expression lines were higher than the wild-type in response to chilling stress, though slightly higher than the wild-type under normal conditions (Figure 7D). Thus, these results strongly suggested that ABF1 promotes trehalose contents to enhance seedlings chilling tolerance.

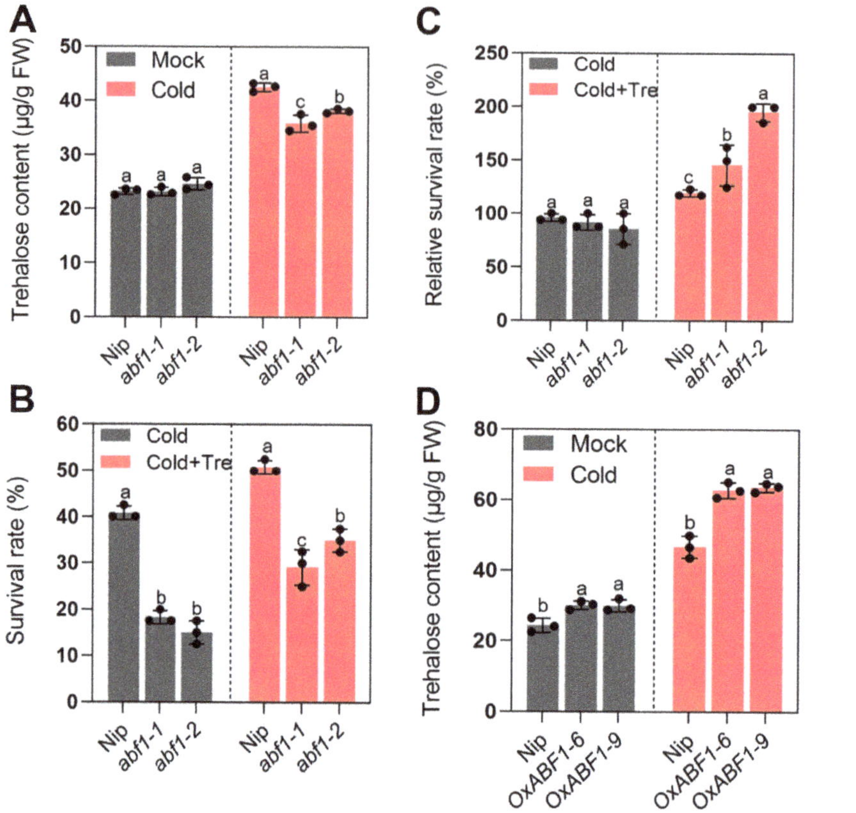

Figure 7. (**A**) Trehalose contents of seedling leaves from wild type and *abf1* mutant lines under chilling stress (4 °C) for 5 days. (**B**) Survival rates of wild-type, *abf1* mutant lines under chilling stress (4 °C) for 5 days. (**C**) Relative survival rates of wild-type, *abf1* mutant lines were expressed as a percentage of those grown under the chilling stress without exogenous trehalose. (**D**) Trehalose contents of seedling leaves from wild type and *ABF1* overexpression lines under chilling stress (4 °C) for 5 days. Error bars indicate SD with biological triplicates (n = 3). Tukey's test with one-way analysis of variance (ANOVA). Different letters indicate statistical differences at $p < 0.05$.

The underlying mechanisms of bZIP transcription factors regulating rice cold tolerance have been reported in the past decades. Liu et al. (2018) have shown that one single nucleotide polymorphism difference in the coding region of *bZIP73* causes the difference in low-temperature tolerance between Indica and Japonica rice. And Japonica type bZIP73^{Jap} interacts with bZIP71 to regulate abscisic acid (ABA) level and active oxygen balance, thus improving rice tolerance to low temperatures [13]. Moreover, the regulatory effect of bZIP73^{Jap} on low-temperature tolerance also occurs during the booting stage. Under natural cold stress conditions, the co-expression of *bZIP73*^{Jap} and *bZIP71* transgenic lines significantly improves seed setting rate and grain yield, not only inhibiting ABA levels in anthers, but also promoting the transportation of soluble sugars from anthers to pollen [17]. This study found that trehalose, except ABA, active oxygen and soluble sugars, also participates in bZIPs-mediated cold tolerance processes. Under low-temperature stress, applying trehalose increases NO levels and stimulates H_2O_2 contents, resulting in enhanced antioxidant activity and reduced membrane lipid peroxide activity [4]. In addition, exogenous trehalose treatment increases the contents of endogenous spermidine, glutathione, and ascorbic acid, promotes the glutathione and ascorbic acid cycle, and finally reduces the

ROS production to improve the cold tolerance of plants [36]. Hence, a possibility exists that trehalose may play a role in ABF1-mediated low-temperature tolerance by affecting the active oxygen homeostasis in vivo.

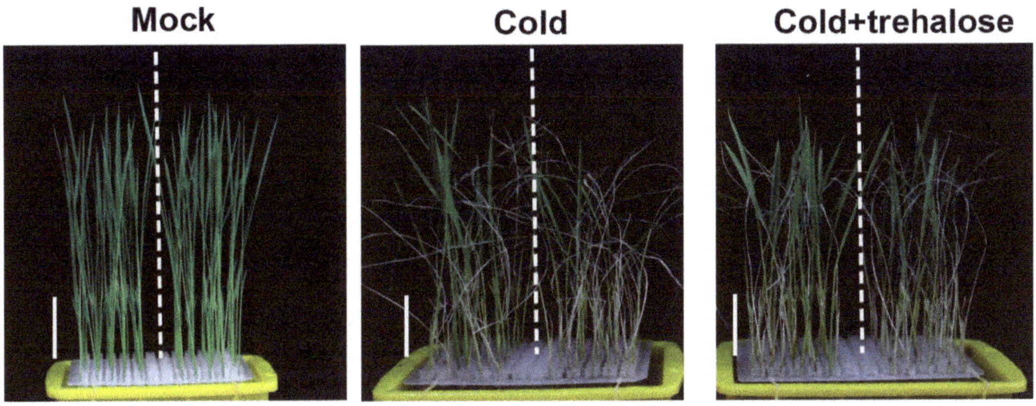

Figure 8. The chilling tolerance of *abf1* mutant lines was enhanced by exogenous trehalose application. Phenotypes of wild type, *abf1* mutant under chilling stress (4 °C) with or without 10 mM exogenous trehalose for 5 days. *abf1* mutant lines are less tolerant to chilling stress with more severe wilting and crimping leaves than the wild type. Applying exogenous trehalose enhanced *abf1* chilling tolerance with less wilting and crimping leaves. Bar = 3 cm.

2.5. ABF1 Binds to and Activates TPS2 Expression

Since ABF1 strengthens rice seedling adaptation by inducing trehalose accumulation, indicating that ABF1 may induce trehalose biosynthesis or suppress trehalose catabolism. This hypothesis triggered us to detect the expression level of a series of crucial enzymes, including TPS (Trehalose-6-phosphate synthase), TPP (Trehalose-6-phosphate phosphatase), and TRE (Trehalase) involved in the trehalose biosynthesis and catabolism pathway [6,7], in *abf1* seedlings under cold treatment. As shown in Figure 9A, *TPS2*, *TPP5* and *TPP6* involved in trehalose biosynthesis were significantly reduced in *abf1* mutant lines (Figure 9A). Among these enzymes, TPS2 attracted our particular interest due to ABF1 binds the G-box (CACGTG) motif of *TPS2* by EMSA (electrophoresis mobility shift assay), and the binding ability disappeared after the unlabeled probe was added (Figure 9C). To our surprise, ABF1 did not interact with other *TPP* or *TPS* members, of which the promoters also harbored G-box cis-elements predicted by PlanCARE (Plant Cis-Acting Regulatory Elements) [37] (Figure S4). Yeast-one hybrid (Y1H) assay further confirmed that ABF1 could bind the promoter of *TPS2* and activate the LacZ expression, though it displayed weak autoactivation activity (Figure 9B). The binding pattern of ABF1 on the promoter of *TPS2* was further validated by ChIP-qPCR (Chromatin immunoprecipitation-quantitative PCR) under chilling stress. ABF1 was notably enriched in the P3 region containing the G-box motif of the promoter of *TPS2*, nearly threefold higher than the other probes (Figure 9D). Finally, a luciferase transient transcriptional activity assay (LUC) was used to detect the regulatory effect of ABF1 on *TPS2*. The luciferase activity of the *proTPS2*: firefly LUC reporter was significantly facilitated by *pro35S*: ABF1: tNOS, which was consistent with the expression pattern of *TPS2* in *abf1* (Figure 9A,E). Intriguingly, the upregulation of *TPS2* by ABF1 was remarkably enhanced when *pro35S*: SAPK10: tNOS was added, indicating that SAPK10-mediated phosphorylation on ABF1 promotes ABF1-mediated *TPS2* expression (Figure 9E). On the other hand, *TPS2* was also induced in response to cold treatment, which was similar to the expression pattern of ABF1 and SAPK10 (Figure S1). These results strongly implied that ABF1 specifically binds to *TPS2* and activates the latter's expression.

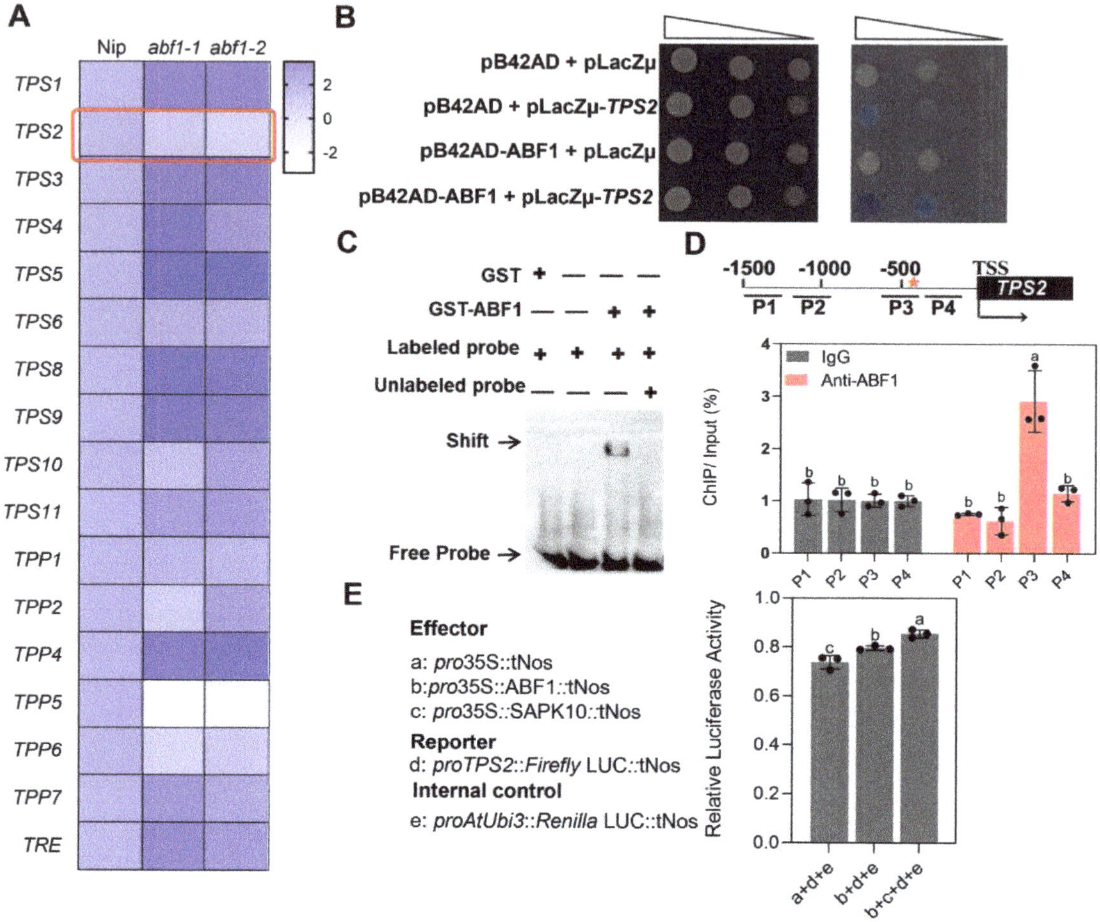

Figure 9. ABF1 directly activates *TPS2* transcription by binding to the G-box in the promoter. (**A**) the expression level of trehalose pathway genes in Nip and *abf1* seedlings under cold stress for 6 h. Data are the means of biological triplicates and were visualized in a heatmap using the log2 fold change of expression ratios relative to Nip. The red box indicated the transcription level of *TPS2* under chilling stress. (**B**) Yeast one-hybrid assay showing the binding of ABF1 to the promoters of *TPS2.* (**C**) Electrophoretic mobility shift assay (EMSA) showing GST-ABF1 specifically binds with the probe 3 (P3) region on the promoter of *TPS2.* (**D**) Probe positions on *TPS2* promoter and genome (**top panel**). Red star: G-box motif (CACGTG) in P3 probe. Black line, untranslated regions; black boxes, exons. Transcription start site (TSS) was set as 0. Numbers indicate the distances (bps) of probes to the TSS. P1–P4, probes 1–4. Chromatin immunoprecipitation quantitative PCR (ChIP-qPCR) assay to show ABF1 binding to the promoter regions of *TPS2* (**bottom panel**). The enrichment values were normalized to the Input. IgG immunoprecipitated DNA was used as a control. Tukey's test with two-way analysis of variance (ANOVA). (**E**) Luciferase transient transcriptional activity assay in rice protoplast. Effectors: 35S:tNOS, 35S:ABF1:tNOS and 35S:SAPK10:tNOS; Reporter: proTPS2:LUC. Error bars indicate SD with biological triplicates (n = 3). Tukey's test with one-way analysis of variance (ANOVA). In (**D**,**E**), different letters indicate statistical differences at $p < 0.05$.

Moreover, we tested the chilling tolerance of *tps2* mutant lines and *TPS2* overexpression lines (*OxTPS2-1* and *OxTPS2-2*) (Figure S5). *tps2* mutant lines harbored a T deletion and a T insertion in the first exon, and shifted the open reading frame of *TPS2*, respectively (Figure S5). Moreover, the trehalose contents were substantially reduced in *tps2* mutant lines, and increased in *TPS2* overexpression lines (Figure S6). In accordance with this, *tps2* mutant lines displayed more severe wilting and crimping, and lower survival rates than the wild type, while *TPS2* overexpression lines exhibited the opposite tendency (Figures 10 and 11A,B). Accordingly, the electrolyte leakage of *tps2* was higher, accompanied by reduced proline content, while *TPS2* overexpression lines showed lower electrolyte leakage and higher proline concentration compared with the wild type in response to low-temperature stress (Figure 11C,D). These findings demonstrated that TPS2 positively participates in rice seedling chilling tolerance, which was up-regulated by ABF1 to enhance trehalose biosynthesis.

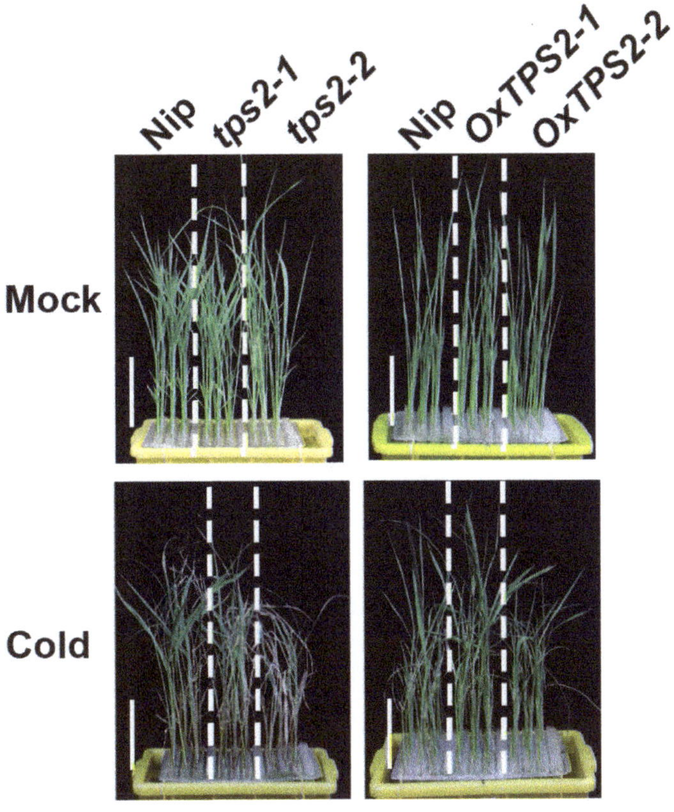

Figure 10. TPS2 enhances rice chilling tolerance. *tps2* mutant lines are less tolerant to chilling stress with more severe wilting and crimping leaves than the wild type. *TPS2* overexpression lines showed the opposite tendency. Seven-day-old seedlings were hydroponically cultured under chilling stress (4 °C) for 5 days and photographed. Bar = 3 cm.

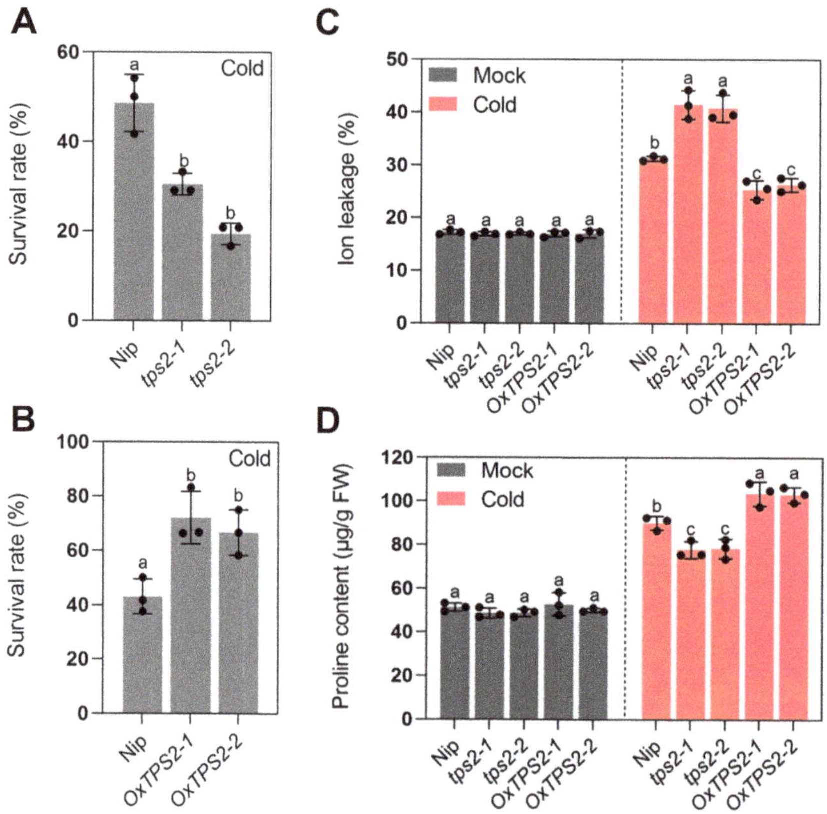

Figure 11. (**A**,**B**) Survival rates of wild type, *tps2* mutant lines and *TPS2* overexpression lines under chilling stress (4 °C) for 5 days. (**C**,**D**) Ion leakage and proline contents of seedling leaves from wild type, *tps2* mutant lines and *TPS2* overexpression lines under chilling stress (4 °C) for 5 days. Error bars indicate SD with biological triplicates ($n = 3$). Tukey's test with one-way analysis of variance (ANOVA). Different letters indicate statistical differences at $p < 0.05$.

As a non-reducing sugar found in bacteria and yeasts, trehalose serves as the source of carbon in higher plants. Trehalose regulates a series of physiological, biochemical, and molecular mechanisms in higher plants. Trehalose supplementation promotes root growth and water absorption, and thus improves plant growth under cold stress [11]. Trehalose supplementation also improves the efficiency of PS-II and the quantum yield of PS-II, while reducing the leakage of electrolytes to adapt to chilling stress [4]. On the other hand, trehalose stimulates the antioxidant defense system, such as CAT, APX, to maintain cellular integrity under cold treatment [38]. The functions and the underlying mechanisms of several TPS or TPP members involved in rice chilling tolerance have been reported [39]. TPS1 enhances rice's abiotic tolerance by increasing the trehalose and proline content and inducing the expression of stress-related genes [9]. Overexpression of *TPS1* increased the tolerance of transgenic rice seedlings to low temperature, salt, and drought treatments, with the concentrations of trehalose and proline higher than the wild type [9].

MAPK3-bHLH002-*TPP1* pathway positively regulates cold tolerance in rice. MAPK3 kinase phosphorylates bHLH002/ICE1 and inhibits the latter's ubiquitination by HOS1, then bHLH002 activates *TPP1* expression and ultimately enhances rice cold tolerance [3]. Recently, Wang et al. (2021) further report that TPP1 controls rice seed germination through crosstalk with the abscisic acid (ABA) catabolic pathway [40]. *tpp1* mutant lines down-regulate ABA catabolism and inhibit seed germination. The transcription factor GAMYB can activate the expression of *TPP1*, increase the trehalose content and induce the transcriptional abundance of the ABA catabolic genes to decrease ABA content in seeds [40]. In this study, we first report the function of cold-induced *TPS2* in strengthening seedling cold adaption by enhancing trehalose and proline contents, similar to TPS1. A previous study has reported auxin and cytokinin regulate root system architecture under low-temperature stress, mainly through modulating root cell division, differentiation, and elongation. Moreover, cytokinin determines root architecture and function by fine-turning auxin transport and signaling [41], indicating the cross-talk of different phytohormones controlling the stem cells of the root in response to chilling stress. Recently, our lab found that ABA-induced SAPK10 can phosphorylate bZIP72 and enhance the latter's binding ability to the promoter of *AOC*, an essential gene in the jasmonic acid (JA) synthesis pathway, thereby inducing *AOC* expression and increasing endogenous JA levels, finally inhibiting seed germination [28]. Based on these regulators were all induced by cold treatment, the reduced trehalose and proline contents in the mutant lines under chilling conditions, and the interaction among the three regulators, we established a novel pathway 'SAPK10-ABF1-*TPS2*' in trehalose-mediated rice cold tolerance in this study (Figure 12). Since ABA and trehalose enhance cold stress tolerance, SAPK10 and ABF1 are ABA signalling positive regulators, and TPS2 is the key enzyme involved in trehalose biosynthesis [25,29,35,39]. Further study will focus on the relationship between trehalose and ABA homeostasis in response to chilling stress, which would deepen our understanding of the regulatory mechanism acclimating to low temperatures.

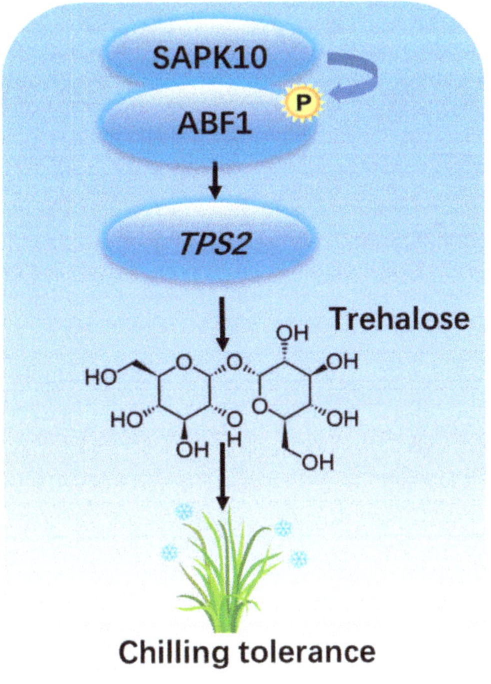

Figure 12. Working model for 'SAPK10-ABF1-*TPS2*' pathway in rice chilling tolerance. P, phosphorylation group. Arrowheads show positive regulation, bent arrow represents phosphorylation.

3. Materials and Methods

3.1. Vector Construction and Plant Transformation

For overexpression of *ABF1*, *SAPK10* and *TPS2*, the coding sequence (CDS) was fused into the binary vector PU1301 driven by the maize ubiquitin promoter using the *Kpn*I and *Bcl*I sites, respectively. *abf1*, *sapk10* and *tps2* mutant lines were generated by CRISPR/Cas9 system as described previously [42]. Briefly, annealed double-strand oligos 20 nt in the length of the gDNA sequence were ligated into the pYLCRISPR/Cas9-MH vector. All the vectors were transformed into Nipponbare (*Oryza sativa* L. ssp. *japonica* cv Nipponbare) callus using the *Agrobacterium*-mediated transformation method [43]. The primers used were listed in Table S1.

3.2. Chilling Stress Treatment

Chilling stress treatment at the seedling stage was conducted as described previously [44] with a few modifications. Briefly, 7-day-old rice seedlings grown using rice nutrient solution [45] were transferred to a growth chamber at 4 °C (14 h: 10 h, light: dark, approximately 200 μmol m^{-2} s^{-1} photon density) for 5 days. Then the seedlings were restored for 7 days in a growth chamber at 28 ± 2 °C (14 h: 10 h light: dark, approximately 200 μmol m^{-2} s^{-1} photon density). For the short-time chilling stress treatment, samples were taken at 0, 1, 3, 6.12 h (light, 4 °C), respectively.

3.3. Yeast Hybrid Assay

Yeast two-hybrid assay was performed according to Matchmaker™ Gold Yeast Two-Hybrid System (Clontech, CA, USA). The CDS of *ABF1* was cloned into the pGADT7 plasmid at the *EcoR*I and *Pst*I sites. The CDS of *SAPK10* was ligated to the pGBKT7 vector at the *EcoR*I and *BamH*I sites. The recombinant vectors were co-transformed into the yeast strain Y2H Gold (Clontech) and selected on a synthetic medium lacking leucine, tryptophan, histidine and Adenine with 0.04 mg mL^{-1} X-α-Gal and 100 ng mL^{-1} Aureobasidin A added.

Yeast one-hybrid assay was performed according to Matchmaker™ One-Hybrid System (Clontech, CA, USA). The CDS of *ABF1* was cloned into the pB42AD plasmid at the *EcoR*I and *Xho*I sites. The 2 kb promoter of *TPS2* was ligated to the pLacZ2μ vector at the *Kpn*I and *Xho*I sites. The recombinant vectors were co-transformed into the yeast strain EGY48 and selected on a synthetic medium lacking tryptophan and uracil with 1% raffinose, 1× BU salts, 80 mg/L X-Gal, and 2% galactose added. The primers used were listed in Table S1.

3.4. BiFC Assay

For Bimolecular Fluorescence Complementation (BiFC) assay, the CDS of *ABF1* was cloned into pDOE plasmid at the *BspE*I site, and the CDS of *SAPK10* was fused to the pDOE at the *BamH*I site as described previously [46]. The recombinant vectors were then transformed into *Agrobacterium* strain EHA105, and infiltrated into tobacco leaves. Fluorescence signals were observed using a Zeiss LSM710 confocal laser-scanning microscope (Carl Zeiss AG, Jena, Germany) after 72 h of infiltration. The primers used were listed in Table S1.

3.5. In Vitro Pull-Down Assay

The CDS of *ABF1* and *SAPK10* were fused into pGEX-4T-1 (G.E. Healthcare, Chicago, IL, USA) and pET-28a (Thermo, Waltham, MA, USA) vectors, respectively. The GST-ABF1, GST and His-SAPK10 were transformed into *Transetta* (DE3) chemically competent cells (Transgen, Beijing, China), and purified using the glutathione S-transferase (GST)-Sefinose™ Kit (Sangon Biotech, Shanghai, China) and 6× His-Tagged Protein Purification Kit (CWBIO, Beijing, China), respectively. The purified proteins were incubated with 50 μL Glutathione High Capacity Magnetic Agarose Beads (Sigma-Aldrich, St. Louis, MO, USA) and 600 μL pull-down buffer (50 mM Tris-HCl, pH 7.5, 5% glycerol, 1 mM EDTA, 1 mM Dithiothreitol (DTT), 1 mM phenylmethylsulfonyl fluoride (PMSF), 0.01% NonidetP-40, and 150 mM KCl) at 4 °C for 2 h. Then the beads were washed five times with pull-down buffer, and suspended in 50 µL of 1× PBS and 10 µL of 6× SDS protein loading buffer, boiled for 5 min, and resolved on 10% acrylamide gels. Individual bands were detected using Supersignal West Pico Chemiluminescent Substrate (Thermo, Waltham, MA, USA) and the ChemDoc™ Touch Imaging system (Bio-Rad, Hercules, CA, USA). The dilution for anti-GST (Yeasen, Shanghai, China) and anti-His (Yeasen, Shanghai, China) was 1: 5000. The primers used were listed in Table S1.

3.6. Coimmunoprecipitation (Co-IP) Assays

The CDS of *SAPK10* was digested with *BamHI* and cloned into the *proUbi*–FLAG vector, and the CDS of *ABF1* was digested with *Kpn*I and *Xba*I and cloned into pCAMBIA1300-35S-GFP vector, respectively. SAPK10-FLAG was transiently co-expressed with an empty green fluorescent protein (GFP) or ABF1-GFP in tobacco leaves by *Agrobacterium* infiltration. The total protein of cotransformed tobacco leaves was extracted by extraction buffer (25 mM Tris-HCl, pH 7.4, 150 mM NaCl,1 mM EDTA, 1% NonidetP-40, 5% glycerol, 1 mM PMSF, 20 µM MG132 and 1× Roche protease inhibitor cocktail (Roche, Basel, Switzerland)). After centrifugation twice (12,000× *g* for 10 min each time), the supernatant was incubated with anti-FLAG M2 magnetic beads (Sigma-Aldrich, St. Louis, MO, USA) at 4 °C for 2 h. Then the beads were washed five times with washing buffer (50 mM Tris, pH 7.5, 150 mM NaCl, 0.2% Triton X-100, 1 mM PMSF, and 1× Roche protease inhibitor cocktail). The immunoprecipitated proteins were suspended in 50 µL of 1× PBS and 10 µL of 6× SDS protein loading buffer, boiled for 5 min, and resolved on 10% acrylamide gels. Individual bands were detected using Supersignal West Pico Chemiluminescent Substrate and the ChemDoc™ Touch Imaging system. The dilution for anti-FLAG (Sigma-Aldrich, St. Louis, MO, USA) and anti-GFP (Yeasen, Shanghai, China) antibodies was 1: 5000. The primers used were listed in Table S1.

3.7. In Vitro Kinase Assay

In vitro kinase assay was performed as described previously [28]. Briefly, His-SAPK10 was co-expressed with GST-ABF1 in *Transetta* (DE3) chemically competent cell (Transgen, Beijing, China), and purified using the GST-Sefinose™ Kit (Sangon Biotech, Shanghai, China) and 6× His-Tagged Protein Purification Kit (CWBIO, Beijing, China), respectively. The purified proteins (100 ng) were incubated with 1 µg calf intestinal phosphatase (CIAP; Takara, Dalian, China) at 37 °C for 30 min, and separated by electrophoresis on 10% acrylamide gels. Phosphorylated bands were detected using biotinylated Phos-tag™ zinc complex BTL111 from Wako (Osaka, Japan). The protein intensities were quantified by ImageJ 1.47V software (ImageJ bundled with 32-bit Java 1.6.0_20).

3.8. Electrophoretic Mobility Shift Assay (EMSA)

GST-ABF1 was expressed in *Transetta* (DE3) chemically competent cell (Transgen, Beijing, China), and purified using the GST-Sefinose™ Kit (Sangon Biotech, Shanghai, China). The probes in a length of 59 bp in the *TPS2* promoter were commercially synthesized by Tsingke (Tsingke Biotech, Hangzhou, China) and labeled by an EMSA Probe Biotin Labeling Kit (Beyotime, Shanghai, China). Briefly, an equal amount of purified recombinant proteins was pre-incubated with EMSA/gel-shift binding buffer, and then incubated with 20 fmol labeled probes or nonlabeled competitive DNA probes. Then the incubated samples were separated by electrophoresis on 6% acrylamide gels. The labeled DNA probes were detected by the LightShift Chemiluminescent EMSA Kit (Thermo, Waltham, MA, USA). The primers used were listed in Table S1.

3.9. Luciferase Transient Transcriptional Activity Assay (LUC)

The CDS of *ABF1* and *SAPK10* were fused into 'None' vectors as effectors by using the *BamHI* and *EcoRI* sites, and the 2 kb promoter of *TPS2* was ligated into 190fLUC vector as a reporter by using the *Hind*III site, respectively. The plasmids were co-transformed into rice protoplasts as described previously [47]. The luciferase activities were detected by the Dual-Luciferase Reporter Gene Assay Kit (Beyotime, Shanghai, China) using the Promega GLOMAX system (Promega, Madison, WI, USA) according to the manufacturer's instructions. The relative luciferase activity was defined as the ratio between fLUC and rLUC (fLUC/rLUC). *AtUbi3*:rLUC was adopted as the internal control. The primers used were listed in Table S1.

3.10. Chromatin Immunoprecipitation-Quantitative PCR (ChIP-qPCR)

The ChIP-qPCR was carried out as described previously [48]. Briefly, chromatin was isolated from 3 g crosslinked leaves of the wild-type under chilling stress at 4 °C for 12 h. Isolated chromatin was sonicated for DNA fragmentation ranging from 200 to 700 bp. Subsequently, sonicated DNA/protein complexes were immune-precipitated with polyclonal rabbit ABF1 antibody against the amino acid residues 149 to 162 (C-KGQEEAPDGSDGPR-COOH) of ABF1, which was commercially synthesized and affinity-purified by Genescript (Genescript, Shanghai, China). After reverse cross-linking and proteinase K treatment, the immunoprecipitated DNA was purified with phenol/chloroform. Then the immunoprecipitated and input DNA were used for quantitative polymerase chain reaction (qPCR), respectively. The primers used were listed in Table S1.

3.11. Trehalose Extraction and Quantification

Briefly, the leaves of 7-day-old seedlings (0.1 g) before or after chilling stress at 4 °C for 72 h were homogenized in 5 mL of 80% (*v*/*v*) hot ethanol for 20 min and centrifuged. Then the supernatants were collected for trehalose extraction and quantification using the Trehalose Content Kit (Grace Biotechnology, Suzhou, China) according to the manufacturer's instructions. The absorbance of trehalose was measured at 620 nm.

3.12. Proline Determination

Briefly, 0.1 g leaves of 7-day-old seedlings before or after chilling stress at 4 °C for 5 days were immersed in 5 mL 3% (*v*/*v*) hot sulfosalicylic acid solution for 30 min, during which the samples were constantly shaken and cooled to room temperature to prepare crude proline extract. A total of 1 mL crude proline extract was added to 2 mL reaction buffer containing 1 mL of 2.5% acidic ninhydrin (60 mL glacial acetic acid, 16.4 mL phosphoric acid, 36.6 mL deionized water, and 2.5 g ninhydrin) and 1 mL glacial acetic acid, then boiled for 30 min with continuous shaking before cooling to room temperature. Finally, 0.5 mL toluene was used to extract 0.75 mL reaction solution with vortexing for 1 min, then incubated for 30 min. The absorbance of proline was measured at 520 nm by a SpectraMax i3x Multi-Mode Microplate Reader (MOLECULAR DEVICES, Sunnyvate, CA, USA).

3.13. Membrane Ion Leakage Measurement

Membrane Ion Leakage was determined as described previously with a few modifications [49]. Leaves of 7-day-old seedlings weighed at 0.3 g before or after chilling stress at 4 °C for 5 days were immersed in 30 mL deionized water and shaken at 120 rpm for 4 h. The membrane ion leakage L_1 was measured using a DDS-307A conductivity meter (LeiCi, Hangzhou, China). Then, the leave samples were incubated in boiling water for 30 min. After cooling down to room temperature, the membrane ion leakage L_2 of the corresponding leaves was measured. The formula $EL = L_1/L_2 \times 100\%$ was used to calculate the ion leakage of leaves.

4. Conclusions

In this study, we characterized the function of ABF1 positively participating in rice chilling tolerance. *ABF1* over-expression lines strengthened chilling tolerance with reduced ion leakage and increased proline contents, while *abf1* mutant exhibited the opposite tendency. Furthermore, SnRK2 protein kinase SAPK10 directly interacts with and phosphorylates ABF1, and strengthens the DNA-binding ability of ABF1 to a member of

Trehalose-6-phosphate synthase *TPS2*, thereby facilitating the transcription of *TPS2* and endogenous trehalose biosynthesis under chilling stress. In a word, this study revealed a novel chilling tolerance strategy mediated by 'SAPK10-ABF1-TPS2' through fine-turning trehalose homeostasis.

Supplementary Materials: The supporting information can be downloaded at: https://www.mdpi.com/article/10.3390/ijms241311082/s1.

Author Contributions: Conceptualization, Y.W. and J.Z.; methodology, X.L. (Xixi Liu), W.L. and X.T.; software, X.L. (Xinyong Liu) and G.L.; investigation, W.Z., J.H. and H.H.; resources, J.Y., L.T. and Z.L.; writing—original draft preparation, Y.S., W.Z. and Y.W.; writing—review and editing, Y.W. and J.Z.; supervision, Y.W. and J.Z.; project administration, J.Z. All the authors read and approved the final manuscript. All authors have read and agreed to the published version of the manuscript.

Funding: This research was funded by Key R&D Program of Zhejiang (Grant No. 2022C02032), Zhejiang Provincial Natural Science Foundation of China (Grant No. LY22C130004 and LY22C130002), National Key R&D Program of China (Grant No. 2022YFE0125600), Natural Science Foundation of China (Grant No. 32071986), CNRRI key research and development project (Grant No. CNRRI-2020-01), and ASTIP program of CAAS.

Institutional Review Board Statement: Not applicable.

Informed Consent Statement: Not applicable.

Data Availability Statement: Data is contained within the article and within Supplementary Material.

Acknowledgments: We thank Babatunde Kazeem Bello (Lianyungang Institute of Agricultural Sciences) for his help in language editing. We also thank Yiting Chen for her help in developing the transgenic plants and managing the hydroponic experiments.

Conflicts of Interest: The authors declare no conflict of interest.

References

1. Guo, X.; Liu, D.; Chong, K. Cold signaling in plants: Insights into mechanisms and regulation. *J. Integr. Plant Biol.* **2018**, *60*, 745–756. [CrossRef]
2. Zhang, Z.; Li, J.; Pan, Y.; Li, J.; Zhou, L.; Shi, H.; Zeng, Y.; Guo, H.; Yang, S.; Zheng, W.; et al. Natural variation in CTB4a enhances rice adaptation to cold habitats. *Nat. Commun.* **2017**, *8*, 14788. [CrossRef]
3. Zhang, Z.; Li, J.; Li, F.; Liu, H.; Yang, W.; Chong, K.; Xu, Y. OsMAPK3 Phosphorylates OsbHLH002/OsICE1 and Inhibits Its Ubiquitination to Activate OsTPP1 and Enhances Rice Chilling Tolerance. *Dev. Cell* **2017**, *43*, 731–743. [CrossRef] [PubMed]
4. Liu, T.; Shi, J.; Li, M.; Ye, X.; Qi, H. Trehalose triggers hydrogen peroxide and nitric oxide to participate in melon seedlings oxidative stress tolerance under cold stress. *Environ. Exp. Bot.* **2021**, *184*, 104379. [CrossRef]
5. Raza, A.; Su, W.; Jia, Z.; Luo, D.; Zhang, Y.; Gao, A.; Hussain, M.A.; Mehmood, S.S.; Cheng, Y.; Lv, Y.; et al. Mechanistic Insights into Trehalose-Mediated Cold Stress Tolerance in Rapeseed (*Brassica napus* L.) Seedlings. *Front. Plant Sci.* **2022**, *13*, 857980. [CrossRef] [PubMed]
6. Avonce, N.; Mendoza-Vargas, A.; Morett, E.; Iturriaga, G. Insights on the evolution of trehalose biosynthesis. *BMC Evol. Biol.* **2006**, *6*, 109. [CrossRef]
7. Van Houtte, H.; Vandesteene, L.; Lopez-Galvis, L.; Lemmens, L.; Kissel, E.; Carpentier, S.; Feil, R.; Avonce, N.; Beeckman, T.; Lunn, J.E.; et al. Overexpression of the trehalase gene AtTRE1 leads to increased drought stress tolerance in Arabidopsis and is involved in abscisic acid-induced stomatal closure. *Plant Physiol.* **2013**, *161*, 1158–1171. [CrossRef] [PubMed]
8. Lin, Q.; Wang, J.; Gong, J.; Zhang, Z.; Wang, S.; Sun, J.; Li, Q.; Gu, X.; Jiang, J.; Qi, S. The Arabidopsis thaliana trehalose-6-phosphate phosphatase gene AtTPPI improve chilling tolerance through accumulating soluble sugar and JA. *Environ. Exp. Bot.* **2023**, *205*, 105117. [CrossRef]
9. Li, H.W.; Zang, B.S.; Deng, X.W.; Wang, X.P. Overexpression of the trehalose-6-phosphate synthase gene OsTPS1 enhances abiotic stress tolerance in rice. *Planta* **2011**, *234*, 1007–1018. [CrossRef]
10. Liu, X.; Fu, L.; Qin, P.; Sun, Y.; Liu, J.; Wang, X. Overexpression of the wheat trehalose 6-phosphate synthase 11 gene enhances cold tolerance in Arabidopsis thaliana. *Gene* **2019**, *710*, 210–217. [CrossRef]
11. Liu, Z.; Ma, L.; He, X.; Tian, C. Water strategy of mycorrhizal rice at low temperature through the regulation of PIP aquaporins with the involvement of trehalose. *Appl. Soil. Ecol.* **2014**, *84*, 185–191. [CrossRef]
12. Nijhawan, A.; Jain, M.; Tyagi, A.K.; Khurana, J.P. Genomic survey and gene expression analysis of the basic leucine zipper transcription factor family in rice. *Plant Physiol.* **2008**, *146*, 333–350. [CrossRef]
13. Liu, C.; Ou, S.; Mao, B.; Tang, J.; Wang, W.; Wang, H.; Cao, S.; Schlappi, M.R.; Zhao, B.; Xiao, G.; et al. Early selection of bZIP73 facilitated adaptation of japonica rice to cold climates. *Nat. Commun.* **2018**, *9*, 3302. [CrossRef]

14. Li, Z.Y.; Fu, D.Y.; Wang, X.; Zeng, R.; Zhang, X.; Tian, J.G.; Zhang, S.S.; Yang, X.H.; Tian, F.; Lai, J.S.; et al. The transcription factor bZIP68 negatively regulates cold tolerance in maize. *Plant Cell.* **2022**, *34*, 2833–2851. [CrossRef]
15. Bai, H.; Liao, X.; Li, X.; Wang, B.; Luo, Y.; Yang, X.; Tian, Y.; Zhang, L.; Zhang, F.; Pan, Y.; et al. DgbZIP3 interacts with DgbZIP2 to increase the expression of DgPOD for cold stress tolerance in chrysanthemum. *Hortic. Res.* **2022**, *9*, uhac105. [CrossRef]
16. Liu, C.; Wu, Y.; Wang, X. bZIP transcription factor OsbZIP52/RISBZ5: A potential negative regulator of cold and drought stress response in rice. *Planta* **2012**, *235*, 1157–1169. [CrossRef]
17. Liu, C.; Schlappi, M.R.; Mao, B.; Wang, W.; Wang, A.; Chu, C. The bZIP73 transcription factor controls rice cold tolerance at the reproductive stage. *Plant Biotechnol. J.* **2019**, *17*, 1834–1849.
18. Zhang, L.; Zhang, L.; Xia, C.; Zhao, G.; Liu, J.; Jia, J.; Kong, X. A novel wheat bZIP transcription factor, TabZIP60, confers multiple abiotic stress tolerances in transgenic Arabidopsis. *Physiol. Plant.* **2015**, *153*, 538–554. [CrossRef] [PubMed]
19. Liang, Y.; Xia, J.; Jiang, Y.; Bao, Y.; Chen, H.; Wang, D.; Zhang, D.; Yu, J.; Cang, J. Genome-Wide Identification and Analysis of bZIP Gene Family and Resistance of TaABI5 (TabZIP96) under Freezing Stress in Wheat (*Triticum aestivum*). *Int. J. Mol. Sci.* **2022**, *23*, 2351. [CrossRef] [PubMed]
20. Liu, X.; Zhou, Y.; Xiao, J.; Bao, F. Effects of Chilling on the Structure, Function and Development of Chloroplasts. *Front. Plant Sci.* **2018**, *9*, 1715. [CrossRef] [PubMed]
21. Li, Z.; Khan, M.U.; Letuma, P.; Xie, Y.; Zhan, W.; Wang, W.; Jiang, Y.; Lin, W.; Zhang, Z. Transcriptome Analysis of the Responses of Rice Leaves to Chilling and Subsequent Recovery. *Int. J. Mol. Sci.* **2022**, *23*, 10739. [CrossRef]
22. Lv, Y.; Yang, M.; Hu, D.; Yang, Z.; Ma, S.; Li, X.; Xiong, L. The OsMYB30 Transcription Factor Suppresses Cold Tolerance by Interacting with a JAZ Protein and Suppressing beta-Amylase expression. *Plant Physiol.* **2017**, *173*, 1475–1491. [CrossRef]
23. Raza, A.; Charagh, S.; Abbas, S.; Hassan, M.U.; Saeed, F.; Haider, S.; Sharif, R.; Anand, A.; Corpas, F.J.; Jin, W.; et al. Assessment of proline function in higher plants under extreme temperatures. *Plant Biol.* **2023**, *25*, 379–395. [CrossRef] [PubMed]
24. Hossain, M.A.; Lee, Y.; Cho, J.-I.; Ahn, C.-H.; Lee, S.-K.; Jeon, J.-S.; Kang, H.; Lee, C.-H.; An, G.; Park, P.B. The bZIP transcription factor OsABF1 is an ABA responsive element binding factor that enhances abiotic stress signaling in rice. *Plant Mol. Biol.* **2010**, *72*, 557–566. [CrossRef]
25. Joo, J.; Lee, Y.H.; Song, S.I. Overexpression of the rice basic leucine zipper transcription factor OsbZIP12 confers drought tolerance to rice and makes seedlings hypersensitive to ABA. *Plant Biotechnol. Rep.* **2014**, *8*, 431–441. [CrossRef]
26. Zhang, C.; Liu, J.; Zhao, T.; Gomez, A.; Li, C.; Yu, C.; Li, H.; Lin, J.; Yang, Y.; Liu, B.; et al. A Drought-Inducible Transcription Factor Delays Reproductive Timing in Rice. *Plant Physiol.* **2016**, *171*, 334–343. [CrossRef]
27. Tang, L.; Xu, H.; Wang, Y.; Wang, H.; Li, Z.; Liu, X.; Shu, Y.; Li, G.; Liu, W.; Ying, J.; et al. OsABF1 Represses Gibberellin Biosynthesis to Regulate Plant Height and Seed Germination in Rice (*Oryza sativa* L.). *Int. J. Mol. Sci.* **2021**, *22*, 12220. [CrossRef]
28. Wang, Y.; Hou, Y.; Qiu, J.; Wang, H.; Wang, S.; Tang, L.; Tong, X.; Zhang, J. Abscisic acid promotes jasmonic acid biosynthesis via a 'SAPK10-bZIP72-AOC' pathway to synergistically inhibit seed germination in rice (*Oryza sativa*). *New Phytol.* **2020**, *228*, 1336–1353. [CrossRef] [PubMed]
29. Kobayashi, Y.; Yamamoto, S.; Minami, H.; Kagaya, Y.; Hattori, T. Differential activation of the rice sucrose nonfermenting1-related protein kinase2 family by hyperosmotic stress and abscisic acid. *Plant Cell* **2004**, *16*, 1163–1177. [CrossRef] [PubMed]
30. Wang, J.; Ren, Y.; Liu, X.; Luo, S.; Zhang, X.; Liu, X.; Lin, Q.; Zhu, S.; Wan, H.; Yang, Y.; et al. Transcriptional activation and phosphorylation of OsCNGC9 confer enhanced chilling tolerance in rice. *Mol. Plant* **2021**, *14*, 315–329. [CrossRef] [PubMed]
31. Jia, M.; Meng, X.; Song, X.; Zhang, D.; Kou, L.; Zhang, J.; Jing, Y.; Liu, G.; Liu, H.; Huang, X.; et al. Chilling-induced phosphorylation of IPA1 by OsSAPK6 activates chilling tolerance responses in rice. *Cell Discov.* **2022**, *8*, 71. [CrossRef] [PubMed]
32. Wang, B.; Xu, B.; Liu, Y.; Li, J.; Sun, Z.; Chi, M.; Xing, Y.; Yang, B.; Li, J.; Liu, J.; et al. A Novel mechanisms of the signaling cascade associated with the SAPK10-bZIP20-NHX1 synergistic interaction to enhance tolerance of plant to abiotic stress in rice (*Oryza sativa* L.). *Plant Sci.* **2022**, *323*, 111393. [CrossRef]
33. Chae, M.J.; Lee, J.S.; Nam, M.H.; Cho, K.; Hong, J.Y.; Yi, S.A.; Suh, S.C.; Yoon, I.S. A rice dehydration-inducible SNF1-related protein kinase 2 phosphorylates an abscisic acid responsive element-binding factor and associates with ABA signaling. *Plant Mol. Biol.* **2007**, *63*, 151–169. [CrossRef] [PubMed]
34. Xiong, L.; Ishitani, M.; Lee, H.; Zhu, J.K. The Arabidopsis LOS5/ABA3 locus encodes a molybdenum cofactor sulfurase and modulates cold stress- and osmotic stress-responsive gene expression. *Plant Cell* **2001**, *13*, 2063–2083.
35. Eremina, M.; Rozhon, W.; Poppenberger, B. Hormonal control of cold stress responses in plants. *Cell Mol. Life Sci.* **2016**, *73*, 797–810. [CrossRef]
36. Liang, Z.; Luo, J.; Wei, B.; Liao, Y.; Liu, Y. Trehalose can alleviate decreases in grain number per spike caused by low-temperature stress at the booting stage by promoting floret fertility in wheat. *J. Agron. Crop. Sci.* **2021**, *207*, 717–732. [CrossRef]
37. Lescot, M.; Dehais, P.; Thijs, G.; Marchal, K.; Moreau, Y.; Van de Peer, Y.; Rouze, P.; Rombauts, S. PlantCARE, a database of plant cis-acting regulatory elements and a portal to tools for in silico analysis of promoter sequences. *Nucleic Acids Res.* **2002**, *30*, 325–327. [CrossRef] [PubMed]
38. Mostofa, M.G.; Hossain, M.A.; Fujita, M. Trehalose pretreatment induces salt tolerance in rice (*Oryza sativa* L.) seedlings: Oxidative damage and co-induction of antioxidant defense and glyoxalase systems. *Protoplasma* **2015**, *252*, 461–475. [CrossRef]
39. Hassan, M.U.; Nawaz, M.; Shah, A.N.; Raza, A.; Barbanti, L.; Skalicky, M.; Hashem, M.; Brestic, M.; Pandey, S.; Alamri, S.; et al. Trehalose: A Key Player in Plant Growth Regulation and Tolerance to Abiotic Stresses. *J. Plant Growth Regul.* **2022**, *40*, 1–23.

40. Wang, G.; Li, X.; Ye, N.; Huang, M.; Feng, L.; Li, H.; Zhang, J. OsTPP1 regulates seed germination through the crosstalk with abscisic acid in rice. *New Phytol.* **2021**, *230*, 1925–1939. [CrossRef]
41. Tiwari, M.; Kumar, R.; Subramanian, S.; Doherty, C.J.; Jagadish, S.V.K. Auxin-cytokinin interplay shapes root functionality under low-temperature stress. *Trends Plant Sci.* **2023**, *28*, 447–459. [CrossRef]
42. Ma, X.; Zhang, Q.; Zhu, Q.; Liu, W.; Chen, Y.; Qiu, R.; Wang, B.; Yang, Z.; Li, H.; Lin, Y.; et al. A Robust CRISPR/Cas9 System for Convenient, High-Efficiency Multiplex Genome Editing in Monocot and Dicot Plants. *Mol. Plant* **2015**, *8*, 1274–1284. [CrossRef]
43. Hiei, Y.; Ohta, S.; Komari, T.; Kumashiro, T. Efficient transformation of rice (*Oryza sativa* L.) mediated by Agrobacterium and sequence analysis of the boundaries of the T-DNA. *Plant J.* **1994**, *6*, 271–282. [CrossRef] [PubMed]
44. Mao, D.; Chen, C. Colinearity and similar expression pattern of rice DREB1s reveal their functional conservation in the cold-responsive pathway. *PLoS ONE* **2012**, *7*, e47275. [CrossRef] [PubMed]
45. Wang, X.F.; Wang, Y.F.; Pineros, M.A.; Wang, Z.Y.; Wang, W.X.; Li, C.Y.; Wu, Z.C.; Kochian, L.V.; Wu, P. Phosphate transporters OsPHT1;9 and OsPHT1;10 are involved in phosphate uptake in rice. *Plant Cell Environ.* **2014**, *37*, 1159–1170. [CrossRef]
46. Gookin, T.E.; Assmann, S.M. Significant reduction of BiFC non-specific assembly facilitates in planta assessment of heterotrimeric G-protein interactors. *Plant J.* **2014**, *80*, 553–567. [CrossRef]
47. Xie, K.; Yang, Y. RNA-guided genome editing in plants using a CRISPR-Cas system. *Mol. Plant* **2013**, *6*, 1975–1983. [CrossRef]
48. Hou, Y.; Wang, L.; Wang, L.; Liu, L.; Li, L.; Sun, L.; Rao, Q.; Zhang, J.; Huang, S. JMJ704 positively regulates rice defense response against Xanthomonas oryzae pv. oryzae infection via reducing H3K4me2/3 associated with negative disease resistance regulators. *BMC Plant Biol.* **2015**, *15*, 286. [CrossRef] [PubMed]
49. He, Y.; Zhang, Z.; Li, L.; Tang, S.; Wu, J.L. Genetic and Physio-Biochemical Characterization of a Novel Premature Senescence Leaf Mutant in Rice (*Oryza sativa* L.). *Int. J. Mol. Sci.* **2018**, *19*, 2339. [CrossRef]

International Journal of
Molecular Sciences

MDPI

Communication

Complete Genomic Sequence of *Xanthomonas oryzae* pv. *oryzae* Strain, LA20, for Studying Resurgence of Rice Bacterial Blight in the Yangtze River Region, China

Yuxuan Hou *, Yan Liang, Changdeng Yang, Zhijuan Ji, Yuxiang Zeng, Guanghao Li and Zhiguo E

State Key Laboratory of Rice Biology and Breeding, China National Rice Research Institute, Hangzhou 311400, China
* Correspondence: houyuxuan@caas.cn

Abstract: *Xanthomonas oryzae* pv. *oryzae* (*Xoo*) is a causative agent of rice bacterial blight (BB). In 2020–2022, BB re-emerged, and there was a break out in the Yangtze River area, China. The pandemic *Xoo* strain, LA20, was isolated and identified from cultivar Quanyou1606 and demonstrated to be the Chinese R9 *Xoo* strain, which is able to override the widely adopted *xa5*-, *Xa7*- and *xa13*-mediated resistance in rice varieties in Yangtze River. Here, we report the complete genome of LA20 by PacBio and Illumina sequencing. The assembled genome consists of one circular chromosome of 4,960,087 bp, sharing 99.65% sequence identity with the traditional representative strain, YC11 (R5), in the Yangtze River. Comparative genome analysis of LA20 and YC11 revealed the obvious variability in *Tal* genes (the uppermost virulence determinants) in numbers and sequences. Particularly, six *Tal* genes were only found in LA20, but not in YC11, among which *Tal1b* (*pthXo1*)/*Tal4* (*pthXo6*), along with the lost one, *pthXo3* (*avrXa7*), might be the major factors for LA20 to overcome *xa5*-, *Xa7*- and *xa13*-mediated resistance, thus, leading to the resurgence of BB. This complete genome of the new pandemic *Xoo* strain will provide novel insights into pathogen evolution, the traits of pathogenicity on genomic level and the epidemic disease status in China.

Keywords: bacterial blight; resurgence; *Xanthomonas oryzae* pv. *oryzae*; genome

1. Introduction

The phytopathogenic *Xanthomonas oryzae* pv. *oryzae* (*Xoo*) causes rice bacterial blight (BB), one of the most devastating diseases of rice worldwide, usually resulting in 10% to 30% yield loss and even 50%, or the loss of the harvest [1]. BB epidemics have been reported in 28 Chinese provinces since it was initially observed in Jiangsu province, the Yangtze River region, China, in the 1930s [2]. By the 1990s, BB was rampant in the vast rice-growing area, especially in the Yangtze River area. Until the early 21st century, the wide utilization of BB resistance (*R*) genes, such as *xa5, Xa7 and xa13,* in rice breeding effectively controlled BB, and the disease barely occurred in rice fields in the Yangtze River region [3,4]. However, due to the co-evolution of *Xoo* and rice [5,6], *R* gene-conferred BB resistance was gradually conquered by emerging *Xoo* strains [7,8]. In 2020–2022, BB had re-emerged, and there were widespread epidemics in the Anhui and Zhejiang provinces in the Yangtze River region.

TALEs (transcription activator-like effectors), the most important virulence factors of *Xoo,* function as transcription activators and bind to the promoters of susceptibility (*S*) or *R* genes of rice to regulate their transcription, which plays crucial roles in pathogenicity and variety discrimination in *Xoo*–rice interaction [9,10]. A typical TALE consists of three parts: the N-terminal domain containing the type III secretion signal, the carboxyl-terminal nuclear localization signal (NLS) and the transcription activation domain (AD) and the central highly conserved repeat units [11]. The promoter-binding specificity of TALEs is determined by repeat units. Each repeat unit has near-perfect 33 to 35 amino acid repeats, with two variable amino acids at position 12 and 13 (termed the 'RVD', for repeat-variable diresidue) [12,13]. TALE-encoding genes (*Tal* genes) are presumed to be highly dynamic and varied in RVDs and have greatly contributed to the production of new toxicity and the escape of plant immunity by *Xoo* [14,15]. Tracking and identifying the variations in *Tal* genes are helpful for accurate pathotype discrimination of the emerging *Xoo* strains and rice-disease-resistance breeding.

In this study, the newly emerging *Xoo* strain, LA20, was isolated from typical diseased rice leaves collected in Anhui province in 2021. We focus on the pathotype, genome sequence and TALEs dissections to uncover the reason for the BB re-outbreak in the Yangtze River region, China.

2. Results

2.1. Pathotype Analysis of Xoo Strain LA20

BB erupted in Anhui province, the Yangtze River region, China, in 2021, and the rate of incidence was up to 85% (Figure 1a). The isolated *Xoo* strain, LA20, was confirmed to be the causative agent by morphology and molecular identification (Figure 1b,c).

Figure 1. Identification and pathotype analysis of the newly emerging *Xoo* strain, LA20. (**a**) The field symptoms of BB in Anhui province, Yangtze, in 2021; rice variety: Quanyou1606. (**b**) colony morphology of LA20. (**c**) PCR amplification of *16SrRNA* and *HP* gene of LA20. (**d–f**) resistance evaluation of IRBB7 to *Xoo* strains LA20 and YC11. (**d**): leaves' phenotype. (**e**,**f**): lesion length and ratio of lesion length to entire leaf length of IRBB7 inoculated with LA20 and YC11 at 21st day, respectively. Error bar: SD from 18 inoculated leaves. Significance of differences was determined by Student's *t* test, ** $p < 0.01$.

Upon pathotype analysis, LA20 clearly showed the "SSSSSS" interaction with six differential rice varieties (near-isogenic lines harboring individual resistance gene) IRBB2 (*Xa2*), IRBB3 (*Xa3*), IRBB5 (*xa5*), IRBB13 (*xa13*), IRBB14 (*Xa14*) and IR24 (*Xa18*) (Table 1). LA20 was demonstrated to be race 9 (R9) according to the pathotype criteria of Chinese *Xoo* strains [16,17]. However, the traditional representative strain, YC11 (R5), from the Yangtze River region showed an "SSRRSS" interaction with different rice varieties, as it failed to overcome *xa5*- and *xa13*-mediated BB resistance [18]. In addition, it was found that IRBB7 (*Xa7*) was also susceptible to LA20 and resistant to YC11 (Figure 1d–f). The results showed that LA20 is a newly discovered hyperpathogenic pandemic strain that is different from

YC11 and is able to overcome widely adopted *xa5*-, *Xa7*- and *xa13*-mediated resistance. LA20 has already been deposited in the China General Microbiological Culture Collection Center (No. 25881).

Table 1. Pathotype of *Xoo* strain, LA20, in six differential rice varieties harboring individual resistance gene.

Differential Variety	Resistance Gene	Mean Lesion Length (cm)	Ratio of Lesion Length to Entire Leaf Length	Resistance Evaluation
IRBB2	*Xa2*	23.3 ± 3.64	0.61 ± 0.10	S
IRBB3	*Xa3*	25.2 ± 5.47	0.65 ± 0.14	S
IRBB5	*xa5*	16.1 ± 4.60	0.46 ± 0.13	S
IRBB13	*xa13*	15.4 ± 2.43	0.41 ± 0.06	S
IRBB14	*Xa14*	20.9 ± 2.47	0.56 ± 0.13	S
IR24	*Xa18*	27.0 ± 4.61	0.73 ± 0. 13	S

S: susceptible (ratio from 0.25 to 1) were measured 21 days after inoculation. ±: SD from 18 inoculated leaves.

2.2. *Complete Genome Characteristics of Xoo Strain, LA20*

To provide better insight into the pathogenicity, adaptation and evolution that caused BB resurgence in the Yangtze River region, the complete genome of LA20 was sequenced. In total, 336,972 reads from PacBio sequencing were obtained, with a mean concordance of 0.89, N50 value of 12,126 bp and average read length of 10,351 bp, consisting of whole 3,488,085,182 bases. High-quality filtered reads were assembled, corrected and annotated. As a result, one unique circular chromosome of 4,960,087 bp was obtained with 63.69% GC content, and the average nucleotide identity (ANI) rates are up to 99.67% with *Xoo* type strain, PXO99A, and 99.65% with the *Xoo* traditional representative strain, YC11, in the Yangtze River region (Table 2). The chromosome contains 3630 protein coding genes, 53 transfer RNAs (tRNAs), 6 ribosomal RNAs (rRNAs), 144 non-coding RNAs (ncRNAs), 811 pseudogenes and 14 *Tal* genes (Table 2). The circular representation of the complete genome of LA20 is shown in Figure S1.

Table 2. Genome comparison between *Xoo* strains, LA20 and YC11.

Genome Features	LA20 (R9)	YC11 (R5)
Accession No.	CP114600	CP031464
Size (bp)	4,960,087	4,867,200
GC content (%)	63.69	63.74
Protein coding genes	3630	3614
tRNAs	53	53
rRNAs	6	6
ncRNAs	144	146
Pseudogenes	811	722
***Tal* genes**	14	12
ANI (%)	99.67	99.65

2.3. *Tal Genes of Xoo Strain LA20*

Despite having a sequence similarity of up to 99.65% with the genome sequences of LA20 and YC11 (Table 2), the data show that there are 10,900 single nucleotide polymorphisms, 188 deletions and 199 insertions (Table S1). These genomic variants may result in the more 16 protein coding genes, 89 pseudogenes and 2 *Tal* genes in the genome of LA20 (Table 2).

The type and number of *Tal* genes are very critical to *Xoo* pathogenicity and rice resistance. It was found that the *Tal* genes between LA20 and YC11 are highly diverse (Figure 2). In LA20, 14 *Tal* genes was identified, which has 2 more than the number of those from YC11; *Tal1b*, *Tal2b*, *Tal3c*, *Tal3b*, *Tal3a* and *Tal4* were found in LA20 and absent in

YC11 (Table 3). Among the above six *Tal* genes, *Tal1b* (*pthXo1*) and *Tal4* (*pthXo6*) have been known to target *OsSWEET11* (*Xa13*) and bZIP transcription factor-encoding gene (*OsTFX1*), inducing rice susceptibility, respectively (Table 3). *Tal2b*- and *Tal3b*-encoding interfering TALEs (iTALEs) are brand new, showing no homology with the reported TALEs (Table 3). *Tal3c* and *Tal3a* in LA20 are identical to *Tal5b* and *Tal5a* in PXO99A (Table 3). Further, *pthXo3* (*avrXa7*), which targets *Xa7* inducing rice resistance, was only found in YC11, but not in LA20 (Table 3). Therefore, we speculated that the diversity of TALEs type in LA20, such as special ones, *Tal1b* (*pthXo1*)/*Tal4* (*pthXo6*), novel ones, *Tal2b* and *Tal3b* (iTALEs) and the lack of *pthXo3* (*avrXa7*), may cause re-epidemics of BB, replacing the traditional representative strain, YC11, in the Yangtze River region, China.

Table 3. *Tal* genes and RVD sequences in LA20 and similar genes in YC11 and PXO99A.

Genes	RVD Sequences	Function	Targeted Genes	YC11	PXO99A
Tal1b	NN-HD-NI-NG-HD-NG-N *-HD-HD-NI-NG-NG-NI-HD-NG-NN-NG-NI-NI-NI-NI-N*-NS-N*	*pthXo1*	*OsSWEET11(Xa13)*	/	*Tal2b*
Tal1a	NI-NG-NN-NG-NK-NG-NI-NN-NI-NN-NI-NN-NS-NG-NS-*NN-NI-N*-NS-NG*	/	/	*Tal4a*	*Tal2a*
Tal2b	NS-NG-NG-NG-NG-HD-H*	*iTal*	/	/	/
Tal2a	NS-HD-NG-NG-HG-NG-HD-HD-NG-HD-NN-HD-NG-HD-NI-NI-NI-N*	*iTal*	/	*Tal3*	/
Tal3c	NI-H*-NI-NN-NN-NN-NN-NN-HD-NI-HD-HG-HD-NI-N*-NS-NI-NI-HG-HD-NS-NS-NG	/	/	/	*Tal5b*
Tal3b	NS-HG-HG-HD-NS-NG-HD-NN-NG-HG-NG-HD-HG-HD-HD-NI-NN-NG	*iTal*	/	/	/
Tal3a	NI-NS-HD-HG-NS-NN-HD-H*-NG-NN-NN-HD-HD-NG-HD-NG	/	/	/	*Tal5a*
Tal4	NI-N*-NI-NS-NN-NG-NN-NS-N*-NS-NN-NS-N*-NI-HG-HD-NI-HD-HD-NG	*pthXo6*	*OsTFX1*	/	*Tal6a*
Tal5a	NN-HD-NS-NG-HD-NN-N*-NI-HD-NS-HD-NN-HD-NN-HD-NN-NN-NN-NN-NN-NN-NN-HD-NG	/	/	*Tal2a*	*Tal9e*
Tal5b	NI-HG-NI-NI-NI-NN-HD-NS-NN-NS-NN-HD-NN-NI-HD-NN-NI-NG-*HD-NG*	/	/	*Tal2b*	*Tal7a*
Tal6a	HD-HD-HD-NG-N*-NN-HD-HD-N*-NI-NI-NN-HD-HI-ND-HD-NI-HD-NG-NG	*pthXo8*	*OsHEN1*	*Tal1a*	*Tal9a*
Tal6b	HD-HD-NN-NN-NI-NG-HD-S*-HG-HD-NG-N*-NG-HD-HD-N*-NI-NI-NN-HD-HI-ND-HD-NG-NN-HG-N*	*avrXa23*	*Xa23*	*Tal1b*	*Tal9b*
Tal6c	NI-NN-N*-NG-NS-NN-NN-NN-NI-NN-NI-NG-HD-HD-NI-HG-N*	*avrXa27*	*Xa27*	*Tal1c*	*Tal9c*
Tal6d	NI-NN-NI-HG-HG-HD-NG-HD-HG-HD-HD-HD-NG	/	/	*Tal1d*	*Tal9d*
/	NI-HG-NI-HG-NI-NI-NI-HD-NN-HD-HD-HD-NG-HD-N*-NI-HD-HD-NN-NS-NI-NN-NN-NG-NN-HD-N*-NS-N*	*pthXo3*	*Xa7*	*Tal4c*	/

* predicted deletion of the thirteen codon in the repeats; / absent. *Tal* genes and RVD sequences are based on AnnoTALE [19].

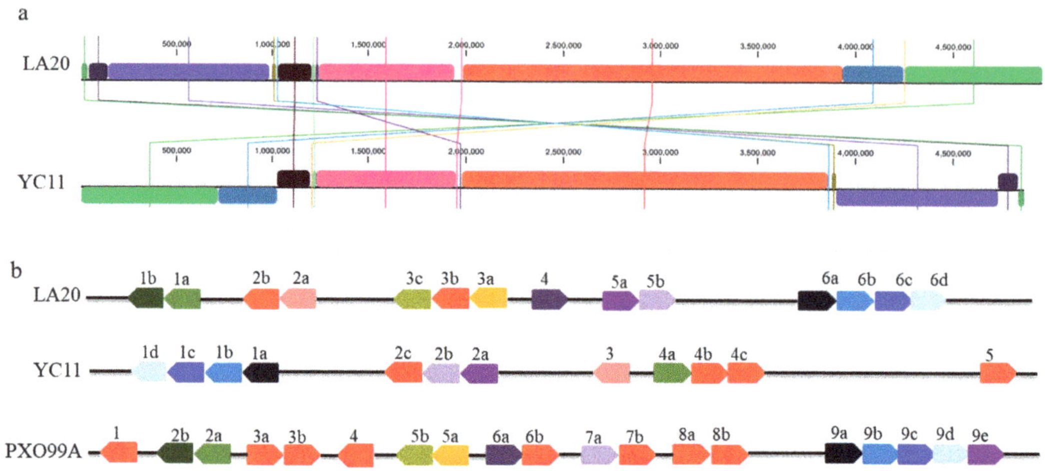

Figure 2. Comparison of whole genome and TALEs in *Xoo* strains. (**a**): Progressive mauve alignment chromosome of LA20 and YC11. The ruler indicates distance from the annotated origin in base pairs. (**b**): The *Tal* genes of *Xoo* strains. The genes are represented as arrows at their relative position in the linearized chromosome. Arabic numbers indicate the serial number of *Tal* gene clusters in a *Xoo* strain. Lowercase letter indicates each *Tal* gene in a certain gene cluster. The identical effectors with identical RVD sequences are lined with the same colors. The strain-specific effectors are indicated in red for LA20, YC11 and PXO99A.

3. Discussion

Recently, BB re-emerged, and there was a break out in the Yangtze River region, which is the major rice-producing region in China. In this study, the representative *Xoo* strain, LA20, was isolated and identified from re-epidemic BB samples. It was demonstrated that LA20 is a Chinese R9 *Xoo* strain that can override *xa5*- and *xa13*-mediated resistance. Moreover, LA20 can also overcome *Xa7*-mediated resistance. The complete genome further revealed variability in *Tal* genes in the pandemic *Xoo* strain, LA20, compared with that of the traditional representative strain, YC11 (R5).

The traditional dominant *Xoo* strain, YC11 (R5), was replaced by the newly emerging one, LA20 (R9), in the Yangtze River region. It is believed that during the long-term co-evolution between pathogen and host, the arms race might be disequilibrated through either the evolution of virulence factors of pathogens or the loss of host resistance genes. TALEs are the most important virulence factors of *Xoo*. An individual *Xoo* strain usually contains 9–21 *Tal* genes [20]. There are two more *Tal* genes in LA20 than there are in YC11. Further, *Tal1b*, *Tal2b*, *Tal3c*, *Tal3b*, *Tal3a* and *Tal4* were found in LA20 and absent in YC11. These variations from TALEs may have contributed to the BB re-outbreak.

BB *R* gene, *xa5*, *xa13* and *Xa7* are usually employed to breed BB-resistant varieties in the Yangtze River region, China. The former YC11 (R5) from the Yangtze River region cannot overcome *xa5*- and *xa13*-mediated BB resistance. However, the special *Tal1b* (*pthXo1*) and *Tal4* (*pthXo6*) of LA20 can recognize and bind to the promoter of the dominant *Xa5* and *Xa13*, overriding *xa5*- and *xa13*-mediated BB resistance [21–24]. Furthermore, *pthXo3* in YC11 can bind to the promoter of *Xa7* and induce its expression to trigger the defense [25]. Conversely, *pthXo3* was not found in LA20, and LA20 broke down the *Xa7*-mediated resistance. Further, it is interesting that two new iTALEs was identified in LA20. iTALEs contain 45 or 129 bp deletions in the sequence encoding the N-terminal region and lack the C-terminal AD domains, unlike typical TALEs [26,27]. It was reported that iTALEs suppressed *Xa1* resistance triggered by typical TALEs [28]. These variations may result in LA20 escaping immunity mediated by *xa5*, *xa13*, *Xa7* and *Xa1*. Analyzing virulence factors will facilitate the further understanding of the co-evolution of rice and *Xoo*, as well as the developing of strategies to clone new disease resistance genes for rice breeding.

In this study, the combination of PacBio and Illumina sequencing quickly, effectively, and accurately conferred the complete gap-free genome of LA20. HiFi long-read sequences with read lengths averaging 10–25 kb and accuracies >99.5% revealed the instability and variability in containing nearly identical tandem-repeat domain *Tal* genes in the epidemic variant, LA20. The approach and genome data will contribute to assessing the epidemic disease status and the valuable resource for a better understanding of pathogen evolution and the molecular pathogenesis of *Xoo* that caused the BB resurgence in the Yangtze River region, China.

4. Materials and Methods

4.1. Isolation and Identification of Xoo

Infected leaves of the rice variety "Quanyou1606" were successively sterilized with 70% ethanol and 1% sodium hypochlorite for 30–60 s, rinsed with distilled water for three times, then cut, crushed and soaked in 1 mL distilled water for 10 min. The suspension was streaked and cultured on peptone sugar agar (PSA) medium for 72 h at 28 °C. Circular, smooth-margined, convex colonies were selected for further purification and molecular identification.

For molecular identification, *Xoo*-specific polymerase chain reaction (PCR) was carried out using *16S rRNA* gene primers 16SrRNAF (5′-AGAGTTTGATCATGGCTCAG-3′)/16SrRNAR (5′-AAGGAGGTGATCCAGCCGC-3′) with target fragment 1539 bp, and *HP* gene primers Xoo163F (5′-CAATGCACACGTGGAAAGGG-3′)/Xoo163R (5′-CTTGCAAG GGATAGAAGCGT-3′) with target fragment 163 bp. Finally, amplified fragments were sequenced, and we performed sequence identity analysis using NCBI Blastn.

4.2. Rice Materials

The differential rice varieties used in this study were a simplified combination of IRBB2, IRBB3, IRBB5, IRBB13, IRBB14 and IR24, which were provided by the International Rice Research Institute (IRRI), Philippines, via National Rice Germplasm Genebank, China. All rice varieties were planted in the experimental field in China National Rice Research Institute. Plants in the booting stage were used for artificial *Xoo* inoculation assays.

4.3. Pathotype Analysis

The booting stage near-isogenic lines harboring individual *Xa*-resistance gene were inoculated with LA20 (OD_{600} = 0.5) by a leaf clipping method [29]. The disease was scored as the percent lesion area (lesion length/leaf length) at 21 days after inoculation, as in our previous study. The ratios of lesion length to entire leaf length of $<1/4$ and $\geq 1/4$ were classified as resistant (R) and susceptible (S), respectively [30]. All experiments were repeated three times. The least significant difference (LSD method) was used for data analysis. A pathotype classification of LA20 was performed according to their reactions on NIL rice (Supplementary Table S2).

4.4. Genome Sequencing, Assembly and Annotation

Genomic DNA of LA20 was extracted using SDS method [31]. The SMRT bell TM Template kit (version 2.0) and NEBNext®Ultra™ DNA Library Prep Kit for Illumina (New England Biolabs Inc., Beverly, MA, USA) were used to construct sequencing library for PacBio Seque1and Illumina NovaSeq PE150 sequencing by Novogene (Beijing, China) [32]. The average fragment length of the sequencing library is about 10 kb for circular consensus sequencing (CCS) to achieve long and high fidelity (HiFi) reads. HiFi reads were de novo assembled with Canu (v2.0) and Racon (v1.4.13) [33]. Reads from Illumina were used to correct genome sequences for improving the quality of assembly using Pilon software (v1.22) [34]. Circos software was used to cyclize and adjust the starting site [35]. Finally, the assembly was annotated using the NCBI Prokaryotic Genome Annotation Pipeline [36] and GeneMarkS (v4.17) [37].

4.5. Comparative Genomics and TALEs Analysis

For structural comparison, complete genomes were aligned using progressive Mauve [38] with default settings. For TALEs analysis, *Tal* genes and RVD sequences were assessed using AnnoTALE tools according to a previous report [19].

5. Conclusions

We demonstrated that the causative agent of newly emerging BB in the Yangtze River region is the Chinese R9 *Xoo* strain, which is able to override the widely adopted *xa5-*, *Xa7-* and *xa13*-mediated resistance in rice varieties in the Yangtze River region. Further, we provide an HiFi long-read gap-free genome sequence of LA20 and identified 14 *Tal* genes. The variability in TALs in their numbers and sequences compared with those of the former representative strain, YC11 (R5), in the Yangtze River region may have caused the re-outbreak of BB.

Supplementary Materials: The following supporting information can be downloaded at: https://www.mdpi.com/article/10.3390/ijms24098132/s1.

Author Contributions: Y.H. conceived the project and designed the research procedure. Y.H., Y.L., C.Y., Z.J., G.L. and Y.Z. performed the experiments. Z.E. analyzed the data. Y.H. wrote the manuscript with input form. All authors have read and agreed to the published version of the manuscript.

Funding: This study was supported by Zhejiang Provincial "San Nong Jiu Fang" Sciences and Technologies Cooperation Program (2022SNJF009), Zhejiang Provincial Natural Science Foundation of China (LY23C130004), and Agricultural Sciences and Technologies Innovation Program of Chinese Academy of Agricultural Sciences.

Institutional Review Board Statement: Not applicable.

Informed Consent Statement: Not applicable.

Data Availability Statement: The whole-genome sequences reported here have been deposited in GenBank (https://www.ncbi.nlm.nih.gov/, accessed on 23 December 2022) under the accession number CP114600 (BioProject: PRJNA913657, BioSample: SAMN32303221).

Acknowledgments: We appreciate Jian Zhang (China National Rice Research Institute) for English polish of this manuscript.

Conflicts of Interest: The authors declare no conflict of interest.

References

1. Nino-Liu, D.O.; Ronald, P.C.; Bogdanove, A.J. *Xanthomonas oryzae* pathovars: Model pathogens of a model crop. *Mol. Plant Pathol.* **2006**, *7*, 303–324. [CrossRef]
2. Xu, X.; Li, Y.; Xu, Z.; Yan, J.; Wang, Y.; Wang, Y.; Cheng, G.; Zou, L.; Chen, G. TALE-induced immunity against the bacterial blight pathogen *Xanthomonas oryzae* pv. *oryzae* in rice. *Phytopathol. Res.* **2022**, *4*, 47. [CrossRef]
3. Chukwu, S.C.; Rafii, M.Y.; Ramlee, S.I.; Ismail, S.I.; Hasan, M.M.; Oladosu, Y.A.; Magaji, U.G.; Akos, I.; Olalekan, K.K. Bacterial leaf blight resistance in rice: A review of conventional breeding to molecular approach. *Mol. Biol. Rep.* **2019**, *46*, 1519–1532. [CrossRef] [PubMed]
4. Yang, Y.; Zhou, Y.H.; Sun, J.; Liang, W.F.; Chen, X.Y.; Wang, X.M.; Zhou, J.; Yu, C.L.; Wang, J.M.; Wu, S.L.; et al. Research Progress on Cloning and Function of Xa Genes Against Rice Bacterial Blight. *Front. Plant Sci.* **2022**, *13*, 847199. [CrossRef] [PubMed]
5. Jones, J.D.G.; Dangl, J.L. The plant immune system. *Nature* **2006**, *444*, 323–329. [CrossRef]
6. An, S.Q.; Potnis, N.; Dow, M.; Vorholter, F.J.; He, Y.Q.; Becker, A.; Teper, D.; Li, Y.; Wang, N.; Bleris, L.; et al. Mechanistic insights into host adaptation, virulence and epidemiology of the phytopathogen *Xanthomonas*. *FEMS Microbiol. Rev.* **2020**, *44*, 1–32. [CrossRef]
7. Mew, T.W.; Vera Cruz, C.M.; Medalla, E.S. Changes in race frequency of *Xanthomonas oryzae* pv *oryzae* in response to rice cultivars planted in the Philippines. *Plant Dis.* **1992**, *76*, 1029–1032. [CrossRef]

8. Quibod, I.L.; Perez-Quintero, A.; Booher, N.J.; Dossa, G.S.; Grande, G.; Szurek, B.; Cruz, C.V.; Bogdanove, A.J.; Oliva, R. Effector Diversification Contributes to *Xanthomonas oryzae* pv. *oryzae* Phenotypic Adaptation in a Semi-Isolated Environment. *Sci. Rep.* **2016**, *6*, 34137. [CrossRef]
9. Kay, S.; Bonas, U. How *Xanthomonas* type III effectors manipulate the host plant. *Curr. Opin. Microbiol.* **2009**, *12*, 37–43. [CrossRef]
10. Moscou, M.J.; Bogdanove, A.J. A Simple Cipher Governs DNA Recognition by TAL Effectors. *Science* **2009**, *326*, 1501. [CrossRef]
11. Boch, J.; Scholze, H.; Schornack, S.; Landgraf, A.; Hahn, S.; Kay, S.; Lahaye, T.; Nickstadt, A.; Bonas, U. Breaking the Code of DNA Binding Specificity of TAL-Type III Effectors. *Science* **2009**, *326*, 1509–1512. [CrossRef] [PubMed]
12. Mak, A.N.S.; Bradley, P.; Cernadas, R.A.; Bogdanove, A.J.; Stoddard, B.L. The Crystal Structure of TAL Effector PthXo1 Bound to Its DNA Target. *Science* **2012**, *335*, 716–719. [CrossRef] [PubMed]
13. Streubel, J.; Blucher, C.; Landgraf, A.; Boch, J. TAL effector RVD specificities and efficiencies. *Nat. Biotechnol.* **2012**, *30*, 593–595. [CrossRef]
14. Erkes, A.; Reschke, M.; Boch, J.; Grau, J. Evolution of Transcription Activator-Like Effectors in *Xanthomonas oryzae*. *Genome Biol. Evol.* **2017**, *9*, 1599–1615. [CrossRef] [PubMed]
15. Booher, N.J.; Carpenter, S.C.D.; Sebra, R.P.; Wang, L.; Salzberg, S.L.; Leach, J.E.; Bogdanove, A.J. Single molecule real-time sequencing of *Xanthomonas oryzae* genomes reveals a dynamic structure and complex TAL (transcription activator-like) effector gene relationships. *Microb. Genom.* **2015**, *1*, e000032. [CrossRef]
16. Liu, H.; Yang, W.; Hu, B.; Liu, F. Virulence analysis and race classification of *Xanthomonas oryzae* pv. *oryzae* in China. *J. Phytopathol.* **2007**, *155*, 129–135. [CrossRef]
17. Chen, X.L.; Yu, L.; Gao, L.L.; Jiang, T.; Li, Q.Y.; Huang, Q. Elevational Variation in Diversity of *Xanthomonas oryzae* pv. *oryzae* in South-West China. *J. Phytopathol.* **2012**, *160*, 261–268. [CrossRef]
18. Li, T.J.; Li, Y.M.; Ma, X.G.; Dan, X.; Huang, X.J.; Li, Q.Y.; Lei, S.M.; Zhang, Z.C.; Huang, S.; Jiang, W.; et al. Comparative Genomic Analysis of Two *Xanthomonas oryzae* pv. *oryzae* Strains Isolated From Low Land and High Mountain Paddies in Guangxi, China. *Front. Microbiol.* **2022**, *13*, 867633. [CrossRef]
19. Grau, J.; Reschke, M.; Erkes, A.; Streubel, J.; Morgan, R.D.; Wilson, G.G.; Koebnik, R.; Boch, J. AnnoTALE: Bioinformatics tools for identification, annotation, and nomenclature of TALEs from *Xanthomonas* genomic sequences. *Sci. Rep.* **2016**, *6*, 21077. [CrossRef]
20. Oliva, R.; Ji, C.H.; Atienza-Grande, G.; Huguet-Tapia, J.C.; Perez-Quintero, A.; Li, T.; Eom, J.S.; Li, C.H.; Nguyen, H.; Liu, B.; et al. Broad-spectrum resistance to bacterial blight in rice using genome editing. *Nat. Biotechnol.* **2019**, *37*, 1344. [CrossRef]
21. Yuan, M.; Ke, Y.G.; Huang, R.Y.; Ma, L.; Yang, Z.Y.; Chu, Z.H.; Xiao, J.H.; Li, X.H.; Wang, S.P. A host basal transcription factor is a key component for infection of rice by TALE-carrying bacteria. *Elife* **2016**, *5*, e19605. [CrossRef]
22. Yang, B.; Sugio, A.; White, F.F. *Os8N3* is a host disease-susceptibility gene for bacterial blight of rice. *Proc. Natl. Acad. Sci. USA* **2006**, *103*, 10503–10508. [CrossRef] [PubMed]
23. Chu, Z.H.; Yuan, M.; Yao, L.L.; Ge, X.J.; Yuan, B.; Xu, C.G.; Li, X.H.; Fu, B.Y.; Li, Z.K.; Bennetzen, J.L.; et al. Promoter mutations of an essential gene for pollen development result in disease resistance in rice. *Gene Dev.* **2006**, *20*, 1250–1255. [CrossRef]
24. Sugio, A.; Yang, B.; Zhu, T.; White, F.F. Two type III effector genes of *Xanthomonas oryzae* pv. *oryzae* control the induction of the host genes *OsTFIIA* gamma 1 and *OsTFX1* during bacterial blight of rice. *Proc. Natl. Acad. Sci. USA* **2007**, *104*, 10720–10725. [CrossRef]
25. Chen, X.F.; Liu, P.C.; Mei, L.; He, X.L.; Chen, L.; Liu, H.; Shen, S.R.; Ji, Z.D.; Zheng, X.X.; Zhang, Y.C.; et al. *Xa7*, a new executor R gene that confers durable and broad-spectrum resistance to bacterial blight disease in rice. *Plant Commun.* **2021**, *2*, 100143. [CrossRef]
26. Read, A.C.; Rinaldi, F.C.; Hutin, M.; He, Y.Q.; Triplett, L.R.; Bogdanove, A.J. Suppression of *Xo1*-Mediated Disease Resistance in Rice by a Truncated, Non-DNA-Binding TAL Effector of *Xanthomonas oryzae*. *Front. Plant Sci.* **2016**, *7*, 1516. [CrossRef]
27. Perez-Quintero, A.L.; Szurek, B. A Decade Decoded: Spies and Hackers in the History of TAL Effectors Research. *Annu. Rev. Phytopathol.* **2019**, *57*, 459–481. [CrossRef] [PubMed]
28. Ji, C.H.; Ji, Z.Y.; Liu, B.; Cheng, H.; Liu, H.; Liu, S.Z.; Yang, B.; Chen, G.Y. Xa1 Allelic *R* Genes Activate Rice Blight Resistance Suppressed by Interfering TAL Effectors. *Plant Commun.* **2020**, *1*, 100087. [CrossRef] [PubMed]
29. Chen, H.L.; Wang, S.P.; Zhang, Q.F. New gene for bacterial blight resistance in rice located on chromosome 12 identified from Minghui 63, an elite restorer line. *Phytopathology* **2002**, *92*, 750–754. [CrossRef] [PubMed]
30. Chen, X.; Wei, S.; Yan, Q.; Huang, F.; Ma, Z.; Li, R. Virulence and DNA fingerprinting analysis of *Xanthomonas oryzae* pv. *oryzae* identify a new pathotype in Guangxi, South China. *J. Basic Microbiol.* **2019**, *59*, 1082–1091. [CrossRef] [PubMed]
31. Lim, H.J.; Lee, E.H.; Yoon, Y.; Chua, B.; Son, A. Portable lysis apparatus for rapid single-step DNA extraction of Bacillus subtilis. *J. Appl. Microbiol.* **2016**, *120*, 379–387. [CrossRef]
32. Reiner, J.; Pisani, L.; Qiao, W.Q.; Singh, R.; Yang, Y.; Shi, L.S.; Khan, W.A.; Sebra, R.; Cohen, N.; Babu, A.; et al. Cytogenomic identification and long-read single molecule real-time (SMRT) sequencing of a Bardet-Biedl Syndrome 9 (BBS9) deletion. *NPJ Genom. Med.* **2018**, *3*, 3. [CrossRef] [PubMed]
33. Ardui, S.; Ameur, A.; Vermeesch, J.R.; Hestand, M.S. Single molecule real-time (SMRT) sequencing comes of age: Applications and utilities for medical diagnostics. *Nucleic Acids Res.* **2018**, *46*, 2159–2168. [CrossRef]
34. Walker, B.J.; Abeel, T.; Shea, T.; Priest, M.; Abouelliel, A.; Sakthikumar, S.; Cuomo, C.A.; Zeng, Q.D.; Wortman, J.; Young, S.K.; et al. Pilon: An Integrated Tool for Comprehensive Microbial Variant Detection and Genome Assembly Improvement. *PLoS ONE* **2014**, *9*, e112963. [CrossRef] [PubMed]

35. Krzywinski, M.; Schein, J.; Birol, I.; Connors, J.; Gascoyne, R.; Horsman, D.; Jones, S.J.; Marra, M.A. Circos: An information aesthetic for comparative genomics. *Genome Res.* **2009**, *19*, 1639–1645. [CrossRef]
36. Tatusova, T.; DiCuccio, M.; Badretdin, A.; Chetvernin, V.; Nawrocki, E.P.; Zaslavsky, L.; Lomsadze, A.; Pruitt, K.D.; Borodovsky, M.; Ostell, J. NCBI prokaryotic genome annotation pipeline. *Nucleic Acids Res.* **2016**, *44*, 6614–6624. [CrossRef] [PubMed]
37. Besemer, J.; Lomsadze, A.; Borodovsky, M. GeneMarkS: A self-training method for prediction of gene starts in microbial genomes. Implications for finding sequence motifs in regulatory regions. *Nucleic Acids Res.* **2001**, *29*, 2607–2618. [CrossRef] [PubMed]
38. Darling, A.E.; Mau, B.; Perna, N.T. Progressive Mauve: Multiple genome alignment with gene gain, loss and rearrangement. *PLoS ONE* **2010**, *5*, e1114. [CrossRef]

International Journal of
Molecular Sciences

MDPI

Article

The Overexpression of *Oryza sativa* L. *CYP85A1* Promotes Growth and Biomass Production in Transgenic Trees

Guodong Li [1,2], Xinzhuan Yao [2], Zhouzhuoer Chen [1,2], Xingyu Tian [1,2] and Litang Lu [1,2,*]

1 The Key Laboratory of Plant Resources Conservation and Germplasm Innovation in Mountainous Region (Ministry of Education), College of Life Sciences/Institute of Agro-Bioengineering, Guizhou University, Guiyang 550025, China
2 College of Tea Sciences, Institute of Plant Health & Medicine, Guizhou University, Guiyang 550025, China
* Correspondence: ltlv@gzu.edu.cn; Tel.: +86-136-3903-1105

Abstract: Brassinosteroids (BRs) are important hormones that play crucial roles in plant growth, reproduction, and responses to abiotic and biotic stresses. CYP85A1 is a castasterone (CS) synthase that catalyzes C-6 oxidation of 6-deoxocastasterone (6-deoxoCS) to CS, after which CS is converted into brassinolide (BL) in a reaction catalyzed by CYP85A2. Here, we report the functional characteristics of rice (*Oryza sativa* L.) *OsCYP85A1*. Constitutive expression of *OsCYP85A1* driven by the cauliflower mosaic virus 35S promoter increased endogenous BR levels and significantly promoted growth and biomass production in three groups of transgenic *Populus tomentosa* lines. The plant height and stem diameter of the transgenic poplar plants were increased by 17.6% and 33.6%, respectively, in comparison with control plants. Simultaneously, we showed that expression of *OsCYP85A1* enhanced xylem formation in transgenic poplar without affecting cell wall thickness or the composition of cellulose. Our findings suggest that *OsCYP85A1* represents a potential target candidate gene for engineering fast-growing trees with improved wood production.

Keywords: *Oryza sativa* L.; *OsCYP85A1*; castasterone (CS); *Populus tomentosa*; biomass production; xylem

1. Introduction

The class of compounds known as plant steroidal hormones contains more than 40 members, which are collectively designated brassinosteroids (BRs). Among BRs, castasterone (CS) and brassinolide (BL) are considered to have the most important biological roles [1,2]. BRs play critical roles in regulating plant developmental processes, including cell division, cell cycle, elongation, morphogenesis, reproduction, senescence, and stress-protective responses [3–5]. Simultaneously, BRs act as regulators of plant vascular structure formation, hypocotyl elongation, root growth, bud growth, xylem formation, and xylem differentiation [6–8].

In general, BL positively contributes to the cellular biochemical, physiological, and morphological processes of taller plants, especially cell division, antioxidant metabolism, gas exchange, and growth [6,9,10]. Previous reports suggest that CS activates BR signal transduction pathways with efficacy similar to that of BL. A rice lamina inclination bioassay revealed that CS induced lamina bending, although it was slightly less effective than BL [11]. CS is biosynthesized from 24-methylcholesterol via two parallel pathways, namely, the early and late C-6 oxidation pathways, after which CS is oxidized by 7-oxalactonation to produce BL. The rice CYP85A enzyme catalyzes C-6 oxidation of 6-deoxocastasterone (6-deoxoCS) to produce CS, a bioactive BR [11]. The tomato *Dwarf* gene encodes *CYP85A1*, which catalyzes the synthesis of CS from 6-deoxoCS [12]. Catabolism of BRs is another important factor that regulates their endogenous levels. In Arabidopsis, the inactivation of BRs is also catalyzed by P450 enzymes, among which the key genes are *CYP734A1* and *CYP72C1*. *CYP72C1* binds to the BL precursor to inactivate it and maintain BR homeostasis, and *CYP734A1* inactivates CS and BL by C-26 hydroxylation [13,14]. The results of a study using a cell-free extract from cultured *Phaseolus vulgaris* cells suggest that P450 enzymes are involved in the conversion of CS to BL [15]. Using reporter gene fusion, *Dwarf* has been shown to be expressed both in vegetative tissues and in fruits. A null allele, extreme dwarf (dx), shows severely reduced stem length, altered leaf morphology, and no detectable CS in vegetative tissues [16]. Previous reports indicate that auxin signaling via *ARF7* directly modulates the expression of *BAS1* by competition with *BZR1*, thereby increasing the level of CS and promoting growth and development in *Arabidopsis thaliana* [17]. Recent reports have demonstrated that higher levels of CS contribute to plant growth and development and increase plant biomass [7,18]. More recently, Song and Duan [19] regulated the activity of antioxidant enzymes by overexpression of *SoCYP85A1*, which increased CS accumulation and enhanced black shank tolerance in tobacco.

One of the most critical steps in the BR pathway is the conversion of 6-deoxoCS to CS [20]. P450 is a class of enzymes involved in oxidative reactions occurring during BR biosynthesis [21]. The *CYP85A1* gene encodes a cytochrome P450 enzyme containing monooxygenase *CYP85A1*, which is a necessary catalyst for several essential reactions required for CS production [22]. The synthesis of CS and BL requires the enzyme *CYP85A2*, which also contributes to the conversion of 6-deoxyCS into CS and finally into BL. Several lines of evidence show that altering CS content affects plant phenotypes. For example, extreme dwarfism in cucumber (*Cucumis sativus* L.) was found to be due primarily to the inhibition of internode elongation caused by mutation in *scp-1/CsCYP85A1* [23]. In another recent study, increasing the abundance of *SoCYP85A1* transcript and endogenous CS in transgenic tobacco resulted in longer main roots and more lateral roots in comparison with control plants, which enhanced drought resistance [24]. Simultaneously, Li et al. [25] found that overexpression of the *CYP85A1* gene in tomatoes increased the CS content in plants while enhancing the quantum yield of photosystem II and the CO_2 assimilation rate. Therefore, increasing the expression of *CYP85A1* will likely increase endogenous CS content, which should enhance plant resistance to stressors, promote growth and increase biomass.

Among trees, poplar (genus *Populus*) has the largest planting area and widest distribution globally due to its strong adaptability. Poplar is an economically important group of species that play an important role in ecological protection. *Populus tomentosa* generally grows upright and has wood that is compact and delicate with very high fiber content, making it particularly useful for plate making, paper making, and other industrial applications. Previous studies have shown that modifying BR biosynthesis genes such as *DWF4* and *CYP85* enhances growth, biomass accumulation, and timber quality [26–28]. *OsCYP85A1/OsBRD1* is a key gene for CS synthesis in rice, which plays an important role in its rapid growth. *OsCYP85A1/OsBRD1* is a bromodomain-containing gene. Brd-containing proteins (alone or as multi-protein complexes) are involved in the regulation of gene expression by different mechanisms viz. chromatin remodeling, histone modifications, and transcriptional machinery regulation [29]. *Populus tomentosa* is an important industrial tree species; however, it took a relatively long time to develop into industrial wood [17]. In this study, the key gene *OsCYP85A1* of rice castasterone (CS) biosynthesis was genetically transformed into *P. tomentosa* to obtain transgenic poplars with the developmental characteristics of gramineous plants. Compared with the WT poplar, transgenic poplars can simultaneously accumulate a large amount of lignin. Their rate of growth is substantially improved so that the trees quickly become woody. Thus, they meet the needs for daily life and industrial production in the future, which have the potential to improve the production of materials and utilize land resources.

2. Results

2.1. Bioinformatic Analysis of CYP85A Gene and Expression Patterns in Poplar

Two CYP85 members, *CYP85A1* and *CYP85A1*-like, were identified in *Oryza sativa* L., and their homologues in *Populus tomentosa, Populus euphratica, Populus trichocarpa, Gossypium hirsutum, Malus domestic, Camellia sinensis, Gossypium arboreum, Nicotiana tabacum,*

Zea mays, *Citrus sinensis*, *Eucalyptus grandis*, *Hibiscus syriacus*, *Morus notabilis*, and *Arabidopsis thaliana* were searched in the NCBI database (Figure 1B). CYP85A in *O. sativa* and *P. tomentosa* is conserved with regard to protein length, molecular weight, isoelectric point (pI), and sequence similarity (Table S1). Interestingly, *OsCYP85A1* and *PtCYP85A*-2 have 67.25% similarity, with the same size and similar pI. *OsCYP85A1* was selected for further analysis and named *OsCYP85A1*-1. *OsCYP85A1*-1 had 65.86% similarity and 83.93% similarity to *PtCYP85A* and *ZmCYP85A*, respectively (the sequence alignment of *P. tomentosa* and *Z. mays* is shown in Figure S1), and the presence of several functional domains indicated that these proteins had similar functions (Figure S5). The *OsCYP85A* protein possesses the typical functional domains of a conserved cytochrome P450 monooxygenase: purine-rich domain, dioxygen-binding domain, steroid-binding domain, and heme-binding domain (Figure 1A). The results also showed that several cis-acting elements, including light and hormone response and circadian rhythm elements, were present in the *OsCYP85A* promoter (Figure S5). To further clarify the role of overexpression of *OsCYP85A1* on *Populus tomentosa*, we isolated total RNA from different organs from 3-month-old *P. tomentosa* and analyzed the transcript abundance of *OsCYP85A1* by quantitative real-time PCR (qRT-PCR) (Table 1). The expression level of *OsCYP85A1* in the stems and leaves was higher than that measured in the roots, which indicated that it played an important role in the growth and development of the aboveground parts of transgenic poplars (Figure 1C). Figure 1C shows the expression analysis and comparison with wild-type controls.

Table 1. List of primers used in this study.

Usage	Primer Name	Primer Sequence
PCR	*OsCYP85A1*	F: AACCTTCCTGGAACCAACTAC R: CGAAGATAACAGCTCGAGTGAA
qRT-PCR	*OsCYP85A1*	F: TGGAGGAGGTAGTCGAATGT R: CCTTCTTCCTCCCATCTGTATTG
qRT-PCR	*PtBRI1*	F: GATGTCAGAGGTGGTCAGAATG R: GGGTGGTGAGTGTGGTTAAA
qRT-PCR	*PtBZR1*	F: GGTTAAGGGCTCAAGGGAATTA R: CTGTGTCCCTTGCGATAAGTAG
qRT-PCR	*PtMYB2*	F: TTGGAGTGATGTAGCAAGGAA R: GATGAAGATGACAGTGACGGAT
qRT-PCR	*PtCYP85A2*	F: GAGAGCAAGGTATGGGAGTATTT R: GCCCTTTCCCTCGTTCATTA
qRT-PCR	*PtCYP85A3*	F: CTATCCAGAGCCTTCAACCTTC R: CCAGTTCCTTTCCAGGACATAG
qRT-PCR	*PtNPR1*	F: GTTGACCTAAATGAGACACC R: TAATCTCAGCCTTGTCCTTG
qRT-PCR	*PtWRKY70*	F: AATCCAAGGAGCTACTAC R: GTTACCATTGTTGTTGTGG
qRT-PCR	*PtMYB115*	F: GCCATTGGAGGTCTTTGCC R: GGTTACCGAGGAGGGAGTGC
qRT-PCR	Actin	F: GCATCCACGAGACTACATACAA R: TCAGCAATACCAGGGAACATAG

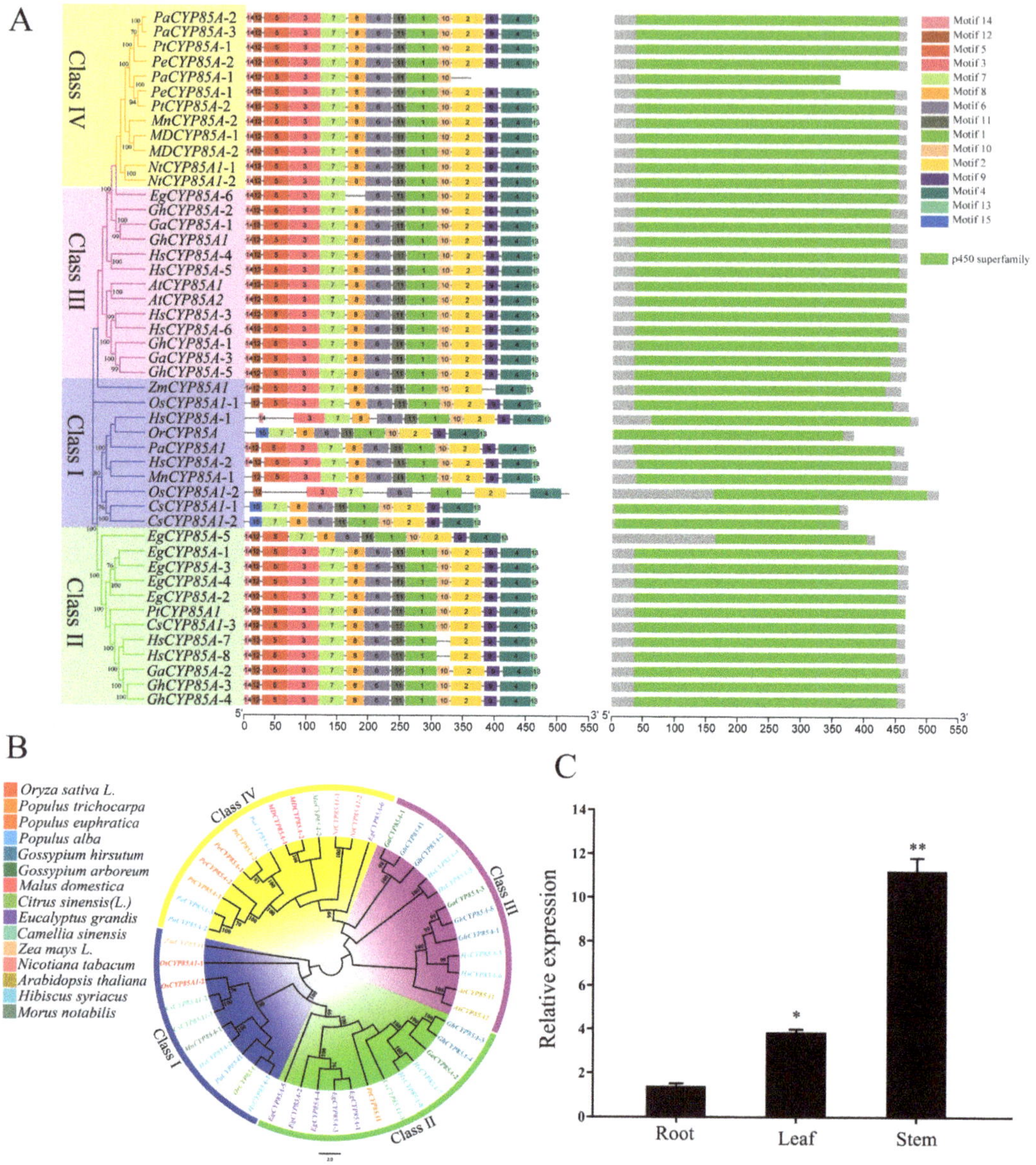

Figure 1. Sequence analysis of CYP85As proteins and expression patterns of *OsCYP85A1* in different plant parts. (**A**) *OsCYP85A* protein has the typical functional domains of a conserved cytochrome P450 monooxygenase. (**B**) Phylogenetic relationships of *OsCYP85A* with CYP85As from *P. euphratica* (Pe), *P. trichocarpa* (Pt), *A. thaliana* (At), *G. hirsutum* (Gh), *Zea mays* (Zm), *Oryza sativa* (Os), *Malus domestica* (MD), *Camellia sinensis* (Cs), *Gossypium arboretum* (Ga), *Nicotiana tabacum* (Nt), *Citrus sinensis* (CS), *Eucalyptus grandis* (Eg), *Hibiscus syriacus* (Hs), and *Morus notabilis* (Mn). The scale bar represents 0.05 substitutions per site. (**C**) Expression of *OsCYP85A1* in different organs of *P. tomentosa*. Error bars represent ± SD from three biological repeats. * Significant at $p < 0.05$ level and ** significant at $p < 0.01$ level according to Student's *t*-test.

2.2. Overexpression of OsCYP85A1 Improves Plant Growth Development

2.2.1. Generation of *OsCYP85A1* Transgenic *P. tomentosa* Lines

To assess the potential value of *OsCYP85A1* in improving biomass production in woody plants, a pCAMBIA130::*OsCYP85A1*::GUS recombinant plasmid was constructed and transformed into *P. tomentosa.* The length of *OsCYP85A1* is 1436 bp. Six transgenic lines were obtained, and three lines with a high transcription level of *OsCYP85A1* were selected for further experiments: TP1, TP2, and TP3 (Figure 2b). PCR analysis confirmed that *OsCYP85A1* was integrated into the poplar genome (Figure 2a). Further analysis by GUS staining (Figure 2c) confirmed the overexpression of *OsCYP85A1* in all selected trans-genic lines. Simultaneously, the overexpression of *OsCYP85A1* in transgenic lines was ver-ified by qRT-PCR. On the 70th day of growth, the height of the transgenic lines was significantly higher than that of WT (Figure 2d).

Figure 2. Molecular analyses of *OsCYP85A1* transgenic trees. (**a**) PCR analysis of wild-type and different independently generated transgenic lines. (**b**) Phenotypes of wild-type and three independent *OsCYP85A1* transgenic lines grown in a glass greenhouse for 7 weeks. (**c**) GUS staining of *OsCYP85A1* transgenic plants. (**d**) The 70 d transgenic plants.

2.2.2. Growth Rate

The plant height and basal stem size affect plant architecture, lodging resistance, yield, and even the optimal plant harvesting strategy [30]. Changes in the height and basal stem growth of transgenic and wild-type *P. tomentosa* plants were recorded between 27 November 2021 and 27 March 2022. No significant differences in plant height were ob-served among the TP1, TP2, TP3, and WT lines for the first 70 days of observation. After 70 days, the TP1, TP2, and TP3 plants were significantly taller than the WT plants (Figure 3C). However, the basal stems of the TP1, TP2, and TP3 plants were significantly thicker than those of the WT plants after day 20 (Figure 3D). The height growth rate (HGR) and basal diameter growth rate (DGR) of the TP1 and TP3 transgenic lines were significantly greater than those of the WT plants (HGR: 1.05 times WT for TP1 and 1.06 times for TP3; DGR: 1.39 times WT for TP1 and 1.17 times WT for TP3). In addition, the HGR and DGR of TP2 were significantly greater than those of the WT plants (HGR: 1.02 times WT; DGR: 1.15 times WT) (Figure 3A,B).

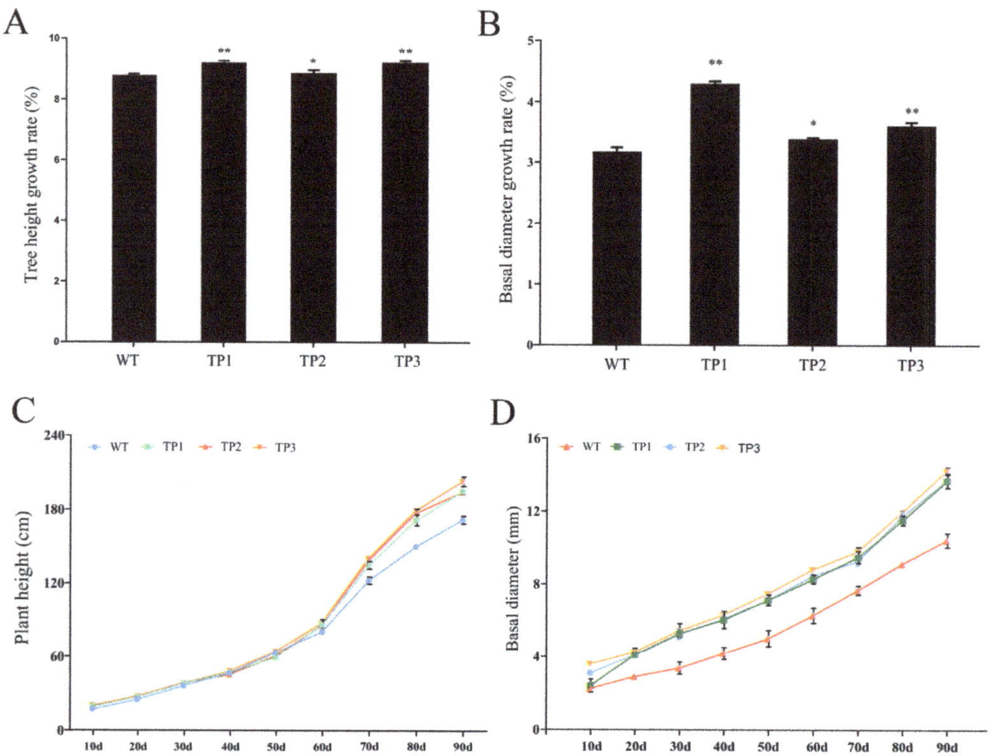

Figure 3. Overexpression of *OsCYP85A1* in poplar increases plant height and stem diameter. (**A**) Tree height growth rate. (**B**) Basal diameter growth rate. (**C**) Plant height. (**D**) Basal diameter. Error bars represent the SDs from three biological replicates. * and ** indicate significant differences in comparison to WT at $p < 0.05$ and $p < 0.01$, respectively (Student's *t*-test).

2.2.3. Effects of *OsCYP85A1* on Chlorophyll Content

Li et al. [25] generated transgenic tomato plants by overexpressing *Dwarf*, a BR biosynthetic gene that encodes *CYP85A1*, and compared the photosynthetic capacity of the BR biosynthetic mutant dim and WT. Overexpression of *dwarf* increased the net photosynthetic rate (*PN*) and photosynthetic capacity by inducing a reduced redox status that maintained the activation states of Calvin cycle enzymes, thereby promoting plant growth and development. The leaves of TP1, TP2, TP3, and WT plants were collected at the same growth stage and from the same part of each plant for analysis of the content of photosynthetic pigments. The analysis revealed a significant difference between the photosynthetic pigment content of TP1, TP2, and TP3 in comparison with that of WT plants (Figure 4A–C). The results of this study were consistent with those of Li et al. [25]. At the same time, the petioles of each group, again collected at the same stage and from the same position, were subjected to anatomical analysis (Figure 4D). The transverse sections of petioles in each group were morphometrically analyzed by Fiji/ImageJ software (Fiji- ImageJ version of Java 6, NIH, Bethesda, MD, USA) (Figure S2A). The results indicated that the TP2 and TP3 petioles had significantly more vessels and sieve tubes in comparison with WT petioles; however, the number of vessels in TP1 was slightly lower than that in WT, but the total number of vessels and sieve tubes in TP1 was also significantly higher than that in WT (Figure S2B,C). Similarly, the number and area of vascular bundles in the whole petiole cross-section of transgenic lines were greater than those of the wild type (Figure S4), which likely facilitated the transportation of water and nutrients in the transgenic plants, promoting increased biomass. Red is the result of safranine staining and decolorization, and green is the result of fixation green staining and decolorization.

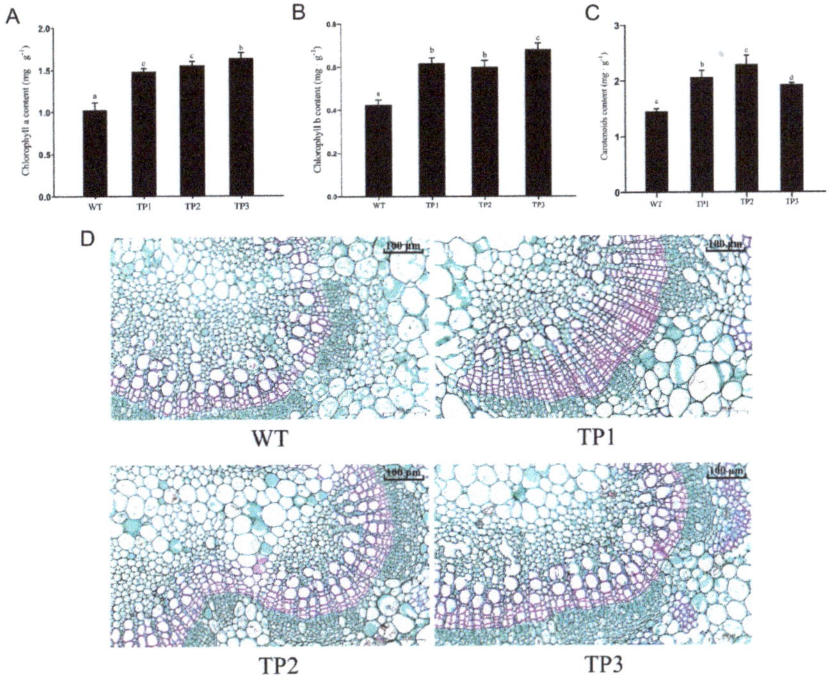

Figure 4. Effects of *OsCYP85A1* on the photosynthetic pigment content and petiole morphological characteristics of transgenic plants. (**A**) Chlorophyll a. (**B**) Chlorophyll b. (**C**) Carotenoid. (**D**) Morphological characteristics of transverse section of petiole per unit area of transgenic plants. Error bars represent mean ± SD of three biological replicates. Different letters indicate a significant difference using the Tukey–Kramer test: $p < 0.05$.

2.3. *Overexpression of OsCYP85A1 Regulates Root Development*

The root development of TP1, TP2, TP3, and WT plants was assessed at 5 weeks and 6 months of age. The roots were grown under the same culture conditions in a tissue culture room (relative humidity of 60–70%, 16 h/8 h light/dark cycle, and 25 °C). In comparison with the WT line, the main roots of the transgenic lines were longer and had a greater number of lateral roots after 5 weeks (Figure 5a). After 6 months, the entire plant was harvested, and all roots were washed. Root segments of the same position and length were selected, and root activity was determined by the TTC method (Figure 5b). A significant difference in root activity was observed between transgenic and WT roots. Higher root activity helps transgenic plants absorb inorganic salts and water from the soil, thus promoting the growth of their aboveground parts [24]. The growth and biomass of transgenic plants were significantly increased in this study, indicating that *OsCYP85A1* has great po-tential for molecular breeding of fast-growing tree species by regulating BR production in poplar and other woody plants.

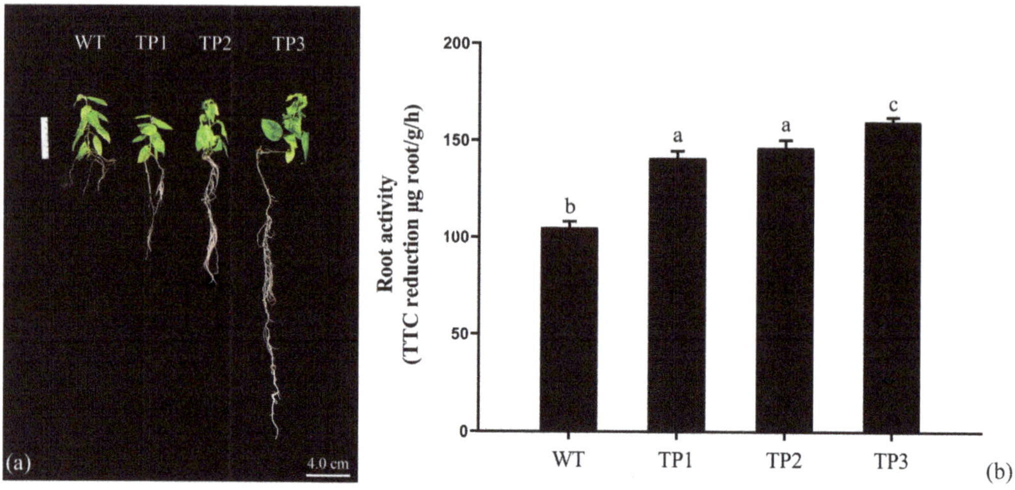

Figure 5. The effect of *OsCYP85A1* overexpression on the root development of transgenic plants. (**a**) Phenotype of 5-week-old seedlings under normal conditions. (**b**) The metabolic activity of transgenic plant roots was estimated by tetrazolium chloride (TTC) reduction. Error bars represent mean ± SD of three biological replicates. Different letters indicate a significant difference using the Tukey–Kramer test: $p < 0.05$.

2.4. Overexpression of OsCYP85A1 Enhanced Plant Disease Resistance

According to previous reports, the contents of disease-resistant metabolites in the leaves of transgenic lines and WT after 90 d of growth were determined, and the expression of disease-resistant genes in plants was analyzed. The results showed that the salicylic acid content of transgenic lines was significantly higher than that of wild type, while the activities of superoxide dismutase (SOD) and peroxidase (POD) were not significantly different from those of the wild type (Figure 6D–F). Interestingly, the expression levels of resistance-related genes *PtNPR1*, *PtMYB115*, and *PtWRKY70* in transgenic lines were significantly higher than those in the wild type (Figure 6D–F). We speculate that this is related to the bromodomain-containing protein encoded by *OsCYP85A1*/*OsBRD1*. This is the same as the previous research results; the increase in BR content will enhance the disease resistance of plants [31].

2.5. OsCYP85A1 Promotes Xylem Differentiation in Transgenic Plants

To study the effect of *OsCYP85A1* on the secondary growth process of woody plants, the stems of WT and transgenic plants were microscopically analyzed. As shown in Figure 7 xylem growth was improved in all tested *OsCYP85A1* transgenic lines. Transgenic plants produced more xylem (Figure S3), but there was no significant difference in the yield of bark between the WT and transgenic plants (Figure 8A–D). There was no significant difference in the content of hemicellulose between the transgenic lines and WT, but the content of pectin was significantly higher than that of the WT (Figure 8E,F). Further studies were carried out on *PtMYB2*, an MYB transcription factor regulating secondary cell wall synthesis in transgenic plants, as well as *PtCYP85A2*, *PtCYP85A3*, *PtBRI1*, and *PtBZR1*, which control CS and BL biosynthesis in *P. tomentosa* (Figure 8G–L). The results of these experiments indicated that *OsCYP85A1* participates in secondary cell division and promotes the anabolism of CS and BL during poplar wood formation, thus promoting the growth and wood formation of transgenic plants.

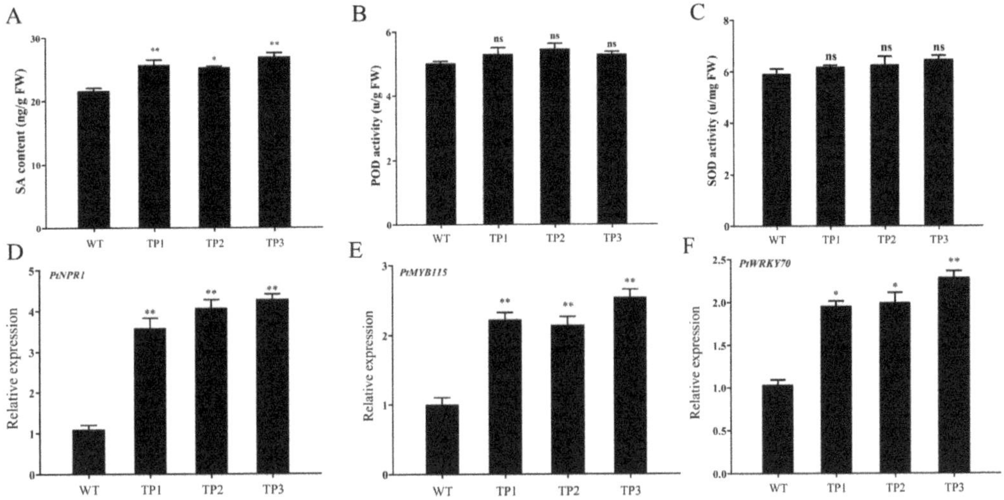

Figure 6. Overexpression of *OsCYP85A1* enhanced plant disease resistance. (**A**) Salicylic acid (SA); (**B**) Superoxide dismutase (SOD); (**C**) Peroxidase (POD). Expression of genes related to poplar disease resistance. The expression level of these genes in the stem of WT was set to 1. ns, nonsignificant; Error bars represent the SDs from three biological replicates. * and ** indicate significant differences in comparison to WT at $p < 0.05$ and $p < 0.01$, respectively (Student's *t*-test).

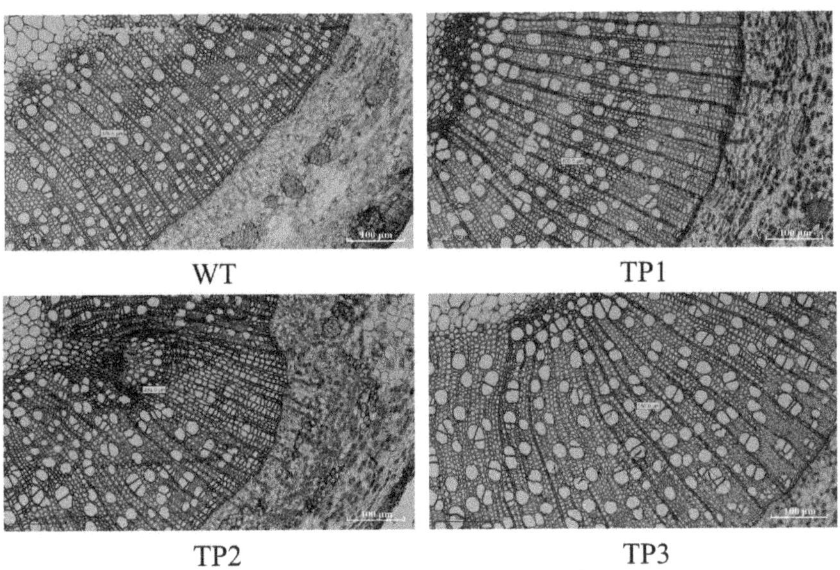

Figure 7. Cross-sections of stems showing increased xylem area in 90 d old wild-type and transgenic plants (lines 1, 2, and 3). Blue is marked as the thickness of xylem per unit area. WT. The thickness of xylem is 376.6 μm. TP1. The thickness of xylem is 450.0 μm. TP1. The thickness of xylem is 434.0 μm. TP1. The thickness of xylem is 550.0 μm.

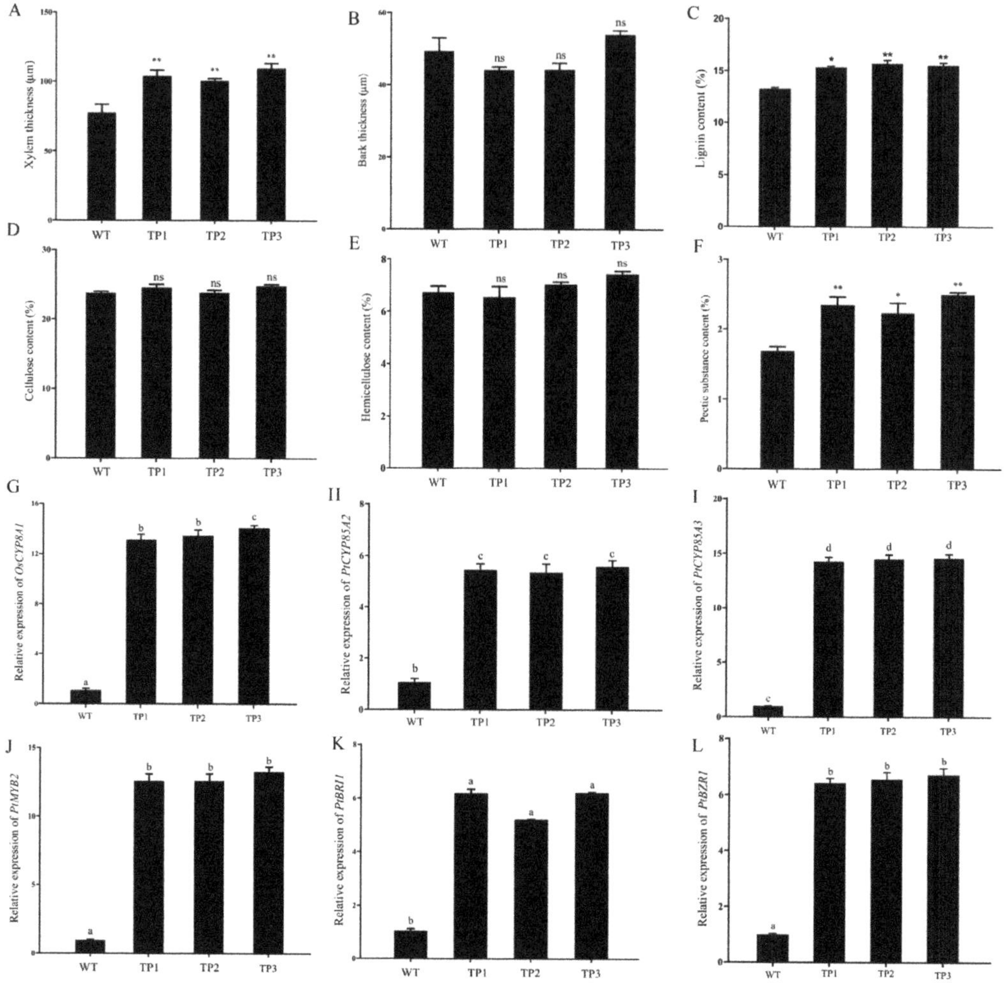

Figure 8. Effects of *OsCYP85A1* on xylem differentiation in transgenic poplar. (**A**,**B**) Measurement of xylem and bark. (**C**,**D**) Cellulose and lignin contents in the xylem tissues. (**E**,**F**) Hemicellulose content and pectic content. (**G**–**L**) Expression analysis of secondary cell wall synthesis-related MYB transcription factors and CS synthesis-related genes. The expression level of these genes in the stem of WT was set to 1. ns, nonsignificant; error bars represent the SDs from three biological replicates. * and ** indicate significant differences in comparison to WT at $p < 0.05$ and $p < 0.01$, respectively (Student's *t*-test). Error bars represent mean ± SD of three biological replicates. Different letters indicate a significant difference using the Tukey–Kramer test: $p < 0.05$.

3. Discussion

Current studies have fully demonstrated that *CYP85A1* is necessary for the production of CS. Until now, functional studies of *CYP85A1* have been carried out in *Arabidopsis*, tomato, and rice, but few studies have assessed the role of this gene in *P. tomentosa* [24]. In this study, *CYP85A1* was observed to have a highly conserved cytochrome P450 domain, indicating that *CYP85A1* belongs to the P450 superfamily. Multiple sequence alignment showed that *OsCYP85A1* has high sequence consistency with *CYP85A1* genes from other plants at the amino acid level. Phylogenetic tree analysis revealed the relationship between

OsCYP85A1 and *CYP85A1* genes from other plants, showing that *OsCYP85A1* is closely related to *CYP85A1* from *Zea mays* (NM_001349820.1, 83.93%) and *P. tomentosa* (EEE85904, 65.86%). Numerous studies have shown that *CYP85A1* regulates biocatalysis, plant growth, and development [21,25]. However, no research has been published on the effects of *CYP85A1* from herbaceous plants on the growth and development of woody plants, which could provide new ideas for the creation of novel industrial tree species.

The qRT-PCR data showed that the *OsCYP85A1* transcript was more abundant in stems and leaves. These results were different from those of previous studies, which reported that CYP85A1 mRNA mainly accumulated in roots or buds [16,32]. The discrepancy between these results may be due to the use of different species and different sample collection times [24]. Alternatively, because the gene is expressed under the control of the constitutive promoter CaMV 35s, the expression in leaves and stems is higher than that in roots and buds [33]. High expression of *OsCYP85A1* in young leaves and stems suggests that it plays a key role in BR metabolism and the normal growth of trees. Under the control of the CaMV 35S promoter, transgenic poplars were generated by Agrobacterium-mediated transformation with *OsCYP85A1*. The agronomic traits of three selected transgenic lines (TP1, TP2, and TP3) were analyzed. In comparison with WT plants, the transgenic lines had longer main roots, more lateral roots, higher plant height, and thicker stems. These findings are consistent with the results of Li et al. [25]. *DWARF* is a key gene involved in BR synthesis in tomatoes. When DWF transcripts and endogenous BR levels are increased, germination, lateral root development, CO2 assimilation, and final plant growth are improved, as demonstrated by elongation and compactness of plant structure. Consistent with Kim et al., *CYP85A2* overexpression enhanced the growth of *Arabidopsis thaliana*; *CYP85A2* overexpression induced higher levels of BL, conferring transgenic plants with significant advantages in morphogenesis, leading to larger rosette leaves and longer petioles in comparison with those of WT plants [34]. Therefore, genes involved in BR biosynthesis can be manipulated to control plant growth, photosynthesis, structure, stress resistance, and other important agronomic traits, thus improving crop yield [35].

Previous reports on the involvement of endogenous BR in promoting wood formation and the abnormal xylem appearance of BR dwarf mutants suggest that BR biosynthesis occurs in vascular tissue [20,36]. Research by Hamasaki et al. [37] showed that light activates the expression of BR biosynthesis genes in the hook region via a phytochrome-signaling pathway and HY5, and BR biosynthesis was shown to be essential for hook opening and petiole development during photomorphogenesis. Genes involved in BR synthesis include *CYP90A1*, *CYP90C1*, and *CYP90D1*, which regulate C-23 hydroxylation shortcuts and camphor-independent pathways, and *BR6ox1/CYP85A1* and *BR6ox2/CYP85A2*, which control the biosynthesis of CS and BL [16,38]. In this study, *OsCYP85A1*, which controls the synthesis of CS in rice, was transformed into *P. tomentosa*. Histological analysis showed that the petioles of transgenic plants had significantly more vessels and sieve tubes in comparison with those of WT plants, which helped transgenic plants transport more organic matter to roots and more inorganic salts and water to leaves.

Physiological, genetic, and molecular studies have revealed the roles of plant hormones in xylogenesis and vascular tissue differentiation. The expression levels of some genes involved in cell wall synthesis were found to be reduced in BR synthetic mutants, such as XETs and MERI5 in *dwf1* [39]. In addition, brassinazole (Brz) was found to be capable of suppressing the development of the secondary xylem of cress plants [40]. Another recent study showed that overexpression of *PtCYP85A3*, a functional allele of *AtCYP85A2* and *SlCYP85A1*, increased the yield of CS and regulated secondary cell division and fiber elongation during poplar wood formation, thereby promoting growth and wood formation [26]. Overexpression of *PtBRI1*, a brassinosteroid-insensitive gene, promotes the accumulation of *PtBZR1* (BRASSINAZOLE RESISTANT1) in the nucleus and then activates *PtWNDs* (*WOOD-ASSOCIATED NAC DOMAIN* transcription factor) to regulate the expression of secondary cell wall biosynthesis genes involved in wood formation; this process plays a crucial role in regulating twig growth and wood formation [17]. In this study, we found that transgenic poplars produced more xylem than wild-type plants, but the increase in the bark was not significant. The expression levels of secondary cell wall-related transcription factor genes, genes controlling BR biosynthesis in poplar, and genes regulating xylem synthesis were significantly up-regulated in transgenic plants. These results indicate that *OsCYP85A1* can participate in both secondary cell division and fiber elongation during poplar wood formation. Overexpression of *OsCYP85A1* increased CS production, thereby promoting the growth and wood formation of transgenic plants.

Previous studies have shown that *CYP85A1* is mainly expressed in roots and hypocotyls, specifically at the root-hypocotyl junction during early development [20]. A common feature of the data obtained from the CYP85 promoter is that the highest expression level occurs early in development, after which expression is down-regulated over time [32]. In this study, it was found that the primary root length of the transgenic plants was 5.2 times that of the wild type, and the number of lateral roots was also significantly higher than that of the wild type. However, when detecting the expression of *OsCYP85A1* in the roots, stems, and leaves of transgenic plants after 90 days of growth, it was found that the expression of *OsCYP85A1* in roots was significantly lower than that in stems and leaves. These results are consistent with the above research. Overexpression of *SoCYP85A1* enhanced drought tolerance in transgenic tobacco with increased CS content, which promoted root development and led to the accumulation of a higher relative water content (RWC) and proline, as well as less malondialdehyde (MDA) and H_2O_2, improved the activities of antioxidant enzymes, and increased the transcript expression of ROS-related and stress-responsive genes under drought stress [24]. In summary, *CYP85A1* has a positive effect on plant root development [41]. In order to verify that *OsCYP85A1* plays an important role in the early stage of plant development, we measured root activity using the TTC method after 90 days of growth. The results showed that the root activity of the transgenic plants was significantly higher than that of the wild type, which likely helped the transgenic plants absorb more nutrients and promote growth and development.

4. Materials and Methods

4.1. Bioinformatics Analysis

The protein sequences of *Oryza sativa* L. CYP85A family members were used as queries to search against the reference protein sequence databases of *Populus tomentosa, Populus euphratica, Populus trichocarpa, Gossypium hirsutum, Malus pumila, Camellia sinensis, Gossypium arboretum, Nicotiana tabacum, Zea mays, Citrus sinensis, Eucalyptus grandis, Hibiscus syriacus, Morus notabilis*, and *Arabidopsis thaliana* (available at the National Center for Biotechnology Information) using BLASTP (http://www.ncbi.nlm.nih.gov/BLAST/, accessed on 31 January 2022). The results were combined and splice variants were removed. Sequence similarity was calculated by Sequencher v4.7. The molecular weight and theoretical isoelectric point (pI) of the proteins were predicted with ProtParam (http://web.expasy.org/protparam/, accessed on 8 June 2022). A phylogenetic tree was built with MEGA 6.0 software according to the neighbor-joining statistical method with 1000 bootstrap replicates. Multiple sequence alignments were generated with DNA-MAN 7.0. Conserved domains and motifs were analyzed by CDD in the NCBI (http://www.ncbi.nlm.nih.gov/Structure/cdd/docs/cdd_search.html, accessed on 8 June 2022). The promoter sequence of the *OsCYP85A* gene was analyzed by the online software Plant CARE (http://bioinformatics.psb.ugent.be/webtools/plantcare/html/, accessed on 27 June 2022) to predict cis-acting elements. The above experiments were carried out in the Tea College of Guizhou University (26°44′ N, 106°65′ E).

4.2. Materials and Transformation

Plant materials: in this study, *P. tomentosa* hybrid poplar (*P. tomentosa* Carr. (♀) × *P. simonii* Carr. (♂)) seedlings were used as the experimental material. Strains and vector plasmids: *Escherichia coli* DH5α, *Agrobacterium tumefaciens* LB4404; CaMV 35S promoter cloning vectors pCAMBIA130::*OsCYP85A1*::GUS. Carrier construction method: the *Os-*

CYP85A1 gene sequence was obtained by PCR amplification using rice cDNA as a template and high-fidelity polymerase PrimeSTAR® Max (Junotec Biotechnology Co., Ltd., Wuhan, China). The plasmid vector pCAMBIA130 was digested with *Xba*I and *BamH*I, and the synthetic target fragment *OsCYP85A1* was digested with *Xba*I and *Kpn*I. pCAMBIA130 and *OsCYP85A1* were ligated by T4 DNA ligase (Covent Biotechnology Co., Ltd., Changsha, China). The pCAMBIA130::*OsCYP85A1*::GUS vector driven by the constitutive promoter CaMV 35S was constructed. The β-glucuronidase (GUS) fusion gene was used as a screening marker and reporter gene. The above materials were stored in the College of Tea Sciences at Guizhou University (26°44′ N, 106°65′ E), Guiyang, China. *P. tomentosa* transformation was implemented by the Agrobacterium-infected leaf disc method. Infected Agrobacterium LB4404-leaf discs were first cultured in MS culture medium (MS) with 1.0 mg·L^{-1} naphthylacetic acid (NAA) and 1.0 mg·L^{-1} zeatin (ZT) in the dark for 2 days, after which they were placed into MS with 1.0 mg·L^{-1} NAA, 1.0 mg ·L^{-1} ZT, 50 mg ·L^{-1} kanamycin, and 200 mg·L^{-1} timentin (TIM) and kept in the dark for 2–3 days until callus formation. Subsequently, the calluses were switched to MS with 1.5 mg·L^{-1} ZT, 25 mg·L^{-1} kanamycin, and 200 mg·L^{-1} TIM and cultured under a 16/8 h light/dark cycle to induce shoot regeneration. Finally, the shoots (about 1 cm in length) were cut down and cultured in 1/2 MS with 0.2 mg·L^{-1} NAA, 50 mg·L^{-1} kanamycin, and 200 mg·L^{-1} TIM. Regenerated roots formed after about 1 week.

4.3. Quantitative PCR (qPCR) and Gus Histochemical Staining

The genomic DNA of rooted plantlets was extracted by the CTAB method [42] to confirm whether the recombinant plasmid had been successfully inserted into the plants using PCR (forward primer *OsCYP5A1*-F and reverse primer *OsCYP5A1*-R; see Table 1). The first leaves of 1-month-old plantlets were subjected to Gus histochemical staining and serial decolorization in 10%, 25%, 50%, 75%, and 95% ethanol at 37 °C until the leaves turned white, after which they were observed and imaged under a stereoscopic microscope.

4.4. Determination of Growth Indices

4.4.1. Growth Measurement

Healthy poplars without pests and diseases that had grown for 90 d were selected as experimental materials [26]. Sapling height was measured using a tape rule from the base of the stem to the terminal bud. Stem basal diameter was measured by a Vernier caliper at the base of stem. Sapling height and stem basal diameter measurements were taken on 27 November 2021 and 27 March 2022, respectively. The relative growth rate of height (HGR) and basal diameter (DGR) were calculated using Equations (1) and (2) [43], where H_1 is initial sapling height (cm), H_2 is final sapling height (cm), D_1 is initial basal diameter (mm), and D_2 is final basal diameter (mm).

$$HGR = (H_2 - H_1)/H_1 \quad (1)$$

$$DGR = (D_2 - D_1)/D_1 \quad (2)$$

4.4.2. Determination of Photosynthetic Pigments

Wild-type (WT) and transgenic plants (TP1, TP2, and TP3) were selected as experimental materials, and the third functional leaf of each plant was collected. Fresh leaves (0.1 g) without the midrib were weighed, ground into a powder, and soaked in 10 mL of 95% alcohol to extract the pigment. The specific process followed the method of Ren et al. [44]. The absorbance values (A_{665}, A_{649}, and A_{470}) of each extract at 665 nm, 649 nm, and 470 nm were measured using a UH5300 spectrophotometer (Hitachi Co., Ltd., Tokyo, Japan). Each treatment was repeated three times. The concentrations and contents of chlorophyll a (*Ca*), chlorophyll b (*Cb*), and carotenoids (*Car*) were calculated using Equations (3)–(5).

$$Ca = 13.95A_{665} - 6.88A_{649} \quad (3)$$

$$Cb = 24.96A_{649} - 7.32A_{665} \tag{4}$$

$$Car = \frac{1000\ A_{470} - 2.05\ Ca\ - 114.8\ Cb}{245} \tag{5}$$

4.4.3. Lignin Content Determination

Quantification of lignin was performed via the acetyl bromide method [45]. Fresh samples of *P. tomentosa* stem segments (1 g) were placed into a screwcap centrifuge tube containing 2 mL of 25% acetyl bromide (v/v in glacial acetic acid) and incubated at 70 °C for 30 min. After complete digestion, the sample was quickly cooled in an ice bath and mixed with 0.9 mL of 2 M NaOH, 0.1 mL of 5 M hydroxylamine-HCl, and a volume of glacial acetic acid sufficient for complete solubilization of the lignin extract. After centrifugation (4500 rpm, 5 min), the absorbance of the supernatant was measured at 280 nm. The absorbance value was used to calculate the relative content of lignin per gram of fresh sample.

4.4.4. Determination of Cellulose, Hemicellulose, and Pectin Content

The method of Peng et al. [46] was used to determine the contents of cellulose, hemicellulose, and pectin in the stem of *P. tomentosa*. Ammonium oxalate was used to extract pectin. The obtained crude cell wall precipitate was added to 5 mL of 0.5% ammonium oxalate, heated in boiling water for 1 h, and centrifuged, after which the supernatant was collected. The precipitate was washed with 5 mL 0.5% ammonium oxalate once, after which it was washed twice with distilled water. All supernatants were collected and diluted to a constant volume, and pentose, hexose, and galacturonic acid were determined by colorimetry. Galacturonic acid reacted with 3-phenyl phenol to form a colored dye, and the absorbance was measured at 530 nm. Pectin content was expressed as galacturonic acid content (%) [47]. Alkali-soluble hemicellulose was extracted by 4 M KOH. First, 5 mL of 4 M KOH was added to the precipitate after the extraction of pectin by ammonium oxalate, and the sample was oscillated at 90× *g* for 1 h, shaken at 150× *g* for 1 h at 25 °C, and centrifuged at 4000× *g* for 5 min. The precipitate was washed once with 5 mL of 4M KOH and twice with distilled water. All supernatants were collected and diluted to a fixed volume of 10 mL. Pentose and hexose were determined by colorimetry. A total of 1 g of residue was put into a 10 mL measuring flask.. Sulfuric acid extraction was used to obtain total cellulose and alkali insoluble hemicellulose. First, 3.0 mL of 67% (v/v) sulfuric acid was added to the above residue and mixed. The sample was shaken at 150× *g* for 1 h at 25 °C, and distilled water was added to a volume of 10 mL. The sample was cooled to room temperature, centrifuged at 4000× *g* for 5 min, and the supernatant was collected. Finally, pentose and hexose were determined by Sulfuric Acid–UV method [48]. The hexose results reported in this step indicate total cellulose, whereas pentose consists of alkali-insoluble hemicellulose, and total hemicellulose is the sum of alkali-soluble hemicellulose and alkali-insoluble hemicellulose. UV spectrophotometer was used to measure the UV absorption at 315 nm. Tamarind xylan was used as a standard for the preparation of calibration curves, and the results were expressed as hemicellulose content in the sample [49]. The UV absorption of hexose at 620 nm was also measured by UV spectrophotometer. Glucose was used as a standard for the preparation of calibration curves, and the results were expressed as the cellulose content in the sample [46].

4.5. Triphenyltetrazolium Chloride (TTC) Mothed Root Activity Determination and Disease Resistance Analysis

Root activity was analyzed by the triphenyl tetrazolium chloride (TTC) method. TTC is a chemical that is reduced by dehydrogenases when added to a tissue. The dehydrogenase activity is regarded as an index of root activity. The specific process followed the method of Zhang et al. [50]. First, 0.5 g fresh *Populus tomentosa* root was immersed in 10 mL of a homogenous mixed solution of 0.4% TTC and phosphate buffer and kept in the dark at 37 °C for 2 h. Subsequently, 2 mL of 1 mol/L H_2SO_4 was added to stop the reaction with the root. The root was dried with filter paper and then extracted with ethyl acetate. The red extractant was transferred into a volumetric flask to reach 10 mL by adding ethyl acetate. The absorbance of the extract at 485 nm was recorded. Root activity was expressed as the TTC reduction intensity. Root activity = amount of TTC reduction (μg)/fresh root weight (g) × time (h). In order to avoid the influence of changes in root microbiome activity on experimental data, in this study, the collected root samples were soaked in 75% ethanol for 5 min, sterile water was used to clear 3~4 times, and the samples were dried before TTC experiment [51].

In accordance with previous studies, physiological parameters such as superoxide dismutase (SOD) and peroxidase (POD) of transgenic poplar and wild-type poplar were measured [52,53]. The total endogenous salicylic acid (SA) in poplar leaves was extracted according to the method described in previous reports [54,55]. The quantification of the total SA content in plant samples was performed with a plant SA ELISA test kit (DREP0874c, Kamai Shu, Shanghai, China), as described by the manufacturer. Each sample had three biological copies. The expression of disease-resistance-related genes *PtNPR1*, *PtMYB115*, and *PtWRKY70* was also analyzed [56].

4.6. Histological Analysis of OsCYP85A1 Transgenic Populus tomentosa and the Wild Type

In order to explore the stem tissue structure of transgenic poplar overexpressing *OsCYP85A1* and wild-type plants, fresh petiole segments of each sample were collected and immediately fixed in a formalin–glacial acetic acid–alcohol (FAA) solution containing 3.8% formalin, glacial acetic acid, and 70% alcohol (*v*:*v*:*v* = 1:1:9) for 24 h. During fixing, the air was extracted through a pump. The paraffin sections were prepared according to the method of Miao et al. [30]. The samples were dehydrated with a series of ethanol solutions and stained with 1% safranine. After they were embedded in paraffin, all cross-sections (20 μm) were cut and stained with 0.1% fast green. The specimens were observed and scanned using Pannoramic DESK (3DHISTECH, Budapest, Hungary). According to the phloroglucinol staining method of Rao et al. [57], the stems of the seventh section were stained, and the sections were decolorized, observed, and photographed under a NIKON Eclipse Ci (Nikon, Tokyo, Japan).

4.7. Analysis of Expression by qRT-PCR

OsCYP85A1 (AB084385.1) encodes the protein responsible for BR C-6 hydroxylation in rice. To understand the possible functions of CYP85A family proteins in wooden plants, we performed a BLAST search in the Joint Genome Initiative poplar database (JGI, *Populus trichocarpa* genome portal v1.1; http://genome.jgi-psf.org/Poptr1_1/Poptr1_1.home.html, accessed on 31 January 2022) and identified two homologues in the Populus genome database in the CDS of *OsCYP85A1*, designated *PtCYP85A2* (EEF10243) and *PtCYP85A3* (EEF02250.2). Total RNA was extracted using a TRIzol-A+ kit (TianGen Biochemical Technology Co., Ltd., Beijing, China), cDNA was synthesized by reverse-transcription using a PrimeScript^{TM} II first strand cDNA Synthesis Kit (Solarbio Technology Co., Ltd., Beijing, China), and the cDNA was diluted by 10-fold for real-time PCR detection. To determine the expression levels of the *OsCYP85A1*, *PtCYP85A2*, *PtCYP85A3*, *PtBRI1* (XM_024597445.1), *PtBZR1* (XM_006375009.2), and *PtMYB2* (XM_002299875) genes, Primer Premier 5.0 was used to design specific primers for *OsCYP85A1*, *PtCYP85A2*, *PtCYP85A3*, *PtBZR1*, *PtBRI1*, and *PtMYB2*. The actin gene from *Populus trichocarpa* was used as an internal reference (Table 1). A real-time quantitative reverse transcription PCR (qRT-PCR) assay was performed on a Bio-Rad CFX Connect^{TM} real-time quantitative PCR instrument (Bio-Rad, Hercules, CA, USA). qRT-PCR was performed with a general-purpose, high-sensitivity, dye-based quantitative PCR detection kit from Nanjing Novozan Biotechnology Co., Ltd. (Nanjing, China). The qRT-PCR reaction system was utilized according to the manufacturer's instructions. The conditions and system of qRT-PCR were as follows: 1 cycle of denaturation (94 °C, 5 min), followed by 40 cycles of amplification (94 °C, 30 s; 50 °C, 45 s), and signal acquisition (72 °C, 113 s). The reaction mixture included 50 μL of the qRT-PCR selection system, which consisted of 20 μL of nuclease-free water, 25 μL of Pfu PCR Mix (Biorun Biotechnology Co., Ltd., Wuhan, China), 1 μL of template, 2 μL of primer (+) (100 μM), and 2 μL of primer (−) (100 μM). The relative levels of expression of the genes were analyzed using the $2^{-\Delta Ct}$ method.

4.8. Data Analysis

The data were subjected to an analysis of variance (ANOVA) using SPSS 18.0 (IBM, Armonk, NY, USA), and the error bars represent the ±standard deviation of data from three independent experiments. The data were normalized, and all samples were normally distributed with homogeneity of variance. Different letters indicate significant differences at $p < 0.05$ and extremely significant differences at $p < 0.01$. The qRT-PCR data were analyzed using SPSS for significant difference analysis ($p < 0.05$). The data were analyzed using Microsoft Excel 2007 (Redmond, WA, USA), and a sequence analysis was performed using DNAMAN 5.0 and a public website (https://www.ncbi.nlm.nih.gov/, accessed on 8 June 2022) database.

5. Conclusions

This study showed that *OsCYP85A1* significantly increased the growth and biomass yield of transgenic plants. First, it affected the vascular tissue in the petiole. The numbers of vessels and sieve tubes in the petioles of transgenic plants were significantly higher than those of the wild type. The content of photosynthetic pigments in the leaves of the transgenic plants at the same position was higher than that of the wild type, which likely helped the transgenic plants to produce more organic matter for their own use. Second, transgenic poplar produced more xylem than wild-type plants, but the amount of bark in each line was similar. *OsCYP85A1* participated in secondary cell division during poplar wood formation. Finally, the effect of *OsCYP85A1* on the root system of transgenic plants showed a promoting effect in the early stage of transgenic poplar development. The main root of transgenic plants was significantly longer than that of WT plants. At the same time, the root activity of transgenic plants after 90 days of growth was also significantly higher than that of WT plants, which was conducive to the absorption of more water and inorganic salts by transgenic *P. tomentosa*, leading to enhanced growth and development. In summary, *OsCYP85A1* has great potential for the molecular breeding of fast-growing tree species due to its capability to regulate BR production in poplar and other woody plants. This study will provide potential candidate genes for economic tree species with long growth cycles and shorten the time to produce finished products for wood or other uses, which meet the needs for daily life and industrial production in the future.

Supplementary Materials: The following supporting information can be downloaded at: https://www.mdpi.com/article/10.3390/ijms24076480/s1.

Author Contributions: Conceptualization, G.L.; methodology, G.L., X.Y. and Z.C.; software, G.L. and X.Y.; validation, G.L. and Z.C.; formal analysis, G.L., X.T. and X.Y.; investigation, G.L. and Z.C.; resources, X.Y. and L.L.; data curation, G.L. and X.Y.; writing—original draft preparation, G.L.; writing—review and editing, G.L. and L.L.; visualization, G.L. and L.L.; supervision, X.Y. and L.L.; project administration, L.L.; funding acquisition, X.Y. and L.L. All authors have read and agreed to the published version of the manuscript.

Funding: This work was supported by the Natural Science Foundation of China (3226180451); the Guizhou Province Outstanding Young Scientific and Technological Talent Cultivation Project (Qiankehe Platform Talent (2019) 5651); and the Guizhou Province Science and Technology Planning Project (Qiankehe Support (2021) General 111).

Institutional Review Board Statement: Not applicable.

Informed Consent Statement: Not applicable.

Data Availability Statement: Not applicable.

Conflicts of Interest: The authors declare no conflict of interest.

References

1. Bajguz, A.; Tretyn, A. The Chemical Characteristic and Distribution of Brassinosteroids in Plants. *Phytochemistry* **2003**, *62*, 20. [CrossRef]
2. Fridman, Y.; Savaldi-Goldstein, S. Brassinosteroids in Growth Control: How, When and Where. *Plant Sci.* **2013**, *209*, 24–31. [CrossRef] [PubMed]
3. Clouse, S.D. Brassinosteroid Signal Transduction: From Receptor Kinase Activation to Transcriptional Networks Regulating Plant Development. *Plant Cell* **2011**, *23*, 1219–1230. [CrossRef] [PubMed]
4. Li, H.; Jiang, L.; Youn, J.; Sun, W.; Cheng, Z.; Jin, T.; Ma, X.; Guo, X.; Wang, J.; Zhang, X.; et al. A Comprehensive Genetic Study Reveals a Crucial Role of *CYP90D2/D2* in Regulating Plant Architecture in Rice (*Oryza sativa*). *N. Phytol.* **2013**, *200*, 1076–1088. [CrossRef]
5. Tanabe, S.; Ashikari, M.; Fujioka, S.; Takatsuto, S.; Yoshida, S.; Yano, M.; Yoshimura, A.; Kitano, H.; Matsuoka, M.; Fujisawa, Y.; et al. A Novel Cytochrome P450 Is Implicated in Brassinosteroid Biosynthesis via the Characterization of a Rice Dwarf Mutant, *Dwarf11*, with Reduced Seed Length. *Plant Cell* **2005**, *17*, 776–790. [CrossRef]
6. Manghwar, H.; Hussain, A.; Ali, Q.; Liu, F. Brassinosteroids (BRs) Role in Plant Development and Coping with Different Stresses. *Int. J. Mol. Sci.* **2022**, *23*, 1012. [CrossRef]
7. Watanabe, B.; Yamamoto, S.; Yokoi, T.; Sugiura, A.; Horoiwa, S.; Aoki, T.; Miyagawa, H.; Nakagawa, Y. Brassinolide-like Activity of Castasterone Analogs with Varied Side Chains against Rice Lamina Inclination. *Bioorg. Med. Chem.* **2017**, *25*, 4566–4578. [CrossRef] [PubMed]
8. Northey, J.G.B.; Liang, S.; Jamshed, M.; Deb, S.; Foo, E.; Reid, J.B.; McCourt, P.; Samuel, M.A. Farnesylation Mediates Brassinosteroid Biosynthesis to Regulate Abscisic Acid Responses. *Nat. Plants* **2016**, *2*, 16114. [CrossRef]
9. Sun, Y.; He, Y.; Irfan, A.R.; Liu, X.; Yu, Q.; Zhang, Q.; Yang, D. Exogenous Brassinolide Enhances the Growth and Cold Resistance of Maize (*Zea mays* L.) Seedlings under Chilling Stress. *Agronomy* **2020**, *10*, 488. [CrossRef]
10. Li, Y.; Wei, K. Comparative Functional Genomics Analysis of Cytochrome P450 Gene Superfamily in Wheat and Maize. *BMC Plant Biol.* **2020**, *20*, 93. [CrossRef]
11. Kim, B.K.; Fujioka, S.; Takatsuto, S.; Tsujimoto, M.; Choe, S. Castasterone Is a Likely End Product of Brassinosteroid Biosynthetic Pathway in Rice. *Biochem. Biophys. Res. Commun.* **2008**, *374*, 614–619. [CrossRef] [PubMed]
12. Nomura, T.; Kushiro, T.; Yokota, T.; Kamiya, Y.; Bishop, G.J.; Yamaguchi, S. The Last Reaction Producing Brassinolide Is Catalyzed by Cytochrome P-450s, CYP85A3 in Tomato and CYP85A2 in Arabidopsis. *J. Biol. Chem.* **2005**, *280*, 17873–17879. [CrossRef]
13. Sakamoto, T.; Kawabe, A.; Tokida-Segawa, A.; Shimizu, B.; Takatsuto, S.; Shimada, Y.; Fujioka, S.; Mizutani, M. Rice CYP734As Function as Multisubstrate and Multifunctional Enzymes in Brassinosteroid Catabolism: Rice Brassinosteroid Catabolic Genes. *Plant J.* **2011**, *67*, 1–12. [CrossRef] [PubMed]
14. Thornton, L.E.; Rupasinghe, S.G.; Peng, H.; Schuler, M.A.; Neff, M.M. Arabidopsis CYP72C1 Is an Atypical Cytochrome P450 That Inactivates Brassinosteroids. *Plant Mol. Biol.* **2010**, *15*, 167–181. [CrossRef] [PubMed]
15. Kim, T.-W.; Chang, S.C.; Lee, J.S.; Hwang, B.; Takatsuto, S.; Yokota, T.; Kim, S.-K. Cytochrome P450-Catalyzed Brassinosteroid Pathway Activation through Synthesis of Castasterone and Brassinolide in *Phaseolus vulgaris*. *Phytochemistry* **2004**, *65*, 679–689. [CrossRef]
16. Shimada, Y.; Goda, H.; Nakamura, A.; Takatsuto, S.; Fujioka, S.; Yoshida, S. Organ-Specific Expression of Brassinosteroid-Biosynthetic Genes and Distribution of Endogenous Brassinosteroids in Arabidopsis. *Plant Physiol.* **2003**, *131*, 287–297. [CrossRef]
17. Jiang, C.; Li, B.; Song, Z.; Zhang, Y.; Yu, C.; Wang, H.; Wang, L.; Zhang, H. PtBRI1.2 Promotes Shoot Growth and Wood Formation through a Brassinosteroid-Mediated PtBZR1-PtWNDs Module in Poplar. *J. Exp. Bot.* **2021**, *72*, 6350–6364. [CrossRef]
18. Youn, J.-H.; Kim, M.K.; Kim, E.-J.; Son, S.-H.; Lee, J.E.; Jang, M.-S.; Kim, T.-W.; Kim, S.-K. ARF7 Increases the Endogenous Contents of Castasterone through Suppression of BAS1 Expression in Arabidopsis Thaliana. *Phytochemistry* **2016**, *122*, 34–44. [CrossRef]
19. Song, W. Fangmeng Duan Overexpression of SoCYP85A1 Increases the Accumulation of Castasterone and Confers Enhanced Black Shank Tolerance in Tobacco through Modulation of the Antioxidant Enzymes' Activities. *Front. Plant Sci.* **2019**, *10*, 10.
20. Castle, J.; Szekeres, M.; Jenkins, G.; Bishop, G.J. Unique and Overlapping Expression Patterns of Arabidopsis CYP85 Genes Involved in Brassinosteroid C-6 Oxidation. *Plant Mol. Biol.* **2005**, *57*, 129–140. [CrossRef]
21. Pérez-España, V.H.; Sánchez-León, N.; Vielle-Calzada, J.-P. *CYP85A1* Is Required for the Initiation of Female Gametogenesis in *Arabidopsis thaliana*. *Plant Signal. Behav.* **2011**, *6*, 321–326. [CrossRef]
22. Shimada, Y.; Fujioka, S.; Miyauchi, N.; Kushiro, M.; Takatsuto, S.; Nomura, T.; Yokota, T.; Kamiya, Y.; Bishop, G.J.; Yoshida, S. Brassinosteroid-6-Oxidases from Arabidopsis and Tomato Catalyze Multiple C-6 Oxidations in Brassinosteroid Biosynthesis. *Plant Physiol.* **2001**, *126*, 770–779. [CrossRef]
23. Wang, H.; Li, W.; Qin, Y.; Pan, Y.; Wang, X.; Weng, Y.; Chen, P.; Li, Y. The Cytochrome P450 Gene CsCYP85A1 Is a Putative Candidate for Super Compact-1 (Scp-1) Plant Architecture Mutation in Cucumber (*Cucumis sativus* L.). *Front. Plant Sci.* **2017**, *8*, 266. [CrossRef]
24. Duan, F.; Ding, J.; Lee, D.; Lu, X.; Feng, Y.; Song, W. Overexpression of SoCYP85A1, a Spinach Cytochrome P450 Gene in Transgenic Tobacco Enhances Root Development and Drought Stress Tolerance. *Front. Plant Sci.* **2017**, *8*, 1909. [CrossRef]

25. Li, X.-J.; Guo, X.; Zhou, Y.-H.; Shi, K.; Zhou, J.; Yu, J.-Q.; Xia, X.-J. Overexpression of a Brassinosteroid Biosynthetic Gene Dwarf Enhances Photosynthetic Capacity through Activation of Calvin Cycle Enzymes in Tomato. *BMC Plant Biol.* **2016**, *16*, 33. [CrossRef] [PubMed]
26. Jin, Y.-L.; Tang, R.-J.; Wang, H.-H.; Jiang, C.-M.; Bao, Y.; Yang, Y.; Liang, M.-X.; Sun, Z.-C.; Kong, F.-J.; Li, B.; et al. Overexpression of *Populus trichocarpa CYP85A3* Promotes Growth and Biomass Production in Transgenic Trees. *Plant Biotechnol. J.* **2017**, *15*, 1309–1321. [CrossRef] [PubMed]
27. Shen, Y.; Li, Y.; Xu, D.; Yang, C.; Li, C.; Luo, K. Molecular Cloning and Characterization of a Brassinosteriod Biosynthesis-Related Gene PtoDWF4 from *Populus tomentosa*. *Tree Physiol.* **2018**, *38*, 1424–1436. [CrossRef] [PubMed]
28. Yu, X.; Du, J.; Dong, Y.; Lu, B.; Yang, M. Cloning and Salt Tolerance Analysis of the PnHB7 Transcription Factor in *Populus nigra* L. *Ind. Crops Prod.* **2020**, *158*, 112943. [CrossRef]
29. Fujisawa, T.; Filippakopoulos, P. Functions of Bromodomain-Containing Proteins and Their Roles in Homeostasis and Cancer. *Nat. Rev. Mol. Cell Biol.* **2017**, *18*, 246–262. [CrossRef]
30. Miao, H.; Li, C.; Duan, Y.; Wei, L.; Ju, M.; Zhang, H. Identification of a Sidwf1 Gene Controlling Short Internode Length Trait in the Sesame Dwarf Mutant Dw607. *Theor. Appl. Genet.* **2020**, *133*, 73–86. [CrossRef]
31. Ali, S.S.; Kumar, G.B.S.; Khan, M.; Doohan, F.M. Brassinosteroid Enhances Resistance to Fusarium Diseases of Barley. *Phytopathology* **2013**, *103*, 1260–1267. [CrossRef] [PubMed]
32. Bancosİ, S.; Nomura, T.; Sato, T.; Molnár, G.; Bishop, G.J.; Koncz, C.; Yokota, T.; Nagy, F.; Szekeres, M. Regulation of Transcript Levels of the Arabidopsis Cytochrome P450 Genes Involved in Brassinosteroid Biosynthesis. *Plant Physiol.* **2002**, *130*, 504–513. [CrossRef] [PubMed]
33. Jiang, P.; Zhang, K.; Ding, Z.; He, Q.; Li, W.; Zhu, S.; Cheng, W.; Zhang, K.; Li, K. Characterization of a Strong and Constitutive Promoter from the Arabidopsis Serine Carboxypeptidase-like Gene AtSCPL30 as a Potential Tool for Crop Transgenic Breeding. *BMC Biotechnol.* **2018**, *18*, 59. [CrossRef] [PubMed]
34. Kim, T.-W.; Hwang, J.-Y.; Kim, Y.-S.; Joo, S.-H.; Chang, S.C.; Lee, J.S.; Takatsuto, S.; Kim, S.-K. Arabidopsis CYP85A2, a Cytochrome P450, Mediates the Baeyer-Villiger Oxidation of Castasterone to Brassinolide in Brassinosteroid Biosynthesis. *Plant Cell* **2005**, *17*, 2397–2412. [CrossRef]
35. Vanstraelen, M.; Benková, E. Hormonal Interactions in the Regulation of Plant Development. *Annu. Rev. Cell Dev. Biol.* **2012**, *28*, 463–487. [CrossRef]
36. Yamamoto, R.; Fujioka, S.; Demura, T.; Takatsuto, S.; Yoshida, S.; Fukuda, H. Brassinosteroid Levels Increase Drastically Prior to Morphogenesis of Tracheary Elements. *Plant Physiol.* **2001**, *125*, 556–563. [CrossRef] [PubMed]
37. Hamasaki, H.; Ayano, M.; Nakamura, A.; Fujioka, S.; Asami, T.; Takatsuto, S.; Yoshida, S.; Oka, Y.; Matsui, M.; Shimada, Y. Light Activates Brassinosteroid Biosynthesis to Promote Hook Opening and Petiole Development in *Arabidopsis thaliana*. *Plant Cell Physiol.* **2020**, *61*, 1239–1251. [CrossRef]
38. Ohnishi, T.; Godza, B.; Watanabe, B.; Fujioka, S.; Hategan, L.; Ide, K.; Shibata, K.; Yokota, T.; Szekeres, M.; Mizutani, M. CYP90A1/CPD, a Brassinosteroid Biosynthetic Cytochrome P450 of Arabidopsis, Catalyzes C-3 Oxidation. *J. Biol. Chem.* **2012**, *287*, 31551–31560. [CrossRef]
39. Clouse, S.D. A History of Brassinosteroid Research from 1970 through 2005: Thirty-Five Years of Phytochemistry, Physiology, Genes, and Mutants. *J. Plant Growth Regul.* **2015**, *34*, 828–844. [CrossRef]
40. Nagata, N.; Asami, T.; Yoshida, S. Brassinazole, an Inhibitor of Brassinosteroid Biosynthesis, Inhibits Development of Secondary Xylem in Cress Plants (*Lepidium sativum*). *Plant Cell Physiol.* **2001**, *42*, 1006–1011. [CrossRef]
41. Kim, S.; Moon, J.; Roh, J.; Kim, S.-K. Castasterone Can Be Biosynthesized from 28-Homodolichosterone in *Arabidopsis thaliana*. *J. Plant Biol.* **2018**, *61*, 330–335. [CrossRef]
42. Feng, Q.; Ou, Y.; Han, Y.; de Dios, V.R.; Wang, L.; Zhang, Q.; Yao, Y. The Brassinosteroid Biosynthesis Enzyme Gene PeCPD Improves Plant Growth and Salt Tolerance in *Populus tomentosa*. *Ind. Crops Prod.* **2021**, *162*, 113218. [CrossRef]
43. Liu, X.; Fu, Z.; Zhang, B.; Zhai, L.; Meng, M.; Lin, J.; Zhuang, J.; Wang, G.G.; Zhang, J. Effects of Sulfuric, Nitric, and Mixed Acid Rain on Chinese Fir Sapling Growth in Southern China. *Ecotoxicol. Environ. Saf.* **2018**, *160*, 154–161. [CrossRef]
44. Ren, Y.; Zhang, W.; Huang, Y.; Zhang, Y.; Wang, J.; Yang, M. Transcriptome Analysis Reveals That Populus Tomentosa Hybrid Poplar 741 Responds to Blue Light Treatment by Regulating Growth-Related Genes and Their Metabolic Pathways. *Ind. Crops Prod.* **2020**, *152*, 112512. [CrossRef]
45. Moreira-Vilar, F.C.; de Siqueira-Soares, R.C.; Finger-Teixeira, A.; de Oliveira, D.M.; Ferro, A.P.; da Rocha, G.J.; Ferrarese, M.; de, L.L.; dos Santos, W.D.; Ferrarese-Filho, O. The Acetyl Bromide Method Is Faster, Simpler and Presents Best Recovery of Lignin in Different Herbaceous Tissues than Klason and Thioglycolic Acid Methods. *PLoS ONE* **2014**, *9*, e110000. [CrossRef] [PubMed]
46. Peng, L.; Hocart, C.H.; Redmond, J.W.; Williamson, R.E. Fractionation of Carbohydrates in Arabidopsis Root Cell Walls Shows That Three Radial Swelling Loci Are Speci®cally Involved in Cellulose Production. *Planta* **2000**, *211*, 406–414. [CrossRef] [PubMed]
47. Cybulska, J.; Zdunek, A.; Kozioł, A. The Self-Assembled Network and Physiological Degradation of Pectins in Carrot Cell Walls. *Food Hydrocoll.* **2015**, *43*, 41–50. [CrossRef]
48. Szymańska-Chargot, M.; Chylińska, M.; Gdula, K.; Kozioł, A.; Zdunek, A. Isolation and Characterization of Cellulose from Different Fruit and Vegetable Pomaces. *Polymers* **2017**, *9*, 495. [CrossRef]
49. Albalasmeh, A.A.; Berhe, A.A.; Ghezzehei, T.A. A New Method for Rapid Determination of Carbohydrate and Total Carbon Concentrations Using UV Spectrophotometry. *Carbohydr. Polym.* **2013**, *97*, 253–261. [CrossRef]

50. Zhang, X.; Huang, G.; Bian, X.; Zhao, Q. Effects of Root Interaction and Nitrogen Fertilization on the Chlorophyll Content, Root Activity, Photosynthetic Characteristics of Intercropped Soybean and Microbial Quantity in the Rhizosphere. *Plant Soil Environ.* **2013**, *59*, 9. [CrossRef]
51. Yamagami, M.; Yanai, M. Effect of rice plant root ttc-reducing activity on the chemical form of iodine in cultivated soil solutions. *Radiat. Prot. Dosim.* **2022**, *198*, 1189–1195. [CrossRef] [PubMed]
52. Wang, L.; Li, Z.; Lu, M.; Wang, Y. ThNAC13, a NAC Transcription Factor from *Tamarix* Hispida, Confers Salt and Osmotic Stress Tolerance to Transgenic *Tamarix* and *Arabidopsis*. *Front. Plant Sci.* **2017**, *8*, 635. [CrossRef] [PubMed]
53. Cheng, Z.; Zhang, X.; Zhao, K.; Yao, W.; Li, R.; Zhou, B.; Jiang, T. Over-Expression of ERF38 Gene Enhances Salt and Osmotic Tolerance in Transgenic Poplar. *Front. Plant Sci.* **2019**, *10*, 1375. [CrossRef] [PubMed]
54. Li, Q.; Wang, G.; Guan, C.; Yang, D.; Wang, Y.; Zhang, Y.; Ji, J.; Jin, C.; An, T. Overexpression of *LcSABP*, an Orthologous Gene for Salicylic Acid Binding Protein 2, Enhances Drought Stress Tolerance in Transgenic Tobacco. *Front. Plant Sci.* **2019**, *10*, 200. [CrossRef]
55. Tsai, W.-A.; Weng, S.-H.; Chen, M.-C.; Lin, J.-S.; Tsai, W.-S. Priming of Plant Resistance to Heat Stress and Tomato Yellow Leaf Curl Thailand Virus with Plant-Derived Materials. *Front. Plant Sci.* **2019**, *10*, 906. [CrossRef]
56. Ullah, C.; Tsai, C.; Unsicker, S.B.; Xue, L.; Reichelt, M.; Gershenzon, J.; Hammerbacher, A. Salicylic Acid Activates Poplar Defense against the Biotrophic Rust Fungus *Melampsora larici-populina* via Increased Biosynthesis of Catechin and Proanthocyanidins. *N. Phytol.* **2019**, *221*, 960–975. [CrossRef]
57. Rao, X.; Chen, X.; Shen, H.; Ma, Q.; Li, G.; Tang, Y.; Pena, M.; York, W.; Frazier, T.P.; Lenaghan, S.; et al. Gene Regulatory Networks for Lignin Biosynthesis in Switchgrass *(Panicum virgatum)*. *Plant Biotechnol. J.* **2019**, *17*, 580–593. [CrossRef]

International Journal of
Molecular Sciences

MDPI

Article

Systematic Analysis of BELL Family Genes in *Zizania latifolia* and Functional Identification of *ZlqSH1a/b* in Rice Seed Shattering

Yan-Ning Xie [1,†], Ting Yang [1,†], Bin-Tao Zhang [2,3], Qian-Qian Qi [1], An-Ming Ding [1], Lian-Guang Shang [2], Yu Zhang [1], Qian Qian [4], Zhong-Feng Zhang [1,*] and Ning Yan [1,*]

1 Tobacco Research Institute of Chinese Academy of Agricultural Sciences, Qingdao 266101, China
2 Shenzhen Branch, Guangdong Laboratory of Lingnan Modern Agriculture, Genome Analysis Laboratory of the Ministry of Agriculture and Rural Affairs, Agricultural Genomics Institute at Shenzhen, Chinese Academy of Agricultural Sciences, Shenzhen 518120, China
3 College of Agronomy, Qingdao Agricultural University, Qingdao 266109, China
4 State Key Laboratory of Rice Biology, China National Rice Research Institute, Chinese Academy of Agricultural Sciences, Hangzhou 310006, China
* Correspondence: zhangzhongfeng@caas.cn (Z.-F.Z.); yanning@caas.cn (N.Y.)
† These authors contributed equally to this work.

Abstract: The loss of seed shattering is an important event in crop domestication, and elucidating the genetic mechanisms underlying seed shattering can help reduce yield loss during crop production. This study is the first to systematically identify and analyse the BELL family of transcription factor-encoding genes in Chinese wild rice (*Zizania latifolia*). *ZlqSH1a* (Zla04G033720) and *ZlqSH1b* (Zla02G027130) were identified as key candidate genes involved in seed shattering in *Z. latifolia*. These genes were involved in regulating the development of the abscission layer (AL) and were located in the nucleus of the cell. Over-expression of *ZlqSH1a* and *ZlqSH1b* resulted in a complete AL between the grain and pedicel and significantly enhanced seed shattering after grain maturation in rice. Transcriptome sequencing revealed that 172 genes were differentially expressed between the wild type (WT) and the two transgenic (*ZlqSH1a* and *ZlqSH1b* over-expressing) plants. Three of the differentially expressed genes related to seed shattering were validated using qRT-PCR analysis. These results indicate that *ZlqSH1a* and *ZlqSH1b* are involved in AL development in rice grains, thereby regulating seed shattering. Our results could facilitate the genetic improvement of seed-shattering behaviour in *Z. latifolia* and other cereal crops.

Keywords: Chinese wild rice (*Zizania latifolia*); rice (*Oryza sativa*); seed shattering; ZlBELL transcription factor; abscission layer

1. Introduction

Seed shattering is an important trait that helps wild plants adapt to the natural environment and maintain population growth. In addition, seed shattering is considered one of the most important events in crop domestication [1]. Crop domestication began approximately 10,000 years ago [2–4] and was accompanied by a reduction in seed shattering, changes in seed shape, reduction of dormancy, increase in grain number, and improvement of plant type and fertility—all of which are important factors in artificial selection [5,6]. Research on the genetic regulatory mechanisms underlying plant seed shattering has important theoretical significance for revealing the molecular mechanisms underlying the domestication and improvement of wild plants. In addition, it has extremely high application value for the rapid domestication of wild plants and the enhancement of their yield [7,8].

In plants, the abscission zone (AZ) consists of plant organs and several adjacent cell layers separated via abscission. During plant organ shedding, the AZ differentiates into one or several layers of parenchyma cells to form an abscission layer (AL) [9]. The AL is located between the glumes and branches and consists of a group of small parenchyma cells. The cell layers adjacent to the AL have thick and lignified cell walls [10] that provide the mechanical force required for separation [11]. With the formation of the AL and protective layers, the shedding signal initiates the degradation and separation of AZ cells, leading to plant organ shedding. From an evolutionary perspective, seed shattering after plant maturation is an adaptive feature that enables wild plants to spread more seeds into the surrounding environment, which is beneficial for self-reproduction and population preservation. In agricultural production, the weakening of seed shattering facilitates crop harvest after maturity, which greatly increases crop yield and improves human welfare [12,13].

Chinese wild rice (*Zizania latifolia*, 2n = 34) is a perennial hydrophyte belonging to the Oryzeae Dum. tribe, Oryzoideae Care subfamily, and Gramineae family [14,15]. It is an important crop that is a close genetic relative of rice, second only to *Leersia* [16]. Its caryopsis is known as Chinese wild rice and has been used as a food grain for more than 3000 years in China [17]. Indeed, Chinese wild rice was one of the six grains eaten by emperors in ancient times; the other five being rice, broomcorn millet, panicled millet, wheat, and beans [18]. *Z. latifolia* grows over a wide geographical range in China, except for Tibet. It grows in lakes, ditches, ponds, rivers, and wetlands, especially in the middle and lower reaches of the Yangtze River and the Huaihe River basin [19]. Chinese wild rice is a type of whole grain that is rich in proteins, essential amino acids, fatty acids, vitamins, and various trace elements [20,21]. Li Shizhen's Compendium of Materia Medica of the Ming Dynasty reported that Chinese wild rice was used as adjuvant therapy for diabetes and gastrointestinal diseases. Its various effects include preventing oxidative damage, improving insulin resistance, reducing lipid toxicity, controlling weight to prevent obesity, and preventing cardiovascular disease and cancer [22]. In addition, *Z. latifolia* has many excellent traits that are considered beneficial in modern rice cultivars. As a species transitioning from wild to domesticated status, it naturally retains a large number of excellent traits lost in domesticated crops [23], such as resistance to rice blasts, sturdy stalks, strong tillering, resistance to low-temperature conditions, resistance to flooding and irrigation, a fast grain filling rate, high biological yield, and higher contents of grain proteins and phenolic compounds than ordinary rice [24]. Notably, Chinese wild rice shows strong seed-shattering behaviour after maturity; as a result, the seeds rupture spontaneously and are difficult to harvest. This results in an extremely low yield, making the commercial cultivation of *Z. latifolia* extremely difficult [25]. In the early 1950s, American breeders began to artificially select Northern wild rice with low seed-shattering behaviour that was suitable for mechanised planting. Thus, they successfully achieved the commercial production of wild rice, and a large amount of wild rice is exported globally every year [26]. With the improvement of people's quality of life, Chinese wild rice's nutritional value and healthcare efficacy are being increasingly recognised. Therefore, *Z. latifolia* deserves further study as an economically important crop with high nutritive and medicinal value.

In recent years, several domestic and international scholars have studied plant seed shattering [10,27–29], providing an important reference for research on the genetic regulation of this behaviour [30,31]. Compared with wild rice, the loss of seed shattering is one of the biggest trait changes in cultivated rice and an important marker of the transition from wild to cultivated rice [8,32]. The form of the AL differs between rice cultivars: in cultivars prone to seed-shattering behaviour, the AL develops from the epidermis to near the vascular bundles; in contrast, cultivars with low seed-shattering behaviour have no AL or have an AL that is difficult to degrade. Cultivars with moderate seed-shattering behaviour either have a partially developed AL on the glumelle side of the branches or form a complete AL on the glumelle side and an irregular AL on the lemma side [11]. In these cultivars, the AL (composed of non-degradable tissue) is wider than that of cultivars whose seeds are easy to shed and narrower than that of cultivars whose seeds are difficult to shed. To date, 12 genes related to rice seed shattering have been cloned [33]. These genes encode transcription factors (TFs) or kinases that regulate seed shattering by affecting the development or degradation of the AL [32]. The *qSH1 (LOC_Os01g62920)* gene was identified by cloning the hybrid of the Indica rice cultivar 'Kasalath' (whose seeds shed easily) and the Japonica rice cultivar 'Nipponbare' (with reduced seed shedding). The *qSH1* gene explains 68.6% of the phenotypic variation in this rice population, and it is considered the main quantitative trait locus/gene regulating AL development and seed abscission in rice. *qSH1* encodes a BELL-type homeotic protein, and its coding sequence is identical between 'Kasalath' and 'Nipponbare'. However, a single-nucleotide polymorphism (SNP; G/T) in the 12-kb regulatory region upstream of the initiator codon of *qSH1* changes the expression pattern of the protein-coding sequence in the AL region. Because of this change, the AL of Japonica rice plants cannot develop normally, resulting in the loss of seed shattering [27].

Gehring (1987) [34] found that mutations in some genes—such as the extremely conserved homeobox genes—can produce homologous phenotypes. Homeobox genes encode a type of transcription regulator called a homeodomain (HD), and TFs regulate the specific expression of plant genes. The TF family in plants includes NAC (NAM, ATAF, and CUC), MADS (MADS-box), ERF (ethylene response factor), and TALE (3-amino-acid-loop extension) [35–38]. The typical homeobox domain consists of 60 amino acids (except the TALE gene family, which encodes an atypical DNA-binding domain consisting of 63 amino acids) [39,40]. Based on differences in gene sequences and evolutionary history, the TALE gene family in plants can be further divided into two subfamilies: BELL (BEL1-like HD) and KNOX (KNOTTED1-like homeobox). With the rapid development of molecular biology, bioinformatics, and next-generation sequencing technology, genes in the KNOX and BELL families have been cloned and verified in different species [35]. The *KNOX* gene family is mainly involved in cell differentiation and maintenance of the shoot apical meristem [36], whereas the *BELL* gene family helps regulate plant development, hormone responses, and stress responses [41–43]. However, the number of members in the *BELL* gene family varies across plant species. For example, there are 4 *BELL* family genes in the bryophyte genome [38], 13 in *Arabidopsis* [44], 15 in maize [45], and 14 in rice and potato [46]. Nevertheless, the proteins encoded by these genes are highly conserved in structure, and all contain the SKY, BEL, and HD domains [47]. Various combinations of heterologous BELL–KNOX dimers with different activities are involved in regulating plant growth and development, including factors such as the maintenance of the shoot apical meristem and its boundaries, leaf development, and flowering [48–50]. In rice, the BELL genes *qSH1* and *SH5* participate in rice grain abscission by promoting the development of the AZ and inhibiting lignin synthesis. In addition, *SHAT1* (shattering abortion 1) and *SH4* (shattering 4) are essential regulators of AZ formation. Yoon et al. (2014) also reported that *SH5* induces the expression of *SHAT1* and *SH4*, promoting AL formation and seed shattering [51].

Yan et al. (2022) [52] performed genome collinearity analysis in *Z. latifolia* and rice to show that the *ZlqSH1a (Zla04G033720)* and *ZlqSH1b (Zla02G027130)* genes in *Z. latifolia* are orthologous to *qSH1* (*LOC_Os01g62920*) in rice. To date, no studies have analysed the BELL family TFs in the *Z. latifolia* genome or reported the function of any BELL genes. This study is the first to perform bioinformatic analysis of BELL TFs in the entire *Z. latifolia* genome using subcellular localisation and RNA-seq. We verified the functions of *ZlqSH1a* and *ZlqSH1b* by over-expressing these genes in transgenic rice. In addition, we measured the breaking tensile strength (BTS) of *ZlqSH1a* and *ZlqSH1b* over-expressing (OE) and wild type (WT) seeds after maturity, observed the structural differences in AL tissue and fracture surfaces in transgenic and WT plants, and performed RNA-seq analysis of transgenic and WT plants to determine the influence of *ZlqSH1a* or *ZlqSH1b* over-expression on gene expression in the rice grain AL. We identified and verified the functions of *ZlqSH1a* and *ZlqSH1b* in rice and discussed their effects on the development of the AL in rice grains. Our results provide theoretical support and new genetic resources for selecting *Z. latifolia* cultivars with appropriate seed-shattering behaviour that makes them suitable for commercial cultivation.

2. Results

2.1. Identification of ZlBELL Family TFs and Chromosome Localisation

A total of 48 *ZlBELL* genes were identified in the *Z. latifolia* genome. Based on their location on the chromosomes, they were sequentially named *ZlBELL01–48* (Figure 1, Table S1). Notably, ZlBELL05 (Zla02G027130) and ZlBELL13 (Zla04G033720) were orthologous to qSH1 (LOC_Os01g62920) in rice and were named ZlqSH1b and ZlqSH1a, respectively [52]. The chromosome localisation analysis of *ZlBELL* genes showed that the 48 *ZlBELL* genes were distributed on 14 chromosomes in the *Z. latifolia* genome. A total of 12 genes (accounting for 25% of all *ZlBELL* genes) were located on chromosome 5. Moreover, 14.58% (7) and 12.50% (6) of the genes were located on chromosomes 9 and 3, respectively. Chromosomes 16 and 17 contained the fewest *ZlBELL* genes (only one gene each, accounting for 2.08% of the total number). The cluster of *ZlBELL* genes was located in the middle of chromosome 12 and at the top of chromosome 9 in *Z. latifolia*.

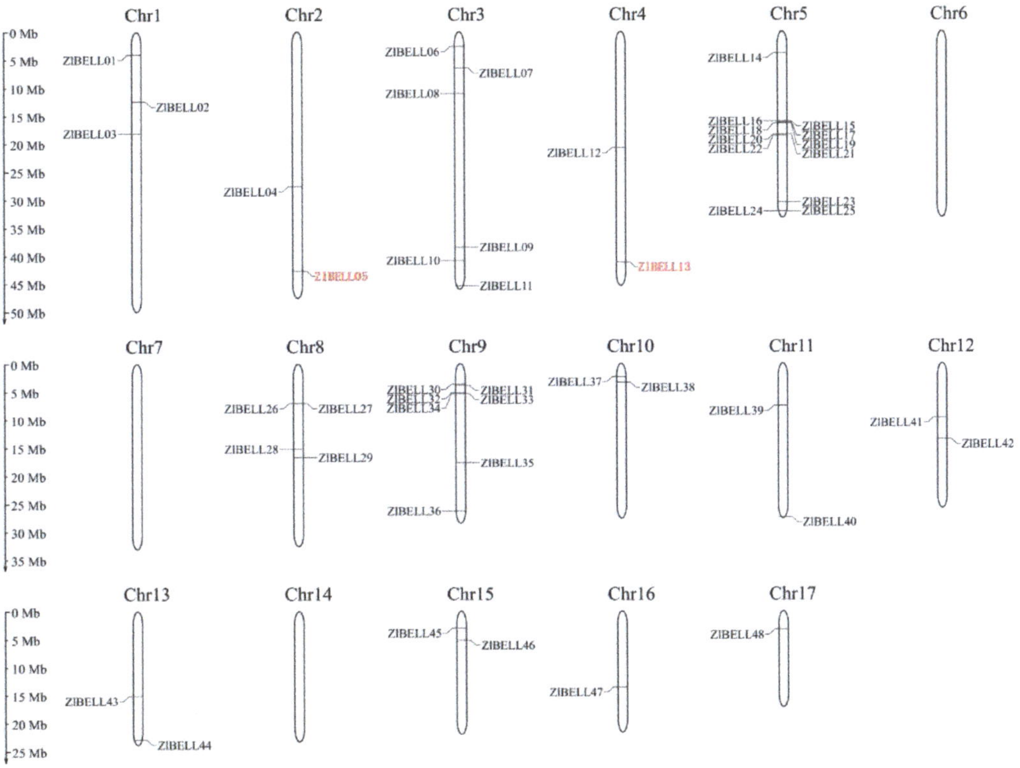

Figure 1. Chromosomal locations of BELL genes in *Zizania latifolia*. The gene IDs of the ZlBELL family are listed in Table S1. ZlBELL05 (ZlqSH1b) and ZlBELL13 (ZlqSH1a) are marked in red.

2.2. Phylogenetic Analysis of ZlBELL Genes

The protein sequences of 48 BELL genes in *Z. latifolia* (Table S1), 13 in *Arabidopsis*, and 14 in rice were used to analyse the BELL phylogenetic relationship between *Arabidopsis*, *Z. latifolia*, and rice. Compared with the six subfamilies of BELL genes in *Arabidopsis* (A, D, E, F, H, and I subfamilies) and seven subfamilies of BELL genes in rice (A, D, E, F, G, H, and J subfamilies), the 48 *ZlBELL* genes in *Z. latifolia* were distributed in nine subfamilies (A, B, C, D, E, F, G, H, and J subfamilies) (Figure 2). The I subfamily had the largest number (five) of *ZlBELL* genes in *Arabidopsis*, and the J subfamily had the largest number (three) in rice. The C subfamily had the largest number (16) of *ZlBELL* genes in *Z. latifolia* but did not contain *Arabidopsis* and rice *BELL* genes. The E subfamilies had fewer *ZlBELL* members (only two), and subfamily I did not contain any *ZlBELL* genes. Notably, ZlBELL05 (ZlqSH1b) and ZlBELL13 (ZlqSH1a) belonged to the D subfamily.

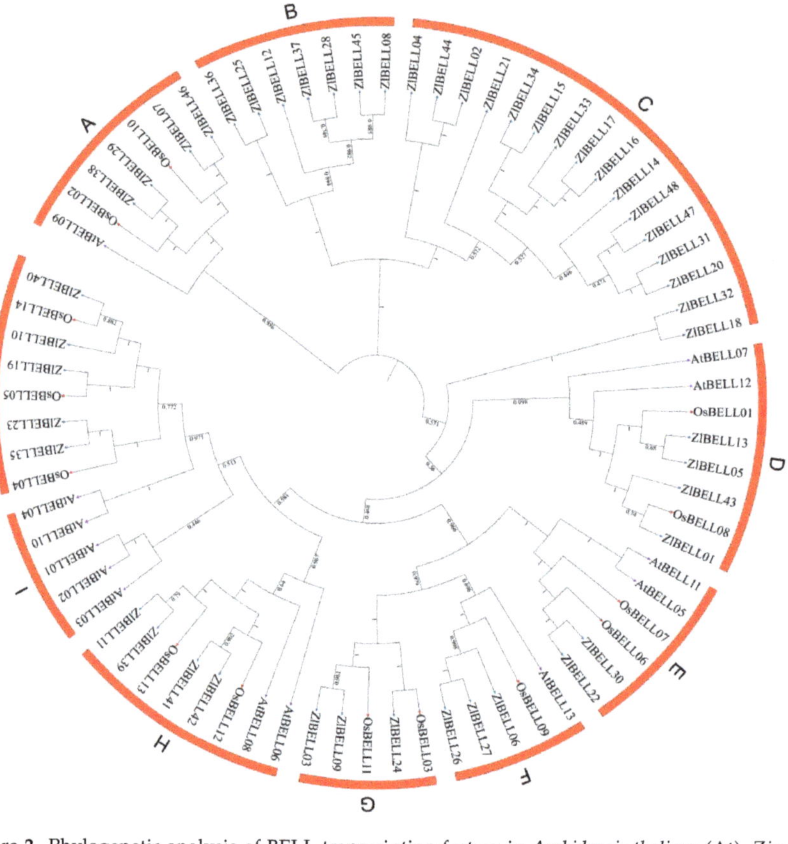

Figure 2. Phylogenetic analysis of BELL transcription factors in *Arabidopsis thaliana* (At), *Zizania latifolia* (Zl), and *Oryza sativa* (Os). The numbers on the major branches indicate bootstrap estimates for 1000 replicate analyses. (**A–J**) represent different subfamilies.

2.3. Gene Structure and Conserved Motif Analysis of Genes in the ZlBELL Family

Phylogenetic analysis of ZlBELL protein sequences revealed that they could be divided into eight subfamilies (Figure 3A). Analysis of the intron/exon structure of each *ZlBELL* gene indicated that the exon number in *ZlBELL* genes varied from four to six; most of the genes (77.08%) contained four to five exons with a length of more than 1.6 kb. However, the intron number ranged from three to five, and did not vary greatly (Figure 3B). Protein domain analysis was performed using the MEME software to elucidate the functional diversity of *ZlBELL* proteins in *Z. latifolia*. The results showed that ZlBELL proteins had a relatively conserved structure. Three relatively conserved motifs (Motifs 1–3) were identified, of which Motifs 1 and 2 were the most conserved and were shared across all subfamilies (Figure 3C). In addition to the shared motifs, each subfamily showed some specificity; for example, Motif 3 was only present in subfamilies I, II, III, IV, and V. Notably, the typical sequence characteristics analysis of BELL family in *Z. latifolia* are shown in Figure S1. It can be seen from Figure 3 and Figure S1 that the conserved motif of the *ZlBELL* family did not completely overlap with its typical sequence.

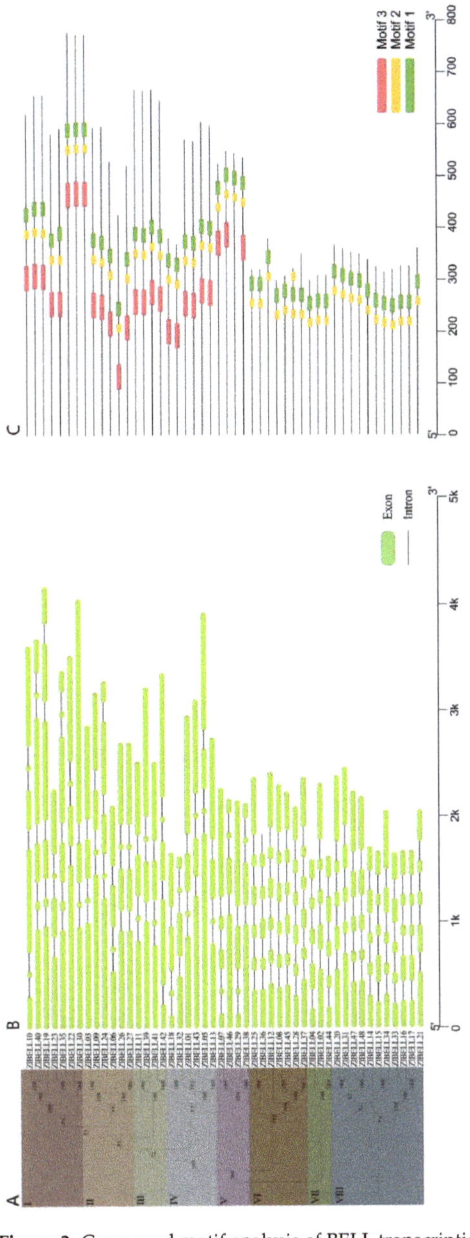

Figure 3. Conserved motif analysis of BELL transcription factors in *Z. latifolia*. (**A**) Phylogenetic tree; the different background colours represent different subfamilies. (**B**) Gene structure; the green boxes and black lines represent the exons and introns, respectively. (**C**) Conserved motifs; different colours represent different conserved motifs.

2.4. Subcellular Localisation of the ZlqSH1a and ZlqSH1b Proteins

Considering the homology of ZlqSH1a (ZlBELL13), ZlqSH1b (ZlBELL05), and qSH1 (LOC_Os01g62920) [52], we chose the ZlqSH1a and ZlqSH1b for subcellular localization. To clarify the specific expression sites of the ZlqSH1a and ZlqSH1b proteins in cells, we used the PSORT program (http://psort.hgc.jp/) to predict whether they contained nuclear localisation signals. To test this prediction, we first constructed the *ZlqSH1a*-GFP and *ZlqSH1b*-GFP fusion protein over-expression constructs. The fusion expression vector, empty vector, and nuclear marker were co-transformed into rice protoplasts and cultured for 8–10 h under low light. Following this, the fluorescent signal was observed using confocal microscopy (Figure 4). The red fluorescent (RFP) regions of the nuclear marker overlapped with the green fluorescent (GFP) regions of *ZlqSH1a*-GFP and *ZlqSH1b*-GFP, indicating that both *ZlqSH1a*-GFP and *ZlqSH1b*-GFP were localised in the nucleus (Figure 4). That is, both proteins were nuclear TFs, which is consistent with the predicted results.

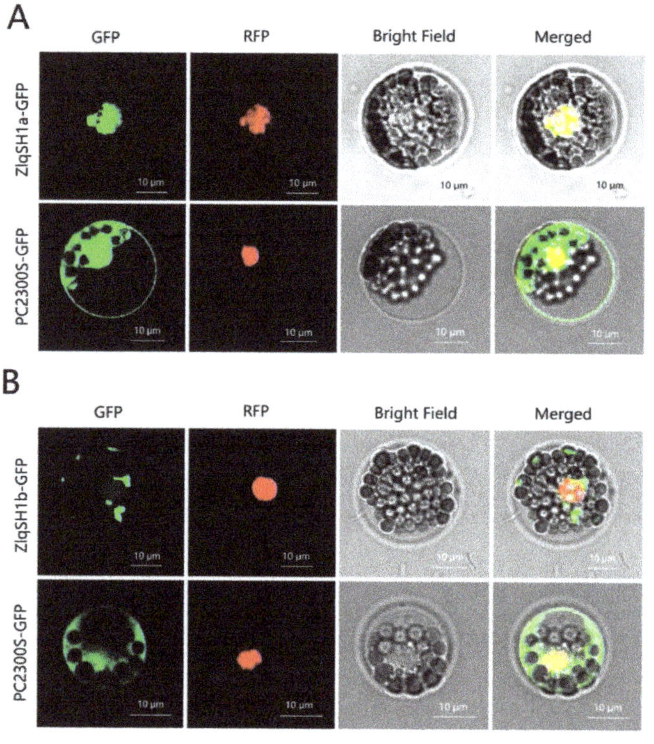

Figure 4. Subcellular localisation of the *ZlqSH1a* and *ZlqSH1b* proteins. (**A**) *ZlqSH1a*-GFP and PC2300S-GFP; (**B**) *ZlqSH1a*-GFP and PC2300S-GFP. GFP: green fluorescence (protein of interest); RFP: red fluorescence (nucleus); Bright: open field channel; Merged: fluorescence integration. Bar: 10 μm.

2.5. Seed Shattering Phenotype and AL-Based Histological Analysis of WT Plants and ZlqSH1a and ZlqSH1b Over-Expressing Rice

To verify the function of the *ZlqSH1a/b* gene, we constructed its over-expression vector and transformed it into a non-shattering rice cultivar, L422, to obtain positive transgenic rice plants. Phenotypic observations showed that at the harvest stage, *ZlqSH1a* and *ZlqSH1b* over-expressing plants had fewer seeds remaining in the rice panicle than WT plants. Thus, *ZlqSH1a/b* are speculated to be key genes for seed shattering in *Z. latifolia*. Thirty days after flowering and pollination, the BTS between the grain and the pedicel were measured in the WT and over-expressing lines using a digital tension meter. The results showed that the maximum tensile strength of the *ZlqSH1a/b* over-expressing lines was lower than that of the WT (Figure 5A). Compared with the WT, the *ZlqSH1a/b* over-expressing lines showed significantly higher natural seed shedding after maturity (Figure 5(B1,C1,D1)). Scanning electron microscopic (SEM) observations of the grain and pedicel fracture surfaces revealed that both *ZlqSH1a* and *ZlqSH1b* were involved in AL development and that the fracture surfaces were relatively smooth in the gene over-expressing plants (Figure 5(C2–C5,D2–D5)). In contrast, the fracture surfaces of the WT were rougher and more fragmented (Figure 5(B2–B5)). The results suggested that seed shattering after seed maturity was higher in *ZlqSH1a* and *ZlqSH1b* over-expressing plants, where the abscission was more complete than in the WT.

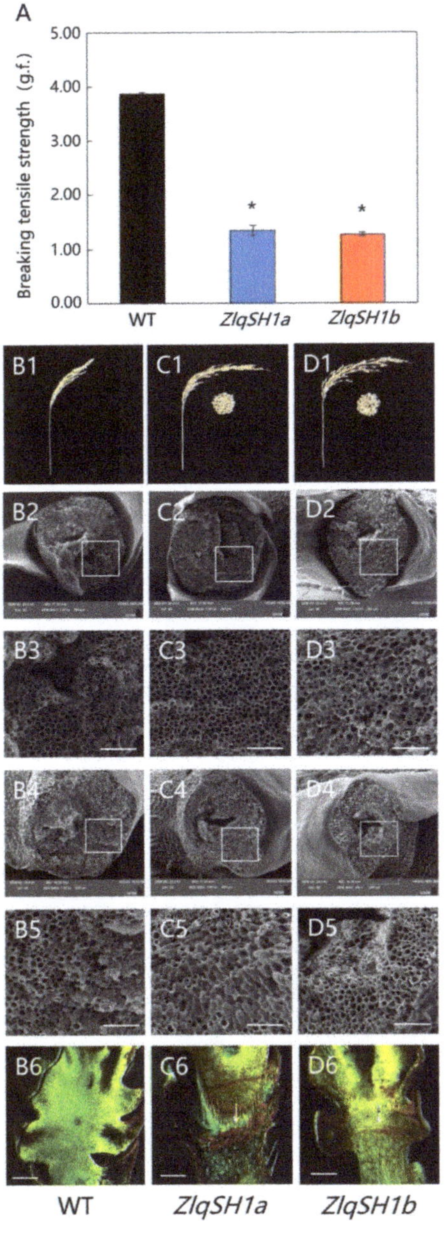

Figure 5. Comparison of the breaking tensile strength (BTS) and results of scanning electron microscopy (SEM) of fracture surfaces and laser confocal microscopy of the longitudinal section of the AL in wild type (WT) and *ZlqSH1a* and *ZlqSH1b* over-expressing plants. (**A**) Comparison of the BTS

of wild type (WT) and *ZlqSH1a* and *ZlqSH1b* over-expressing plants 30 days after pollination. The values are means ± SD (n = 50 grains). The g.f. is the gravitational unit of force. Different letters denote significant differences (* $p < 0.05$) determined using Tukey's honestly significant difference analysis. (**B1,C1,D1**) represent the panicle shape and natural seed-shattering condition of WT and *ZlqSH1a* and *ZlqSH1b* over-expressing plants, respectively. The middle part of the photo of the transgenic plants shows the seeds that were spontaneously shed that were collected in a bag. Bars = 10 cm. (**B2,C2,D2,B4,C4,D4**) show the scanning electron micrographs of the pedicel–seed junction after grain separation. Bars = 200 µm. (**B3,C3,D3,B5,C5,D5**) show close-up SEM images of regions enclosed in the white boxes. Bars = 50 µm. (**B6,C6,D6**) show fluorescence images of longitudinal sections of the pedicel–seed junction stained with acridine orange. The white arrows point to areas corresponding to the abscission layers in WT and the *ZlqSH1a* and *ZlqSH1b* over-expressing plants. Bars = 100 µm.

To accurately determine the histological differences in the AL between *ZlqSH1a/b* over-expressing plants and the WT, we used laser scanning confocal microscopy (LSCM) to compare the longitudinal sections of spikelets at the flowering stage. Samples from the *ZlqSH1a* and *ZlqSH1b* over-expressing plants showed a complete AL composed of small flat parenchymal cells of equal diameter between the pedicel and grain. The longitudinal section showed continuous lines, indicating cell abscission between the pedicel and the grain (Figure 5(C6,D6)). In contrast, no abscission was formed in the WT, and no single obvious line indicated cell abscission (Figure 5(B6)). These anatomical features of the AL were observed 3–5 days after the rice plant flowered, and similar features were maintained in the mature grain. Considering the natural grain shedding behaviour of *ZlqSH1a* and *ZlqSH1b* over-expressing plants, the results of BTS and SEM analysis indicated that seed shattering was significantly higher in the transgenic plants than in the WT. This was consistent with the observations of the longitudinal AL sections, indicating that *ZlqSH1a* and *ZlqSH1b* were involved in AL tissue development.

2.6. Effects of ZlqSH1a and ZlqSH1b Over-Expression on the Transcriptome of the AL in Rice

To analyse the gene functions of *ZlqSH1a* and *ZlqSH1b*, we performed RNA-seq experiments using AZ tissues from WT and *ZlqSH1a* and *ZlqSH1b* over-expressing plants. The unit structure of the AL consists of a 1 mm pedicel region and a 1.5 mm grain. We analysed the differential expression of genes in the two groups of samples and visualised the statistically significant differences using a volcano plot (Figure 6A,B). Hierarchical cluster analysis was performed using the selected differentially expressed genes (DEGs), such that genes with similar expression patterns were clustered (Figure 6C,D). Compared with the WT, 327 DEGs were identified in *ZlqSH1a* over-expressing plants, of which 155 were up-regulated and 172 were down-regulated (Table S2). Compared with the WT, 696 DEGs were identified in *ZlqSH1b* over-expressing plants, out of which 323 were up-regulated and 373 were down-regulated (Table S3).

The two sets of DEGs shared 172 genes. Gene Ontology (GO) and Kyoto Encyclopaedia of Genes and Genomes (KEGG) enrichment analyses were performed on these DEGs. The DEGs were searched against the GO database to classify the normalised gene function and to elucidate the cell components, biological processes, and molecular functions involved in AL development. Pathway annotation and enrichment analysis of DEGs helped further interpret the gene function. A total of 562 DEGs were assigned to three classes (Biological Process, Molecular Function, and Cellular Component) and 44 subclasses of GO (Figure 6E,F). The main subclasses were "cell part", "cell", and "organelle" in the Cellular Component; "metabolic process", "cellular process", and "single-organism process" in Biological Process; and "binding", "catalytic activity", and "transporter activity" in Molecular Function. All DEGs were subjected to KEGG pathway enrichment analysis to understand the complex biological processes of the transcriptome (Figure 6G,H). The results showed that 411 DEGs were annotated and enriched in several KEGG pathways, including "MAPK signalling pathway-plant", "Starch and sucrose metabolism", "Carbon metabolism", and "Plant hormone signal transduction". In this study, we mainly focused on "Carbon metabolism" and "Plant hormone signal transduction".

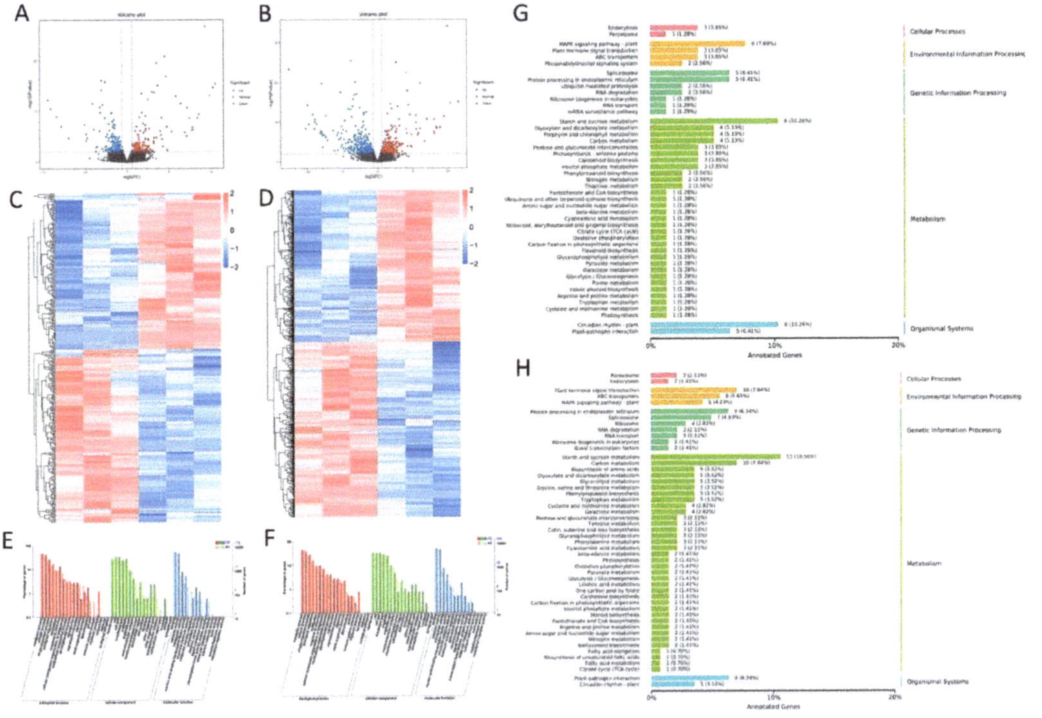

Figure 6. Analysis of RNA-seq results of AL tissues in wild type (WT) plants and *ZlqSH1a* and *ZlqSH1b* gene over-expressing plants. (**A**,**B**) Volcano plots of differential expression between WT and the *ZlqSH1a* and *ZlqSH1b* over-expressing plants, respectively. The green dots represent down-regulated differentially expressed genes (DEGs), the red dots represent up-regulated DEGs, and the black dots represent non-DEGs. (**C**,**D**) DEG cluster diagrams between WT tissues and tissues from *ZlqSH1a* and *ZlqSH1b* over-expressing plants, respectively. The abscissa shows the sample name and the clustering patterns of samples, and the ordinate represents the clustering patterns of DEGs and other genes. Different columns represent different samples, and different rows represent different genes. The colour represents the level of gene expression (log10-transformed) in the sample (FPKM + 0.000001). (**E**,**F**) Statistics related to the Gene Ontology (GO) annotation of DEGs in WT vs. *ZlqSH1a* and *ZlqSH1b* over-expressing plants, respectively. The abscissa shows the GO classification, the left side of the ordinate is the percentage of the number of genes, and the right side is the number of genes. (**G**,**H**) Statistics related to the Kyoto Encyclopaedia of Genes and Genomes (KEGG) annotation of DEGs in WT vs. *ZlqSH1a* and *ZlqSH1b* over-expressing plants, respectively. The ordinate shows the name of the KEGG metabolic pathway, and the abscissa shows the number (and proportion) of genes annotated under the corresponding pathway.

2.7. Expression of Seed Shattering-Related Genes and Validation by qRT-PCR

As an important plant hormone, ethylene can regulate the abscission of seeds and flowers [53]. Moreover, increased polygalacturonase activity is related to cell separation during fruit shedding [54]. The previous results showed that *ZlqSH1a* and *ZlqSH1b* over-expressing plants were significantly different from WT plants in terms of seed-shattering-related morphology and histology. Based on the annotations of the two sets of DEGs, we selected three seed-shattering-related genes for qRT-PCR validation. The three candidate genes—including two down-regulated genes and one up-regulated gene—were regulated in the same direction in the *ZlqSH1a* and *ZlqSH1b* over-expressing plants. The following candidate genes were included: polygalacturonase 1 (*PG1*), polygalacturonase 2 (*PG2*), and ethylene-responsive transcription factor (*ERF*). All candidate genes were expressed in the AL of WT and transgenic plants. We used the rice housekeeping gene *UBQ5* as the housekeeping gene to normalise the gene expression levels. Compared with the WT, the expression levels of *PG1* and *PG2* were significantly down-regulated in *ZlqSH1a* and *ZlqSH1b* over-expressing plants, whereas *ERF* was significantly up-regulated ($p < 0.05$) (Figure 7). This was consistent with the transcriptome sequencing results (Figure 6, Tables S2 and S3).

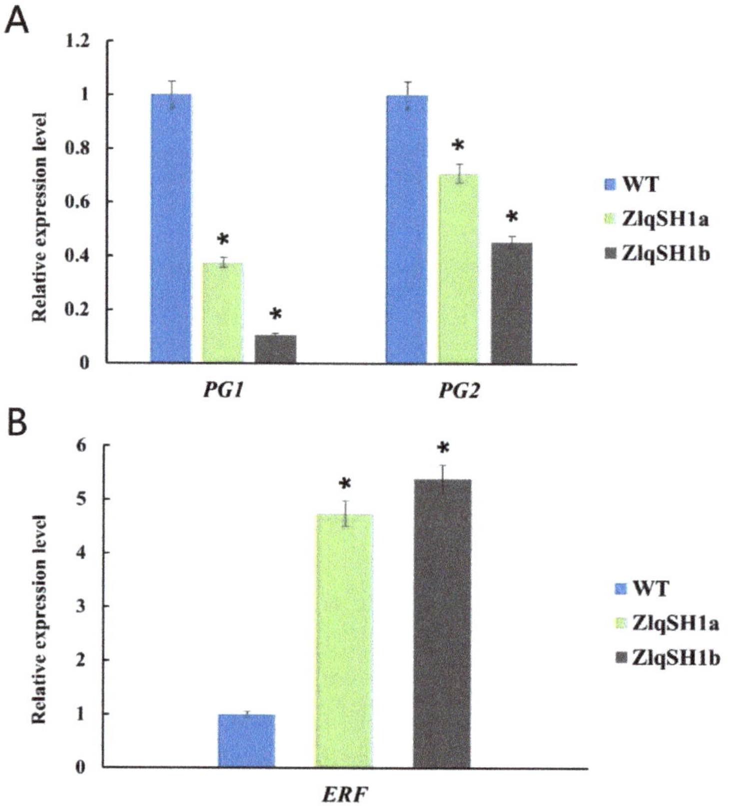

Figure 7. Expression of seed-shattering related genes ((**A**) PG1 and PG2; (**B**) ERF) and validation by qRT-PCR. The rice housekeeping gene UBQ5 was used as an internal control to normalize gene expression data. Two-tailed Student's *t*-tests were used to compare the expression levels between wild-type plants and the *ZlqSH1a/b* over-expressing plants (* $p < 0.05$). *PG1*, polygalacturonase 1; *PG2*, polygalacturonase 2; *ERF*, ethylene-responsive transcription factor.

3. Discussion

3.1. Full-Genome Analysis of BELL Family TFs in *Z. latifolia*

Studies have shown that approximately 300 million years ago, at least one *BELL* gene was already present in the common ancestor of extant spermatophytes and that this gene domain is highly conserved [55]. To date, members of the BEL1-like family of TFs have been identified in every plant species studied. With the advent of full-genome sequences, the breadth and potential functions of this key family of DNA-binding proteins can now be fully understood. Genes in the *BELL* family are involved in plant vegetative growth, fruit development, and seed abscission. In apples, the BELL1-like gene *MDH1* is mainly expressed in developing and mature flowers and fruits. In *Arabidopsis* lines, the ectopic expression of *MDH1* leads to dwarfing, reduced fertility, deformation of the carpel, and deformed silique folds [56]. In rice, the *qSH1* and *SH5* genes (homologs of the Arabidopsis *PNY*) participate in seed shattering by promoting AZ development and inhibiting lignin synthesis [51].

This study is the first to systematically identify BELL TFs in *Z. latifolia*. We identified 48 *ZlBELL* genes (Figure 1, Table S1). The chromosome localisation analysis revealed that the *ZlBELL* genes were unevenly distributed on the chromosomes of *Z. latifolia*. Some *ZlBELL* genes were located at the top and bottom of the chromosome (Figure 1), which is consistent with the findings of Sharma et al. (2014) [57]. This indicates that the *ZlBELL* genes underwent a certain degree of contraction and expansion. Introns and exons can evolve through gain, loss, insertion, or deletion [58]. Here, we found that the number of exons in the *ZlBELL* gene family ranged from four to six. Most of the genes (77.08%) contained four to five exons with a length of more than 1.6 kb, whereas the number of introns ranged from three to five and did not vary greatly. Analysis of the structure of this gene family revealed the length of the introns and exons encoding key functional domains and showed the degree of conservation of their splicing mode (Figure 3). The results indicated that in *ZlBELL* genes, the insertion or loss of introns and exons was relatively stable during the evolutionary process, resulting in relatively small differences in the number of exons and introns. This ensures stability in the biological functions of these genes to a certain extent. The conserved motifs of 48 *ZlBELL* genes in *Z. latifolia* were analysed, revealing three highly conserved amino acid motifs. Most of the conserved motifs in the same subfamily were similar, indicating stability in the function of the coding protein in each subfamily. Almost all ZlBELL proteins contained Motifs 1 and 2, which were always adjacent to each other and together constituted the ZlBELL domain. The uniqueness and conserved nature of these motifs in the same subfamily also supports the evolutionary classification of the ZlBELL gene family (Figure 3C). The 48 *ZlBELL* genes in *Z. latifolia* were distributed in 9 subfamilies (Figure 3). The C subfamily contained the largest number of members (16), whereas the E subfamily contained only two members. This indicates that the members of this subfamily evolved relatively slowly, and we speculate that the function of this gene family is relatively conserved.

3.2. Histological Comparison of the AL between ZlqSH1a/ZlqSH1b Over-Expressing Plants and WT

Rice grains are shed at the AL, a junction between small branches. The AL consists of one or several layers of dense cells with similar morphology. Seed shattering in rice is a complex biological trait mainly controlled by the formation and fracture of the AL [28]. *qSH1* is one of the major genes controlling seed shattering in rice [27], and the locus containing this gene (*LOC_Os01g62920*) explains 68.6% of the phenotypic variation in this species. Continuous expression of *qSH1* (encoding a BELL-type homeoprotein) at the base of spikelets promotes the formation of the AL. As explained above, an SNP causes changes in the expression pattern of *qSH1* in the abscission region of 'Nipponbare' plants, resulting in the loss of seed shattering. Yan et al. (2022) [52] performed genome collinearity and homology analysis in *Z. latifolia* and rice and found that the *ZlqSH1a* (*Zla04G033720*) and *ZlqSH1b* (*Zla02G027130*) genes in *Z. latifolia* were homologs of the *qSH1* (*LOC_Os01g62920*) gene in rice. Based on this, we speculated that *ZlqSH1a* and *ZlqSH1b* might be related to seed shattering in *Z. latifolia*.

In the present study, we transformed *ZlqSH1a* and *ZlqSH1b* into the rice cultivar L422 (which does not exhibit seed-shattering behaviour) and constructed *ZlqSH1a* and *ZlqSH1b* gene over-expressing rice plants. We then used a digital tension meter to measure the BTS between the grain and pedicel of WT and *ZlqSH1a* and *ZlqSH1b* over-expressing plants at the same developmental period. The results showed that the maximum tensile strength was lower in *ZlqSH1a/b* over-expressing lines than in the WT (Figure 5A). In addition, the number of naturally shed seeds after maturity was significantly higher in the *ZlqSH1a/b* over-expressing lines than in the WT (Figure 5(B1,C1,D1)). SEM analysis showed that the phenotype of the AL was very different between *ZlqSH1a* and *ZlqSH1b* over-expressing lines and the WT. In particular, the fracture surfaces of the grain and pedicel were relatively smooth in *ZlqSH1a* and *ZlqSH1b* over-expressing plants (Figure 5(C2–C5,D2–D5)) and rougher and more broken in the WT (Figure 5(B2–B5)). LSCM observation of the longitudinal section of the AL revealed a complete AL composed of small and flat parenchymal cells of equal diameter between the pedicel and grain of the *ZlqSH1a* and *ZlqSH1b* gene over-expressing plants. (Figure 5(C6,D6)). In contrast, no AL had been formed in the WT plants, and no obvious line indicated cell abscission (Figure 5(B6)). These results suggest that *ZlqSH1a* and *ZlqSH1b* affect seed shattering in rice by controlling the formation of the AL. To verify the soundness of this inference, we used RNA-seq to compare and analyse the gene expression levels of ALs in *ZlqSH1a* and *ZlqSH1b* gene over-expressing plants and the WT.

3.3. *ZlqSH1a and ZlqSH1b Are Involved in the Regulation of AL Growth and Development*

We selected two groups of DEGs based on RNA-seq analysis, which revealed that 172 genes were shared among the DEGs (Tables S2 and S3). GO and KEGG enrichment analysis of these DEGs can shed light on the cellular components, biological processes, and molecular functions involved in the development of the AL. Pathway annotation and analysis of DEGs can also help interpret the functions of the genes. We selected three genes related to seed shattering for qRT-PCR validation. Phytohormones are signal molecules produced by plants that play important roles in the regulation of plant growth and development. Ethylene—a phytohormone—regulates the shedding of flowers and seeds, and elevated levels of ethylene are often associated with tissue ageing and cellular stress [53]. Ethylene is an effective inhibitor of auxin, and the auxin content of the AZ regulates the sensitivity of this site to ethylene [59]. In the present study, expression levels of the *ERF* gene were significantly higher in *ZlqSH1a* and *ZlqSH1b* over-expressing plants than in the WT (Figure 7), suggesting that *ERF* may be related to enhanced seed shattering in the transgenic plants. In addition to plant hormones, some cell wall hydrolases also influence plant organ abscission. Some studies suggest that the increased activity of polygalacturonase is related to cell separation during fruit shedding. This is because polygalacturonase promotes the hydrolysis of pectin in the cell wall, thus changing its texture and hardness [54,60,61]. The AL of several plant species show increased polygalacturonase activity before the shedding of leaves, flowers, and fruits [62], and polygalacturonase-induced cell wall degradation is an important step in the loss of cell–cell adhesion in the abscission layer [59]. In the present study, GO functional annotation revealed that the candidate genes *PG1* and *PG2* were involved in regulating polygalacturonase activity. The expression levels of *PG1* and *PG2* were significantly lower in *ZlqSH1a* and *ZlqSH1b* over-expressing plants than in the WT, indicating that these genes are regulated by *ZlqSH1a* and *ZlqSH1b* and are involved in the development of abscission. The findings suggest that *ZlqSH1a* and *ZlqSH1b* are involved in AL development in rice grains and thereby regulate seed shattering. Hence, the *ZlqSH1a* and *ZlqSH1b* genes can be exploited as key target genes in molecular breeding to cultivate new *Z. latifolia* varieties resistant to seed shattering.

4. Materials and Methods

4.1. *Samples*

The *Z. latifolia* plant used in this study was collected from Baimahu Village in Jinhu County, Huai'an City, Jiangsu Province (33°11′9″ N; 119°9′37″ E). *Oryza sativa* cv. L422 was provided by Hefei Jian Gu Biotechnology Co., Ltd. *Escherichia coli*-competent DH5α and T vector were purchased from Sangon Biotech (Shanghai, China) Co., Ltd. The over-expression vector 1390-UBI was provided by Hefei Jian Gu Biotechnology Co., Ltd. (Hefei, China), and the pCAMBIA1390-ubiquitin over-expression vector was provided by Wuhan Edgene Biotechnology Co., Ltd. Equal amounts of soil and water were added to each plant box, and the experiment was repeated three times.

4.2. *Full-Genome Analysis of BELL TFs in Z. latifolia*

4.2.1. Identification of the BELL TFs Family in *Z. latifolia*

The protein sequence of rice BELL was downloaded from PlantTFDB (v5.0) (http://planttfdb.gao-lab.org). The full genome data of *Z. latifolia* (genomic sequence, coding sequence, and protein sequence) were downloaded from the Genome Warehouse accessed on 4 December 2021 (https://ngdc.cncb.ac.cn/gwh/Assembly/22880/show). Using the rice BELL sequence as the query sequence, the candidate *BELL* gene in the *Z. latifolia* genome was identified using BLAST. Subsequently, the hidden Markov model (HMM) of the BELL family gene was downloaded from the PFAM database (http://pfam.xfam.org). Finally, the PFAM, NCBI conserved domains (http://www.ncbi.nlm.nih.gov/Structure/cdd/wrpsb.cgi) and SMART (http://smart.embl-heidelberg.de/) databases were consulted to verify the presence of the BELL conserved domain.

4.2.2. Chromosome Localisation Analysis and Phylogenetic Analysis

The location information of all *BELL* genes in the *Z. latifolia* genome was extracted, and the online tool Map Gene 2 Chromosome (v2.0) (http:/mg2c.iask.in/mg2c_v2.0) was used to draw the chromosome localisation map of the *ZlBELL* family gene. The neighbour-joining method in MEGA7 was used to construct the phylogenetic tree for the protein sequences of *Arabidopsis*, *Z. latifolia* and rice with default parameters (Test of Phylogeny: Bootstrap method; No. of Bootstrap Replications: 1000; Gaps/Missing Data Treatment: Pairwise deletion; Multiple Sequence Alignment: MAFFT v 7.5).

4.2.3. Gene Structure and Conservative Motif Analysis

The MEME (v.5.5.0) online tool (http://meme-suite.org/tools/meme; best match length: 6–50, maximum number of motifs: 10) was used to analyse the conserved motifs in the extracted BELL protein sequence. The GSDS (v2.0) system (http://gsds.gao-lab.org/index.php) was used to visualise the gene structure.

4.3. *Gene Cloning of ZlqSH1a and ZlqSH1b*

4.3.1. Total RNA Extraction and Reverse Transcription

The total RNA of *Z. latifolia* leaves was extracted with the FastPure® Universal Plant Total RNA Isolation Kit (RC-401, Vazyme, Nanjing, China) according to the manufacturer's instructions. The concentration of the extracted RNA was measured using an ultramicro spectrophotometer (OSE-260, Tiangen, Beijing, China). The extracted RNA was used as a substrate for reverse transcription using the PrimeScript™ II 1st Strand cDNA Synthesis Kit. The cDNA products obtained were stored at −20 °C for later use.

4.3.2. Amplification, Transformation, and Sequencing of ZlqSH1a and ZlqSH1b

The prepared cDNA was used as a template for PCR amplification with the following primers: ZlqSH1a-F and ZlqSH1a-R for ZlqSH1a; and ZlqSH1b-F and ZlqSH1b-R for ZlqSH1b (Table S4). The PCR product was subjected to agarose gel electrophoresis (AGE) to analyse the fragment sizes and to verify whether the presence of a heteroband. Subsequently, the PCR product was recovered by DNA precipitation; 10 μL of the mixture was added to 10 μL of *Taq* enzyme, heated at 72 °C for 15 min, and ligated to a ploy(A) tail. After AGE, the product was cut from the gel and recovered. The target gene was ligated to a T vector as follows: 0.5 μL of T vector, 4.5 μL of DNA, and 5 μL of the reaction buffer were added to a centrifuge tube and incubated at 4 °C overnight. The *ZlqSH1a* and *ZlqSH1b* genes were ligated to the T vector and transformed into *E. coli*-competent DH5α cells. Positive clones were screened and sequenced.

4.4. *Subcellular Localisation of ZlqSH1a/ZlqSH1b*

The *ZlqSH1a/ZlqSH1b* constructs and the empty vector PC2300S-GFP were propagated in a bacterial solution, and the plasmids were extracted. The plasmids of *ZlqSH1a/ZlqSH1b* and the empty vector PC2300S-GFP were used to co-transform the rice protoplast with the nuclear marker. After 16 h of incubation at 28 °C in the dark, GFP and RFP fluorescence was detected using an Olympus FV1000 laser scanning microscope (Olympus, Tokyo, Japan) at 488 and 543 nm.

4.5. Construction of ZlqSH1a/ZlqSH1b Over-Expressing Rice

4.5.1. Construction of ZlqSH1a/ZlqSH1b Over-Expressing Carriers

DNA fragments containing the full-length *ZlqSH1a/ZlqSH1b* genes were amplified with primers for seamless cloning and amplification (*ZlqSH1a/ZlqSH1b-FP, ZlqSH1a/ZlqSH1b-RP*) (Table S4). The amplification products were detected by AGE, and the target fragments were cut under UV light and purified. The 1390-ubi vector was digested with the restriction enzyme *Bam*HI, the digested product was separated by AGE, and the linearised 1390-ubi large fragment was recovered. After recovery, the digested *ZlqSH1a/ZlqSH1b* gene fragment was ligated with the 1390-ubi vector. A 5 μL aliquot of the product was used to transform DH5α-competent cells. The next day, monoclonal colonies were selected, and the PCR primers 1390-FP and ZlqSH1a/ZlqSH1b-RP were designed for the bacterial solution (Table S4). The monoclonal colonies with positive PCR results were sent for sequencing-based verification. The single colonies with correct sequencing results were placed in LB culture medium containing kanamycin and shaken overnight at 37 °C at 200 r/min. The over-expression vector was extracted for vector transformation.

4.5.2. Agrobacterium-Mediated Transformation of Rice

The activated *Agrobacterium* strain EHA105 and previously prepared rice calluses were co-cultured to facilitate infection with *Agrobacterium*. A 25 mL aliquot of the *Agrobacterium* resuspension was added to the callus and the solution was incubated for 15 min, during which the solution was shaken gently from time to time. After soaking, the resuspension was poured out, and the callus was placed in a Petri dish lined with several sterile filter papers (which absorbed the residual bacterial liquid on the surface of the callus). Next, the disposable Petri dish was lined with three sterile filter papers, and 2.5 mL of inoculating culture medium was added. The dried callus was spread evenly on the filter paper and cultured at 23 °C for 48 h in the dark. The co-cultured callus was spread evenly and sparsely in the recovery culture medium. After approximately 5 days, the callus was transferred to the screening culture medium. After 3–4 weeks, 3–5 small, fresh, resistant callus particles in good growth condition were selected from each independent transformant and transferred to the regenerative culture medium. When the seedlings grew to 2–5 cm in length, one seedling in good growth condition from each independent transformant was transferred to a root-based culture medium. The culture conditions from recovery to root-based culture were as follows: incubation at 30 °C under a light incubator with a 16 h light/8 h dark photoperiod.

4.5.3. Identification of ZlqSH1a/ZlqSH1b Over-Expressing Rice Plants

The following primers were designed for each promoter region to identify positive transformation by the over-expression vectors: *ZlqSH1a-F* and *ZlqSH1a-R* for *ZlqSH1a* and *ZlqSH1b-F* and *ZlqSH1b-R* for *ZlqSH1b* (Table S4). The PCR reaction mixture was prepared as follows: KOD One PCR Master Mix, 25 μL; 1390-FP, 2 μL; RP-1390, 2 μL; template, 1 μL; ddH_2O, 20 μL. The PCR program was as follows: 94 °C for 5 min; followed by 30 cycles of 94 °C for 30 s, 60 °C for 30 s, and 72 °C for 30 s; followed by 72 °C for 5 min (Figure S2).

4.6. Measurement of Seed Shattering

The flowering time of the rice was set after the heading had occurred. This was done to accurately quantify the changes in the seed-shattering phenotype of T2 generation of the *ZlqSH1a/ZlqSH1b* over-expressing rice plants A digital tension meter was used to measure the BTS between the grains and pedicels of the WT and over-expressing lines at the same period 30 days after flowering and pollination. A small clip was suspended on the small hook of the tension meter, and the clip was clamped to the middle part of the seed glume and pulled horizontally so that the seed and the branch separated. At this time, the tension meter showed the maximum tensile strength required in the process, and this value was recorded. The measurement was repeated 50 times, and the average value was recorded as the tensile strength of the seed.

4.7. Laser Scanning Confocal Microscopy

Approximately 20 spikelets were sampled from each plant at the flowering stage to observe the structure of the AZ. A longitudinal section was made manually through the junction between the flower and pedicel, and the sections were stained with acridine orange. The sections were observed using a Leica SP8 laser scanning microscope (Leica SP8; Leica Biosystems, Nussloch, Germany) at 488 and a 543 nm.

4.8. Scanning Electron Microscopy

Thirty days after pollination, the florets (grains) in the middle area of the inflorescence were selected. The grains (approximately 2–4 mm in length) were cut from the pedicel near the branches and stored at 4 °C in the fixative solution used for electron microscopy. The fixed samples were taken out, and the connection between the grain and the pedicel was gently broken with tweezers. The samples were glued to the metal block and observed under a JEOL JSM-840 scanning electron microscope (JEOL Ltd., Tokyo, Japan).

4.9. RNA-seq Analysis

The AL of WT and *ZlqSH1a/ZlqSH1b* over-expressing plants were collected 3–5 days after flowering, and RNA was extracted from the AL tissue. This step was performed for three biological replicates, with three plants in each replicate. Paired-end libraries were constructed and sequenced using Illumina HiSeq 2500. The raw reads were mapped to the reference genome (Os-Nipponbare-Reference-IRGSP-1.0, MSU7) using TopHat2 with the default parameters [63]. Cuffdiff was used to calculate the fragments per kilobase of exon of the mapped reads per million genes, and the DEGs (fold change ≥ 2) between WT and *ZlqSH1a/ZlqSH1b* over-expressing plants were identified [64]. The DEGs were functionally analysed using the agriGO and KEGG databases [65,66].

4.10. Data Validation by qRT-PCR

Three seed-shattering-related genes were analysed in the *ZlqSH1a/ZlqSH1b* over-expressing and WT plants using qRT-PCR analysis. cDNAs were reverse transcribed using residual RNA from the RNA-seq analysis. The synthesised cDNAs were amplified using SYBR Premix Ex Taq (TaKaRa Biotechnology Co., Ltd., Dalian, China). Quantitative RT-PCR was performed using the SYBR Green QPCR mix (Bio-Rad Laboratories, Hercules, CA, USA), and the *UBQ5* gene (NCBI accession number: LOC4341684) was used as an endogenous control to normalise gene expression. The cycling conditions included incubation for 30 s at 95 °C, followed by 40 cycles of amplification (95 °C for 5 s and 60 °C for 30 s). All samples were repeated at least three times. The primers used in the qRT-PCR are listed in Table S4.

4.11. Statistical Analysis

The experiments were arranged in a randomised block design. Data were collected from each experimental group containing different numbers of plants, and the mean $\pm$ SD was calculated for the total number of plants in each experiment. Data were analysed with analysis of variance (ANOVA) in IBM SPSS 18 IIBM Corp., Armonk, NY, USA) statistical software. Tukey's multiple comparison test was used at a significance level of $p < 0.05$ to determine any statistically significant differences between treatments.

5. Conclusions

The present study explored the functions of two genes in the BELL family of TFs (*ZlqSH1a* and *ZlqSH1b*) in *Z. latifolia*. We found that over-expression of *ZlqSH1a* and

ZlqSH1b was responsible for seed-shattering behaviour after the maturation of rice grains. Our findings shed light on the mechanisms underlying seed shattering in a crop plant and can facilitate the genetic improvement of *Z. latifolia* and other cereal crops.

Supplementary Materials: The following supporting information can be downloaded at: https://www.mdpi.com/article/10.3390/ijms232415939/s1.

Author Contributions: N.Y., Z.-F.Z., Y.-N.X., T.Y. and Q.Q. designed the study. Y.-N.X., T.Y., B.-T.Z., Q.-Q.Q., A.-M.D., L.-G.S., Y.Z. and Q.Q. performed the experiments. Y.-N.X. and B.-T.Z. performed data analyses. Y.-N.X. and N.Y. wrote the manuscript. N.Y., Z.-F.Z., Y.-N.X. and B.-T.Z. critically reviewed the manuscript. All authors have read and agreed to the published version of the manuscript.

Funding: This work was supported by the Youth Innovation Program of the Chinese Academy of Agricultural Sciences (Y2023QC), the Central Public-interest Scientific Institution Basal Research Fund (1610232020008, 1610232021006, and 1610232022004), the Postdoctoral Applied Research Project Fund of Qingdao, and the Agricultural Science and Technology Innovation Program (No. ASTIP-TRIC05).

Institutional Review Board Statement: Not applicable.

Informed Consent Statement: Not applicable.

Data Availability Statement: Not applicable.

Conflicts of Interest: The authors declare no conflict of interest.

References

1. Zhang, L.B.; Zhu, Q.; Wu, Z.Q.; Ross-Ibarra, J.; Gaut, B.S.; Ge, S.; Sang, T. Selection on grain shattering genes and rates of rice domestication. *New Phytol.* **2009**, *184*, 708–720. [CrossRef] [PubMed]
2. Tanksley, S.D.; McCouch, S.R. Seed banks and molecular maps: Unlocking genetic potential from the wild. *Science* **1997**, *277*, 1063–1066. [CrossRef] [PubMed]
3. Salamini, F.; Ozkan, H.; Brandolini, A.; Schafer-Pregl, R.; Martin, W. Genetics and geography of wild cereal domestication in the near east. *Nat. Rev. Genet.* **2002**, *3*, 429–441. [CrossRef]
4. Doebley, J. The genetics of maize evolution. *Annu. Rev. Genet.* **2004**, *38*, 37–59. [CrossRef]
5. Sweeney, M.; McCouch, S. The complex history of the domestication of rice. *Ann. Bot.* **2007**, *100*, 951–957. [CrossRef]
6. Vaughan, D.A.; Balazs, E.; Heslop-Harrison, J.S. From crop domestication to super-domestication. *Ann. Bot.* **2007**, *100*, 893–901. [CrossRef] [PubMed]
7. Jiang, L.Y.; Ma, X.; Zhao, S.S.; Tang, Y.Y.; Liu, F.X.; Gu, P.; Fu, Y.C.; Zhu, Z.F.; Cai, H.W.; Sun, C.Q.; et al. The APETALA2-like transcription factor SUPERNUMERARY BRACT controls rice seed shattering and seed size. *Plant Cell* **2019**, *31*, 17–36. [CrossRef]
8. Wu, W.G.; Liu, X.Y.; Wang, M.H.; Meyer, R.S.; Luo, X.J.; Ndjiondjop, M.N.; Tan, L.B.; Zhang, J.W.; Wu, J.Z.; Cai, H.W.; et al. A single-nucleotide polymorphism causes smaller grain size and loss of seed shattering during African rice domestication. *Nat. Plants* **2017**, *3*, 17064. [CrossRef]
9. Li, C.B.; Zhou, A.L.; Sang, T. Genetic analysis of rice domestication syndrome with the wild annual species, oryza nivara. *New Phytol.* **2006**, *170*, 185–193. [CrossRef]
10. Li, C.B.; Zhou, A.L.; Sang, T. Rice domestication by reducing shattering. *Science* **2006**, *311*, 1936–1939. [CrossRef]
11. Ji, H.S.; Chu, S.H.; Jiang, W.Z.; Cho, Y.I.; Hahn, J.H.; Eun, M.Y.; McCouch, S.R.; Koh, H.J. Characterization and mapping of a shattering mutant in rice that corresponds to a block of domestication genes. *Genetics* **2006**, *173*, 995–1005. [CrossRef] [PubMed]
12. Li, L.F.; Olsen, K.M. To have and to hold: Selection for seed and fruit retention during crop domestication. *Curr. Top. Dev. Biol.* **2016**, *119*, 63–109. [PubMed]
13. Yan, H.X.; Ma, L.; Wang, Z.; Lin, Z.M.; Su, J.; Lu, B.R. Multiple tissue-specific expression of rice seed-shattering gene SH4 regulated by its promoter pSH4. *Rice* **2015**, *8*, 12. [CrossRef] [PubMed]
14. Yu, X.; Yang, T.; Qi, Q.; Du, Y.; Shi, J.; Liu, X.; Liu, Y.; Zhang, H.; Zhang, Z.; Yan, N. Comparison of the contents of phenolic compounds including flavonoids and antioxidant activity of rice (*Oryza sativa*) and Chinese wild rice (*Zizania latifolia*). *Food Chem.* **2021**, *344*, 128600. [CrossRef] [PubMed]
15. Lu, R.; Chen, M.; Feng, Y.; Yuan, N.; Zhang, Y.; Cao, M.; Liu, J.; Wang, Y.; Hang, Y.; Sun, X. Comparative plastome analyses and genomic resource development in wild rice (*Zizania* spp., Poaceae) using genome skimming data. *Ind. Crops Prod.* **2022**, *186*, 115244. [CrossRef]
16. Xu, X.W.; Wu, J.W.; Qi, M.X.; Lu, Q.X.; Lee, P.F.; Lutz, S.; Ge, S.; Wen, J. Comparative phylogeography of the wild-rice genus *Zizania* (Poaceae) in eastern Asia and North America. *Am. J. Bot.* **2015**, *102*, 239–247. [CrossRef]
17. Yu, X.; Qi, Q.; Li, Y.; Li, N.; Xie, Y.; Ding, A.; Shi, J.; Du, Y.; Liu, X.; Zhang, Z.; et al. Metabolomics and proteomics reveal the molecular basis of colour formation in the pericarp of Chinese wild rice (*Zizania latifolia*). *Food Res. Int.* **2022**, *162*, 112082. [CrossRef]

18. Yu, X.; Chu, M.; Chu, C.; Du, Y.; Shi, J.; Liu, X.; Liu, Y.; Zhang, H.; Zhang, Z.; Yan, N. Wild rice (*Zizania* spp.): A review of its nutritional constituents, phytochemicals, antioxidant activities, and health-promoting effects. *Food Chem.* **2020**, *331*, 127293. [CrossRef]
19. Fan, X.R.; Ren, X.R.; Liu, Y.L.; Chen, Y.Y. Genetic structure of wild rice *Zizania latifolia* and the implications for its management in the Sanjiang Plain, Northeast China. *Biochem. Syst. Ecol.* **2016**, *64*, 81–88. [CrossRef]
20. Yan, N.; Du, Y.; Liu, X.; Chu, C.; Shi, J.; Zhang, H.; Liu, Y.; Zhang, Z. Morphological characteristics, nutrients, and bioactive compounds of *Zizania latifolia*, and health benefits of its seeds. *Molecules* **2018**, *23*, 1561. [CrossRef]
21. Chen, Y.; Liu, Y.; Fan, X.; Li, W.; Liu, Y. Landscape-scale genetic structure of wild rice *Zizania latifolia*: The roles of rivers, mountains and fragmentation. *Front. Ecol. Evol.* **2017**, *5*, 17. [CrossRef]
22. Sumczynski, D.; Kotaskova, E.; Orsavova, J.; Valasek, P. Contribution of individual phenolics to antioxidant activity and in *vitro* digestibility of wild rices (*Zizania aquatica* L.). *Food Chem.* **2017**, *218*, 107–115. [CrossRef] [PubMed]
23. Chen, Y.Y.; Chu, H.J.; Liu, H.; Liu, Y.L. Abundant genetic diversity of the wild rice *Zizania latifolia* in central China revealed by microsatellites. *Ann. Appl. Biol.* **2012**, *161*, 192–201. [CrossRef]
24. Shan, X.H.; Liu, Z.L.; Dong, Z.Y.; Wang, Y.M.; Chen, Y.; Lin, X.Y.; Long, L.K.; Han, F.P.; Dong, Y.S.; Liu, B. Mobilization of the active MITE transposons *mPing* and *Pong* in rice by introgression from wild rice (*Zizania latifolia* Griseb.). *Mol. Biol. Evol.* **2005**, *22*, 976–990. [CrossRef] [PubMed]
25. Zhai, C.K.; Tang, W.L.; Jang, X.L.; Lorenz, K.J. Studies of the safety of Chinese wild rice. *Food Chem. Toxicol.* **1996**, *34*, 347–352. [CrossRef] [PubMed]
26. Guo, H.B.; Li, S.M.; Peng, J.; Ke, W.D. *Zizania latifolia* Turcz. cultivated in China. *Genet. Resour. Crop Evol.* **2007**, *54*, 1211–1217. [CrossRef]
27. Konishi, S.; Izawa, T.; Lin, S.Y.; Ebana, K.; Fukuta, Y.; Sasaki, T.; Yano, M. An SNP caused loss of seed shattering during rice domestication. *Science* **2006**, *312*, 1392–1396. [CrossRef]
28. Simons, K.J.; Fellers, J.P.; Trick, H.N.; Zhang, Z.C.; Tai, Y.S.; Gill, B.S.; Faris, J.D. Molecular characterization of the major wheat domestication gene Q. *Genetics* **2006**, *172*, 547–555. [CrossRef]
29. Pourkheirandish, M.; Hensel, G.; Kilian, B.; Senthil, N.; Chen, G.; Sameri, M.; Azhaguvel, P.; Sakuma, S.; Dhanagond, S.; Sharma, R.; et al. Evolution of the grain dispersal system in barley. *Cell* **2015**, *162*, 527–539. [CrossRef]
30. Fu, Z.Y.; Song, J.C.; Zhao, J.Q.; Jameson, P.E. Identification and expression of genes associated with the abscission layer controlling seed shattering in *Lolium perenne*. *AoB Plants* **2019**, *11*, ply076. [CrossRef]
31. Bull, S.E.; Seung, D.; Chanez, C.; Mehta, D.; Kuon, J.E.; Truernit, E.; Hochmuth, A.; Zurkirchen, I.; Zeeman, S.C.; Gruissem, W.; et al. Accelerated ex situ breeding of GBSS- and PTST1-edited cassava for modified starch. *Sci. Adv.* **2018**, *4*, eaat6086. [CrossRef] [PubMed]
32. Zhou, Y.; Lu, D.F.; Li, C.Y.; Luo, J.H.; Zhu, B.F.; Zhu, J.J.; Shangguan, Y.Y.; Wang, Z.X.; Sang, T.; Zhou, B.; et al. Genetic control of seed shattering in rice by the APETALA2 transcription factor *SHATTERING ABORTION1*. *Plant Cell* **2012**, *24*, 1034–1048. [CrossRef] [PubMed]
33. Lin, Z.W.; Griffith, M.E.; Li, X.R.; Zhu, Z.F.; Tan, L.B.; Fu, Y.C.; Zhang, W.X.; Wang, X.K.; Xie, D.X.; Sun, C.Q. Origin of seed shattering in rice (*Oryza sativa* L.). *Planta* **2007**, *226*, 11–20. [CrossRef] [PubMed]
34. Gehring, W.J. Homeo boxes in the study of development. *Science* **1987**, *236*, 1245–1252. [CrossRef]
35. Bian, Z.Y.; Gao, H.H.; Wang, C.Y. NAC transcription factors as positive or negative regulators during ongoing battle between pathogens and our food crops. *Int. J. Mol. Sci.* **2021**, *22*, 81. [CrossRef]
36. Dreni, L.; Kater, M.M. MADS reloaded: Evolution of the *AGAMOUS* subfamily genes. *New Phytol.* **2014**, *201*, 717–732. [CrossRef]
37. Mizoi, J.; Shinozaki, K.; Yamaguchi-Shinozaki, K. AP2/ERF family transcription factors in plant abiotic stress responses. *Biochim. Biophys. Acta* **2012**, *1819*, 86–96. [CrossRef]
38. Hamant, O.; Pautot, V. Plant development: A TALE story. *C. R. Biol.* **2010**, *333*, 371–381. [CrossRef]
39. Chen, H.; Rosin, F.M.; Prat, S.; Hannapel, D.J. Interacting transcription factors from the three-amino acid loop extension superclass regulate tuber formation. *Plant Physiol.* **2003**, *132*, 1391–1404. [CrossRef]
40. Bertolino, E.; Reimund, B.; WildtPerinic, D.; Clerc, R.G. A novel homeobox protein which recognizes a TGT core and functionally interferes with a retinoid-responsive motif. *J. Biol. Chem.* **1995**, *270*, 31178–31188. [CrossRef]
41. Kanrar, S.; Onguka, O.; Smith, H.M.S. *Arabidopsis* inflorescence architecture requires the activities of KNOX-BELL homeodomain heterodimers. *Planta* **2006**, *224*, 1163–1173. [CrossRef] [PubMed]
42. Cole, M.; Nolte, C.; Werr, W. Nuclear import of the transcription factor SHOOT MERISTEMLESS depends on heterodimerization with BLH proteins expressed in discrete sub-domains of the shoot apical meristem of *Arabidopsis thaliana*. *Nucleic Acids Res.* **2006**, *34*, 1281–1292. [CrossRef] [PubMed]
43. Bellaoui, M.; Pidkowich, M.S.; Samach, A.; Kushalappa, K.; Kohalmi, S.E.; Modrusan, Z.; Crosby, W.L.; Haughn, G.W. The *Arabidopsis* BELL1 and KNOX TALE homeodomain proteins interact through a domain conserved between plants and animals. *Plant Cell* **2001**, *13*, 2455–2470. [CrossRef] [PubMed]
44. Arnaud, N.; Pautot, V. Ring the BELL and tie the KNOX: Roles for TALEs in gynoecium development. *Front. Plant Sci.* **2014**, *5*, 93. [CrossRef]
45. Niu, X.; Fu, D. The roles of BLH transcription factors in plant development and environmental response. *Int. J. Mol. Sci.* **2022**, *23*, 3731. [CrossRef]

46. Mukherjee, K.; Brocchieri, L.; Burglin, T.R. A comprehensive classification and evolutionary analysis of plant homeobox genes. *Mol. Biol. Evol.* **2009**, *26*, 2775–2794. [CrossRef]
47. Becker, A.; Bey, M.; Burglin, T.R.; Saedler, H.; Theissen, G. Ancestry and diversity of BEL1-like homeobox genes revealed by gymnosperm (*Gnetum gnemon*) homologs. *Dev. Genes Evol.* **2002**, *212*, 452–457. [CrossRef]
48. Hackbusch, J.; Richter, K.; Muller, J.; Salamini, F.; Uhrig, J.F. A central role of *Arabidopsis thaliana* ovate family proteins in networking and subcellular localization of 3-aa loop extension homeodomain proteins. *Proc. Natl. Acad. Sci. USA* **2005**, *102*, 4908–4912. [CrossRef]
49. Muller, J.; Wang, Y.M.; Franzen, R.; Santi, L.; Salamini, F.; Rohde, W. In vitro interactions between barley TALE homeodomain proteins suggest a role for protein-protein associations in the regulation of *Knox* gene function. *Plant J.* **2001**, *27*, 13–23. [CrossRef]
50. Kanrar, S.; Bhattacharya, M.; Arthur, B.; Courtier, J.; Smith, H.M.S. Regulatory networks that function to specify flower meristems require the function of homeobox genes PENNYWISE and POUND-FOOLISH in *Arabidopsis*. *Plant J.* **2008**, *54*, 924–937. [CrossRef]
51. Yoon, J.; Cho, L.H.; Kim, S.L.; Choi, H.; Koh, H.J.; An, G. The BEL1-type homeobox gene SH5 induces seed shattering by enhancing abscission-zone development and inhibiting lignin biosynthesis. *Plant J.* **2014**, *79*, 717–728. [CrossRef] [PubMed]
52. Yan, N.; Yang, T.; Yu, X.T.; Shang, L.G.; Guo, D.P.; Zhang, Y.; Meng, L.; Qi, Q.Q.; Li, Y.L.; Du, Y.M.; et al. Chromosome-level genome assembly of *Zizania latifolia* provides insights into its seed shattering and phytocassane biosynthesis. *Commun. Biol.* **2022**, *5*, 36. [CrossRef] [PubMed]
53. Sexton, R.; Roberts, J.A. Cell biology of abscission. *Annu. Rev. Plant Physiol.* **1982**, *33*, 133–162. [CrossRef]
54. Rui, Y.; Xiao, C.; Yi, H.; Kandemir, B.; Wang, J.Z.; Puri, V.M. Polygalacturonase involved in expansion3 functions in seedling development, rosette growth, and stomatal dynamics in *Arabidopsis thaliana*. *Plant Cell* **2017**, *29*, 2413–2432. [CrossRef]
55. Champagne, C.E.; Ashton, N.W. Ancestry of KNOX genes revealed by bryophyte (*Physcomitrella patens*) homologs. *New Phytol.* **2001**, *150*, 23–36. [CrossRef]
56. Dong, Y.H.; Yao, J.L.; Atkinson, R.G.; Putterill, J.J.; Morris, B.A.; Gardner, R.C. MDH1: An apple homeobox gene belonging to the BEL1 family. *Plant Mol. Biol.* **2000**, *42*, 623–633. [CrossRef]
57. Sharma, P.; Lin, T.; Grandellis, C.; Yu, M.; Hannapel, D.J. The BEL1-like family of transcription factors in potato. *J. Exp. Bot.* **2014**, *65*, 709–723. [CrossRef]
58. Xu, G.; Guo, C.; Shan, H.; Kong, H. Divergence of duplicate genes in exon-intron structure. *Proc. Natl. Acad. Sci. USA* **2012**, *109*, 1187–1192. [CrossRef]
59. Taylor, J.E.; Whitelaw, C.A. Signals in abscission. *New Phytol.* **2001**, *151*, 323–340. [CrossRef]
60. Xiao, C.; Barnes, W.J.; Zamil, M.S.; Yi, H.; Puri, V.M.; Anderson, C.T. Activation tagging of *Arabidopsis POLYGALACTURONASE INVOLVED IN EXPANSION2* promotes hypocotyl elongation, leaf expansion, stem lignification, mechanical stiffening, and lodging. *Plant J.* **2017**, *89*, 1159–1173. [CrossRef]
61. Swain, S.; Kay, P.; Ogawa, M. Preventing unwanted breakups: Using polygalacturonases to regulate cell separation. *Plant Signal. Behav.* **2011**, *6*, 93–97. [CrossRef] [PubMed]
62. Gonzalez-Carranza, Z.H.; Lozoya-Gloria, E.; Roberts, J.A. Recent developments in abscission: Shedding light on the shedding process. *Trends Plant Sci.* **1998**, *3*, 10–14. [CrossRef]
63. Kim, D.; Pertea, G.; Trapnell, C.; Pimentel, H.; Kelley, R.; Salzberg, S.L. TopHat2: Accurate alignment of transcriptomes in the presence of insertions, deletions and gene fusions. *Genome Biol.* **2013**, *14*, R36. [CrossRef]
64. Trapnell, C.; Williams, B.A.; Pertea, G.; Mortazavi, A.; Kwan, G.; van Baren, M.J.; Salzberg, S.L.; Wold, B.J.; Pachter, L. Transcript assembly and quantification by RNA-Seq reveals unannotated transcripts and isoform switching during cell differentiation. *Nat. Biotechnol.* **2010**, *28*, 511–515. [CrossRef] [PubMed]
65. Xin, Y.; Zhou, D.; Zhen, S. PlantGSEA: A gene set enrichment analysis toolkit for plant community. *Nucleic Acids Res.* **2013**, *41*, 98–103.
66. Tian, T.; Liu, Y.; Yan, H.; You, Q.; Yi, X.; Du, Z.; Xu, W.; Su, Z. agriGO v2.0: A GO analysis toolkit for the agricultural community. *Nucleic Acids Res.* **2017**, *45*, 122–129. [CrossRef] [PubMed]

International Journal of
Molecular Sciences

MDPI

Article

CRISPR/Cas9 Gene Editing of *NtAITRs*, a Family of Transcription Repressor Genes, Leads to Enhanced Drought Tolerance in Tobacco

Guimin Li [1,†], Yanxing Ma [2,†], Xiaoping Wang [1], Nini Cheng [1], Deyu Meng [1], Siyu Chen [1], Wei Wang [1], Xutong Wang [1], Xiaojun Hu [1], Li Yan [1] and Shucai Wang [1,*]

1 Laboratory of Plant Molecular Biology & Crop Gene Editing, School of Life Sciences, Linyi University, Linyi 276000, China
2 Key Laboratory of Molecular Epigenetics of MOE, Northeast Normal University, Changchun 130024, China
* Correspondence: wangshucai@lyu.edu.cn
† These authors contributed equally to this work.

Abstract: Tobacco is a cash crop throughout the world, and its growth and development are affected by abiotic stresses including drought stress; therefore, drought-tolerant breeding may help to improve tobacco yield and quality under drought stress conditions. Considering that the plant hormone ABA (abscisic acid) is able to regulate plant responses to abiotic stresses via activating ABA response genes, the characterization of ABA response genes may enable the identification of genes that can be used for molecular breeding to improve drought tolerance in tobacco. We report here the identification of NtAITRs (*Nicotiana tabacum* ABA-induced transcription repressors) as a family of novel regulators of drought tolerance in tobacco. Bioinformatics analysis shows that there are a total of eight *NtAITR* genes in tobacco, and all the NtAITRs have a partially conserved LxLxL motif at their C-terminus. RT-PCR results show that the expression levels of at least some *NtAITRs* were increased in response to ABA and drought treatments, and NtAITRs, when recruited to the *Gal4* promoter via a fused GD (*Gal4* DNA-binding domain), were able to repress transcription activator LD-VP activated expression of the *LexA-Gal4-GUS* reporter gene. Roles of NtAITRs in regulating drought tolerance in tobacco were analyzed by generating CRISPR/Cas9 gene-edited mutants. A total of three Cas9-free *ntaitr12356* quintuple mutants were obtained, and drought treatment assays show that drought tolerance was increased in the *ntaitr12356* quintuple mutants. On the other hand, results of seed germination and seedling greening assays show that ABA sensitivity was increased in the *ntaitr12356* quintuple mutants, and the expression levels of some ABA signaling key regulator genes were altered in the *ntaitr12356-c3* mutant. Taken together, our results suggest that *NtAITRs* are ABA-responsive genes, and that NtAITRs function as transcription repressors and negatively regulate drought tolerance in tobacco, possibly by affecting plant ABA response via affecting the expression of ABA signaling key regulator genes.

Keywords: NtAITRs; transcription factors; tobacco; ABA; drought stress; abiotic stress; gene editing

1. Introduction

Environmental stresses including abiotic stresses and biotic stresses reduce the yield of plants, including crops [1,2]. It is estimated that biotic stresses such as pathogens and insects cause ~20% whereas abiotic stresses such as drought and salt cause ~50% of global yield loss for most major crops [3,4]. On the other hand, to survive the abiotic stress conditions they are constantly exposed to, plants have evolved mechanisms in response to abiotic stresses [5], and dissection of the mechanisms underlining plant responses to abiotic stresses may lead to the breeding of crops with enhanced abiotic stress tolerance, therefore increasing crop yield under abiotic stress conditions.

It is proposed that plants respond to abiotic stresses via two different signaling pathways, i.e., the plant hormone ABA (abscisic acid)-dependent and -independent pathways [5]. As a key stress hormone, ABA accumulates dramatically in response to abiotic stresses such as drought and salt, leading to the activation/repression of ABA response genes via signaling transduction through several key regulators, including the PYR1/PYL/RCAR (Pyrabactin resistance 1/PYR1-like/regulatory component of ABA receptor) ABA receptors, the ABA signaling negative regulators A-group PP2C (PROTEIN PHOSPHATASE 2C) phosphatases, the ABA signaling positive regulators SnRK2 (NON-FERMENTING 1 (SNF1)-RELATED PROTEIN KINASES) protein kinases, and downstream ABF/AREB/ABI5 type bZIP (basic region leucine zipper) transcription factors, thereby affecting plant responses to abiotic stresses [6–13].

Consistent with this, changes in expression levels of the ABA signaling key regulator genes including the *PYR1/PYL/RCAR* receptor genes, the *PP2C* phosphatase genes, the *SnRK2* protein kinase genes and the *ABF/AREB/ABI5* transcription factor genes affected plant responses to ABA, resulting in changes in abiotic stress tolerance in plants [14–17]. Similarly, changes in expression levels of protein stability regulator genes of some ABA signaling key regulators, including the E2 ligase gene *VPS23A* and E3 ligase genes *CUL4*, *RSL1*, *KEG* (*KEEP ON GOING*), *DWA1*, *DWA2*, *PUB12* and *PUB13*, are also able to affect plant response to ABA and thereby plant responses to abiotic stresses [18–26]. In addition, changes in expression levels of ABA response genes such as the MYB transcription factor genes *MYB44* and *MYB71*, the bHLH transcription factor gene bHLH112, the WD40 protein genes *AIW1* (*ABA Induced WD40-repeat 1*) and *AIW2*, and heat shock factor gene *HSFA6b* also affected plant responses to abiotic stresses [27–31]. Considering that there are hundreds of thousands of ABA-responsive genes remained uncharacterized, it is very likely that some of the genes may be involved in the regulation of plant responses to abiotic stresses.

In an attempt to identify novel regulators of plant abiotic stress responses from unknown-function ABA response genes, we identified AITRs (ABA-induced transcription repressors), a novel family of transcription repressors in Arabidopsis as negative regulators of plant abiotic stress responses [12,13]. We found that the Arabidopsis *aitr* mutants showed enhanced tolerance to abiotic stresses including drought and salt [12], and knockout of the entire family of *AITRs* did not affect plant growth and development in Arabidopsis [13]. Considering that AITRs are conserved in angiosperms [12], *AITRs* may be good candidate genes for molecular breeding to improve abiotic stress tolerance in crops. Indeed, we found that AITRs in several different crops including tomato, rice, soybean and cotton shared similar features with Arabidopsis AITRs, i.e., their gene expression is induced by ABA, and they function as transcription repressors [12,32,33]. In addition, expression of cotton *GhAITR-A1* restored the enhanced abiotic stress tolerance in *aitr1* mutants [32], indicating that GhAITRs may also function as negative regulators of plant abiotic stress tolerance. Most importantly, knockout of several *GmAITR* genes by CRISPR/Cas9 gene editing moderately enhanced salt tolerance in soybean [33]. However, we also note that *gmaitr* mutants showed slightly enhanced ABA sensitivity, a phenotype opposite to that of the Arabidopsis *aitr* mutants, which showed greatly decreased ABA sensitivity [12,33]. These results suggested that AITRs in different plants may have different functions at least in some aspects.

Tobacco (*Nicotiana tabacum*) is an economically important crop throughout the world, which is usually grown for smoking, but its biomass can also be a good source of advanced biofuels [34,35]. Similar to other plants, tobacco's growth and development is also affected by environment stresses including abiotic stresses [36–41], and some studies have found that under abiotic stress conditions, numerous physiological and molecular changes in tobacco including accumulation of ABA were observed [39–41]. Studying the mechanisms of stress tolerances and mining regulatory genes involved in regulating abiotic stress responses, including the ABA response genes in tobacco, are essential for molecular breeding to improve abiotic stress tolerance in tobacco.

In this study, we provide evidence that tobacco AITRs (NtAITRs) are involved in the regulation of plant drought tolerance in tobacco. We found that the expression levels of at least some of the *NtAITRs* increased in response to ABA and drought treatment, and NtAITRs function as transcription repressors. By generating *ntaitr* mutants using CRISPR/Cas9 gene editing and examining their responses to ABA and abiotic stresses, we found that NtAITRs negatively regulate drought tolerance in tobacco via affecting the plant responses to ABA; therefore, *NtAITRs* may be targeted for CRISPR/Cas9 gene editing to improve drought tolerance in tobacco.

2. Results

2.1. Identification of AITRs in Tobacco

By using full-length amino acid sequences of Arabidopsis AITRs for *N. tabacum* Protein BLASTing on NCBI (https://blast.ncbi.nlm.nih.gov/Blast.cgi, accessed on 30 October 2018), we identified a total of eight AITRs in tobacco, i.e., NtAITR1 (LOC107760464), NtAITR2 (LOC107761277), NtAITR3 (LOC107767518), NtAITR4 (LOC107770126), NtAITR5 (LOC107791551), NtAITR6 (LOC107804873), NtAITR7 (LOC107815750) and NtAITR8 (LOC107829869). The coding sequences of the *NtAITRs* were obtained by using the locus numbers for gene searching on the KEGG GENES Database (https://www.genome.jp/kegg/genes.html, accessed on 30 October 2018), and corresponding genome sequences were then obtained by BLASTing the *N. tabacum* genome with the coding sequences on the Sol Genomics Network (https://solgenomics.net/organism/Nicotiana_tabacum/genome, accessed on 30 October 2018). We found that two of the *NtAITRs*, i.e., *NtAITR2* and *NtAITR7*, have two exons, and the other six *NtAITRs* have a single exon (Figure S1), whereas all the AITRs in Arabidopsis and soybean have only one exon [12,33].

Amino acid sequence alignment shows that NtAITRs shared high identity and similarity, especially between each pair of the proteins, i.e., NtAITR1 and NtAITR3, NtAITR2 and NtAITR7, NtAITR4 and NtAITR8, and NtAITR5 and NtAITR6, but all of them have only a partially conserved LxLxL motif at the C-terminus (Figure 1).

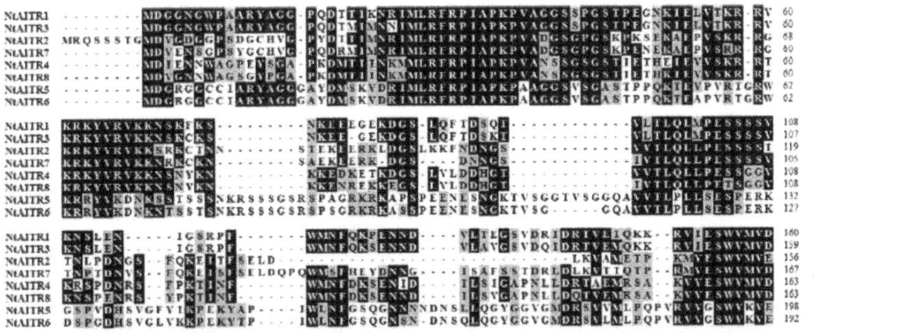

Figure 1. Amino acid sequence alignment of NtAITRs. Full-length amino acid sequences of the NtAITRs were obtained on NCBI (https://blast.ncbi.nlm.nih.gov/Blast.cgi, accessed on 30 October 2018) and used for multiple sequence alignment by using BioEdit. Identical amino acids are shaded in black and similar amino acids are shaded in gray. Underline indicates the partially conserved LxLxL transcriptional repression motif.

Protein domain assays show that there are five conserved domains in all the NtAITRs, including one with the partially conserved LxLxL motif (Figure S2). Phylogenic analysis of AITRs from tobacco, Arabidopsis and soybean shows that NtAITRs formed one clade with three Arabidopsis AITRs (AITR1, AITR3 and AITR4) and two soybean AITRs (GmAITR3 and GmAITR6), whereas the other three Arabidopsis AITRs (AITR2, AITR5 and AITR6) and the other four soybean AITRs (GmAITR1, GmAITR2, GmAITR4 and GmAITR5) formed another clade (Figure 2). In addition, the phylogenic analysis also shows that NtAITR5 and NtAITR6 are closely related, and the other six NtAITRs are closely related (Figure 2).

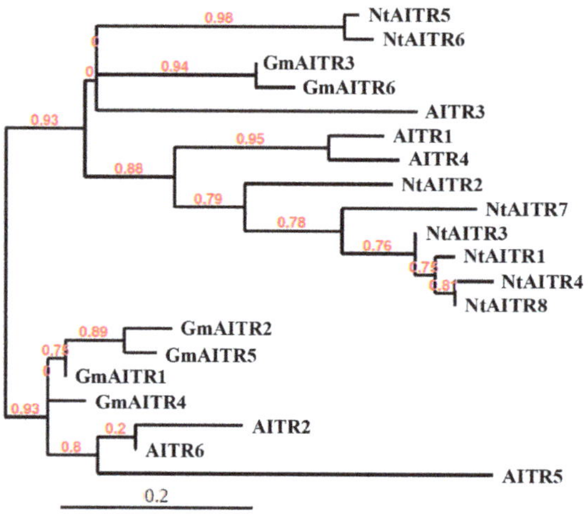

Figure 2. Phylogenetic analysis of NtAITRs, GmAITRs and Arabidopsis AITRs. Full-length amino acid sequences of GmAITRs and Arabidopsis AITRs were obtained from Phytozome (https://phytozome-next.jgi.doe.gov, accessed on 30 October 2018) and subjected to phylogenetic analysis with NtAITRs on phylogeny.fr (http://www.phylogeny.fr, accessed on 1 September 2022) by using the "One Click" mode with default settings. Branch support values are indicated above the branches.

2.2. Expression Levels of NtAITRs Are Increased in Response to ABA and Drought Treatments

Thus far, it has been shown that all the *AITRs* characterized from different plants including the model plant Arabidopsis, and crops including tomato, rice, soybean and cotton, are all inducible by ABA treatment, and all the AITRs function as transcription repressors [12,32,33]. To examine if that is also the case with *NtAITRs*, we first examined their expression in response to ABA treatment. Tobacco seedlings if the K326 wild type were treated with ABA for a short time, and the expression of NtAITRs was examined by RT-PCR. As shown in Figure 3A, it is clear that the expression levels of all the *NtAITRs* but *NtAITR2* were increased in the ABA-treated samples, as a very faint band was obtained for *NtAITR2* in the ABA-treated sample, which indicates that it may respond slightly to ABA treatment. We also examined the expression of *NtAITRs* in response to drought treatment by using PEG treatment to mimic drought condition, and we found that the expression levels of at least some *NtAITRs* including *NtAITR1*, *NtAITR2*, *NtAITR3* and *NtAITR5* were increased in response to drought treatment (Figure 3B).

We then examined the transcription activities of NtAITRs by using protoplast transfection assays. We successfully cloned full-length CDS for *NtAITR1*, *NtAITR3*, *NtAITR4*, *NtAITR6*, *NtAITR7* and *NtAITR8*, and generated constructs with a fused N-terminal GD tag. Plasmids of the *GD-NtAITR* constructs were co-transfected with plasmids of the transcription activator gene *LD-VP* and the *LexA-Gal4:GUS* reporter gene into protoplasts isolated from Col wild type Arabidopsis leaves. As shown in Figure 4, LD-VP-activated GUS expression was repressed when *GD-AITRs* were co-transfected, suggesting that all the NtAITRs examined function as transcription repressors.

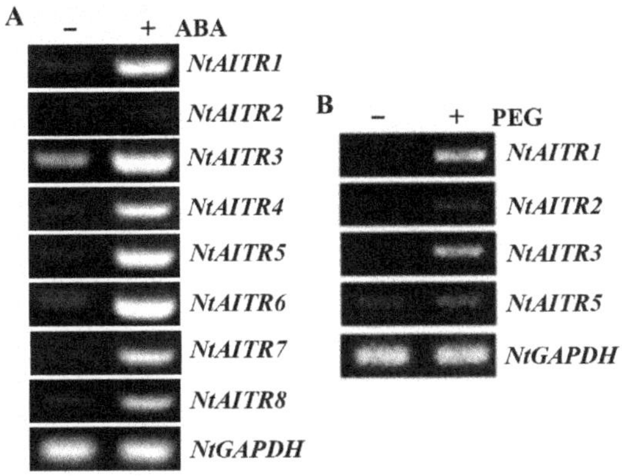

Figure 3. Expression of *NtAITRs* in response to ABA and drought treatments. (**A**) Expression of *NtAITRs* in response to ABA treatment. Ten-day-old K326 wild type tobacco seedlings were treated with 50 μM ABA for 4 h, then RNA was isolated and used for RT-PCR analysis to examine the expression of *NtAITRs*. The expression of *NtGAPDH* gene was used as a control. (**B**) Expression of *NtAITRs* in response to drought treatment. Twenty-day-old soil-grown K326 wild type seedlings were watered with 20% PEG6000, seedlings were collected 6 h after and then RNA was isolated and used for RT-PCR analysis to examine the expression of *NtAITRs*. The expression of *NtGAPDH* gene was used as a control.

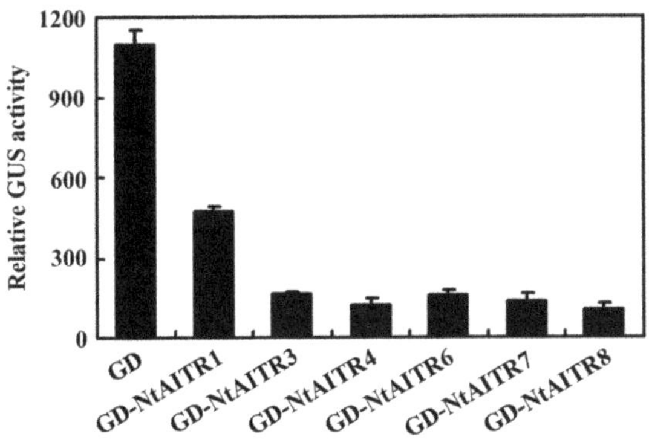

Figure 4. Transcription activities of NtAITRs. Plasmids of the *LexA-Gal4:GUS* reporter and the *LD-VP* activator were co-transfected with the *GD-NtAITR* effectors into protoplasts isolated from leaves of the Col wild type Arabidopsis. The transfected protoplasts were incubated in darkness for 20–22 h, and then GUS activity was assayed. Co-transfection of *GD* was used as a control. Data represent the mean ± SD of three replicates.

2.3. Generation of Transgene-Free *ntaitr* Mutants

The above results show that NtAITRs shared similar features with AITRs characterized from other plants including Arabidopsis, soybean, tomato, cotton and rice, i.e., their gene expression is induced by ABA, and they function as transcription repressors. These results indicate that NtAITRs may also be involved in the regulation of ABA and abiotic stress responses in tobacco. To examine if this is indeed the case, we decided to generate *ntaitr* mutants by using CRISPR/Cas9 gene editing.

Considering that *NtAITRs* shared high nucleotide sequence identity, especially between each pair of the genes, i.e., *NtAITR1* and *NtAITR3*, *NtAITR2* and *NtAITR7*, *NtAITR4* and *NtAITR8*, and *NtAITR5* and *NtAITR6* (Figure S3), we selected four different target sequences that may simultaneously target several different *NtAITRs*, including *NtAITR1*,

NtAITR2, *NtAITR3*, *NtAITR5* and *NtAITR6* (Figure 5A), and made an *FT* expression cassette containing the CRISPR/Cas9 construct with four sgRNAs (Figure 5B). The containing of the *FT* expression cassette enables the reduction in the time required to generate gene-edited mutants and facilitates the selection of Cas9-free mutants [42]. By using the construct to transform wild-type tobacco, we successfully obtained some transgenic plants with early flowering phenotypes. Gene editing status in these early flowering plants was examined, and transgene-free mutants were obtained in next generation based on flowering phenotype, then confirmed by PCR amplification of the *Cas9* fragment, and gene editing status was examined again.

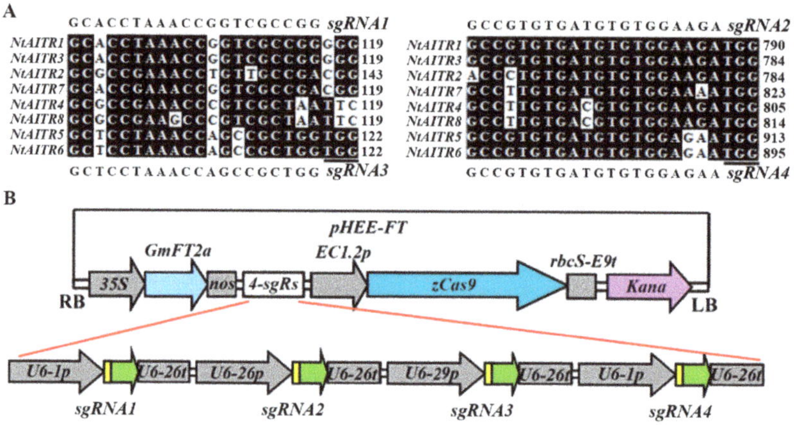

Figure 5. Generation of an *FT* expression cassette containing *pHEE* CRISPR/Cas9 construct for *NtAITRs* editing. (**A**) Target sequences inserted into the sgRNAs and alignment of the sequences may be targeted in *NtAITRs* by using Bioedit. Numbers indicate the nucleotide position relative to the first nucleotide in the coding sequence of *NtAITRs*, underlines indicate PAM sites. (**B**) Diagram of the *pHEE* vector with an *FT* expression cassette and four sgRNAs. Target sequences that can target *NtAITR1*, *NtAITR2*, *NtAITR3*, *NtAITR5* and *NtAITR6* were introduced into the sgRNA expression cassettes by PCR amplification, followed by Golden Gate reaction with the *pHEE-FT* vector.

Three transgene-free homozygous quintuple mutants, i.e., *ntaitr1 ntaitr2 ntaitr3 ntaitr5 ntaitr6-c1* (*ntaitr12356-c1*), *ntaitr12356-c2* and *ntaitr12356-c3* were obtained. In all the mutants, either a single-nucleotide insertion/deletion or the deletion of a few nucleotides occurred at the target sites of the *NtAITRs*, resulting in a frame shift after the edited sites of the corresponding NtAITRs (Figure 6A). It should be noted that *NtAITR2* in *ntaitr12356-c2* and *NtAITR3* in *ntaitr12356-c2* were edited in a bioallelic manner (Figure 6B), but either way resulted in a frame shift (Figure 6A); therefore, seeds collected from the mutants were directly used for the following experiments.

2.4. The ntaitr12356 Mutant Plants Show Enhanced Tolerance to Drought

Expression level changes of *AITRs* in both Arabidopsis and soybean affected plant response to abiotic stresses [12,13,33]. After showing that NtAITRs shared similar features with AITRs from other plants examined so far, we examined drought tolerance in the *ntaitr12356* quintuple mutants. The K326 wild type tobacco and the wild type-I were used as controls. The wild type-I is a non-edited plant segregated from the offspring of the T0 plant from which the *ntaitr12356-c3* quintuple mutant was isolated. Seeds of the wild type, the wild type-I and the *ntaitr12356* quintuple mutants were germinated in soil pots and grown in a growth room. Ten days after germination, the plants were watered thoroughly, and then watering was withheld for drought treatment. Twenty days after withholding watering, watering was resumed. As shown in Figure 7, the seedlings of the wild type, the wild type-I and the *ntaitr12356* quintuple mutants were largely similar before the drought treatment. All the plants continued growing during the treatment, but most of the leaves of the plants nearly dried after the drought treatment, with the three *ntaitr12356* quintuple mutants to a less severe degree (Figure 7). Five days after watering was resumed, at least four-fifths of each of the three *ntaitr12356* quintuple mutants survived, whereas at least four-fifths of the wild type and the wild type-I plants died, and no big difference was observed among the three *ntaitr* quadruple mutants or between the wild type and the wild type-I plants (Figure 7), indicating that drought tolerance was enhanced in the *ntaitr12356* quintuple mutants.

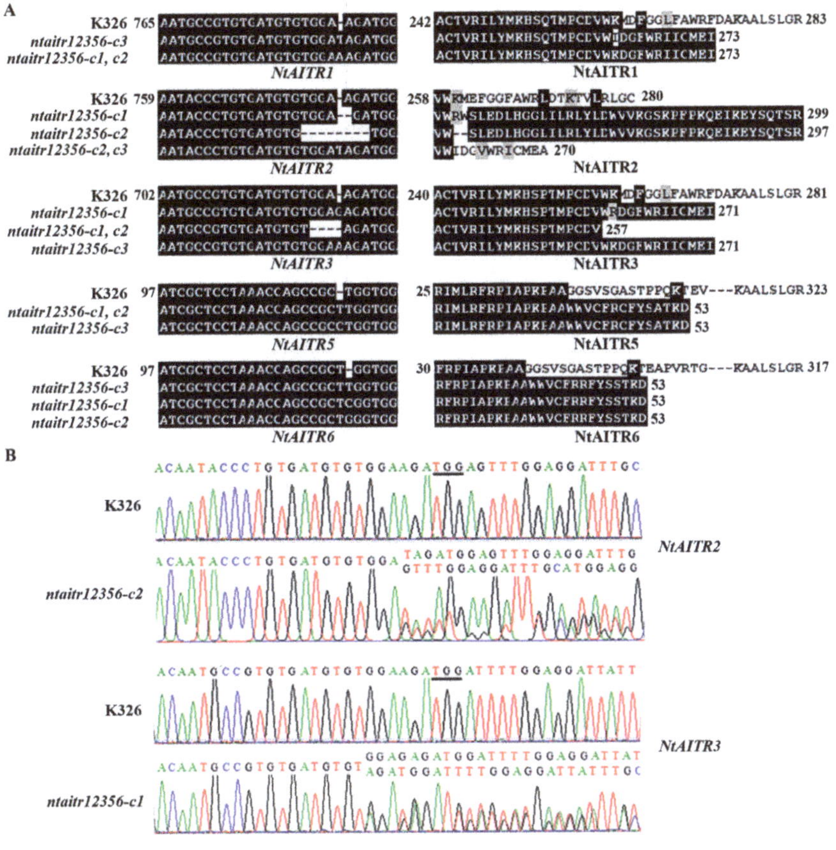

Figure 6. Generation of the *ntaitr12356* quintuple mutants. (**A**) Alignment of DNA sequences (**left**) and amino acid sequences (**right**) at the target sites of NtAITR1, NtAITR2, NtAITR3, NtAITR5 and NtAITR6, respectively, in the K326 wild type and the *ntaitr12356* quintuple mutants. The CRISPR/Cas9 construct was transformed into the K326 wild type tobacco, gene editing status was examined in the early flowering T0 plants and transgene-free homozygous mutants were isolated in T1 or T2 generation of the edited T0 plants. Numbers indicate the nucleotide position relative to the first nucleotide of *NtAITRs*, or the first amino acid of NtAITRs. (**B**) Editing status of *NtAITR2* in the *ntaitr12356-c2* mutant and *NtAITR3* in the *ntaitr12356-c1* mutant. Underlines indicate PAM sites.

2.5. The ntatir12356 Mutants Are Hypersensitive to ABA

After showing that drought tolerance was enhanced in *ntaitr12356* quintuple mutants, we further examined if this may be caused by alternated ABA response in the mutants. Seed germination and cotyledon greening assays were used to examine ABA sensitivity of the *ntaitr12356* quintuple mutants, and the K326 wild type tobacco and the wild type-I were used as controls. As shown in Figure 8A, in the seed germination assays, all three *ntaitr12356* quintuple mutants showed a clearly increased sensitivity to ABA compared to the wild type and the wild type-I. We also note that little if any difference on the germination rate was observed among the three *ntaitr12356* quintuple mutants, and the seed germination rate of the wild type-I was also largely indistinguishable from that of the wild type (Figure 8A).

Figure 7. Drought tolerance of the *ntaitr12356* quintuple mutants. Seeds of the K326 wild type, the wild type-I and the *ntaitr12356* mutants were germinated in soil trays and grown for 10 days before drought treatment. Drought treatment was performed by withholding watering for 20 days, and then watering was resumed. Pictures were taken before drought treatment, after drought treatment and 5 days after watering was resumed. Plants that survived were counted 5 days after watering was resumed, and survival rate was calculated. Data represent the mean ± SD of five replicates.

Clearly increased ABA sensitivity for all three *ntaitr12356* quintuple mutants was also observed in the seedling greening assays (Figure 8B). Quantitative assays show that in the presence of ABA, less than 20% of the seedlings of all the three *ntaitr12356* quintuple mutants produced green cotyledons, compared with ~60% for wild type and wild type-I seedlings (Figure 8C). Again, no difference was observed among the three *ntaitr12356* quintuple mutants, or between the wild type and the wild type-I (Figure 8C).

It has been shown that both the Arabidopsis AITRs and GmAITRs regulate ABA response via regulating the expression of some ABA signaling key regulator genes [12,13,33]. We thus examined if the expression levels of the ABA signaling key regulator genes may be changed in the *ntaitr12356* quintuple mutants. As the three *ntaitr12356* quintuple mutants showed a similar response to ABA in both the seed germination and seedling greening assays, we used only the *ntaitr12356-c3* mutant for the experiment. As shown in Figure 8D, the expression level of *NtPYL7*, an ABA receptor gene [43], was increased, whereas that of *NtPP2CA*, a PP2C phosphatase gene, was decreased in the *ntaitr12356-c3* mutant compared to the wild type seedlings.

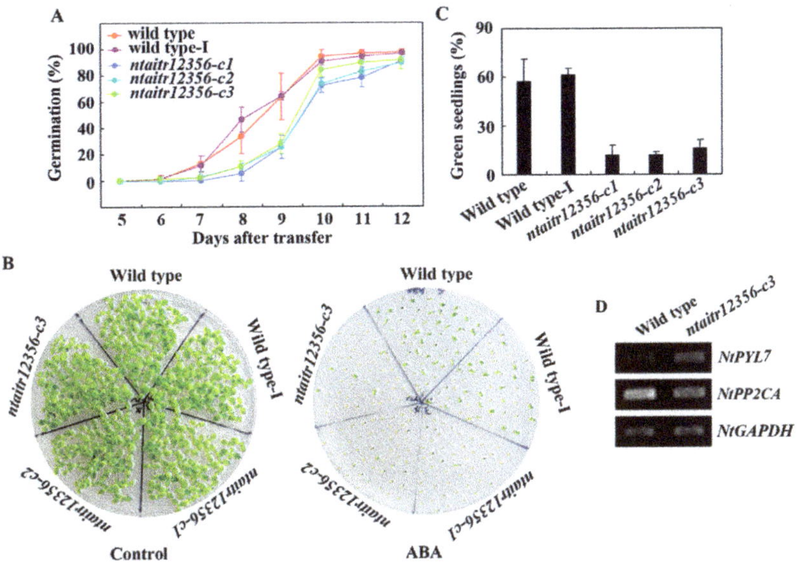

Figure 8. ABA sensitivity of the *ntaitr12356* quintuple mutants. (**A**) Seed germination of the K326 wild type, the wild type-I and the *ntaitr12356* quintuple mutants in response to ABA. Sterilized seeds the K326 wild type, the wild type-I and the *ntaitr12356* mutants were kept at 4 °C in darkness for 7 days and then sown on 1/2 MS plates in the presence or absence of 2 μM ABA. The plates were transferred to a growth room and seeds germinated were counted daily after the plates were transferred. Data represent the mean ± SD of three replicates. (**B**) Seedling greening of the K326 wild type, the wild type-I and the *ntaitr12356* quintuple mutants in response to ABA. Pictures were taken 12 days after the plates were transferred. (**C**) Percentage of green seedlings of the K326 wild type, wild type-I and the *ntaitr12356* quintuple mutants in response to ABA. Green seedlings were counted 12 days after the transfer, and the percentage of green seedlings was calculated. Data represent means ± SD of three replicates. (**D**) Expression of *NtPYL7* and *NtPP2CA* in the K326 wild type and the *ntaitr12356-c3* quintuple mutant seedlings in response to ABA. Ten-day-old seedlings were treated with 50 μM ABA or solvent methanol as a control for 4 h. Total RNA was isolated, and RT-PCR was used to examine the expression of the ABA signaling component genes. The expression of *NtGAPDH* was used as a control.

3. Discussion

ABA signaling via several key regulators activates/represses ABA responses genes, thereby regulating plant abiotic stress tolerance [6–13]. By characterizing unknown-function ABA response genes, we previously identified AITRs as a novel family of transcription repressors that negatively regulate abiotic stress tolerance in Arabidopsis [12,13]. As AITRs are conserved in angiosperms [12], we then characterized AITRs in several different crops to see if *AITRs* may be targeted for CRISPR/Cas9 gene-editing-based molecular breeding to improve abiotic stress tolerance in crops. Thus far, we found that *gmaitr* mutants generated by CRISPR/Cas9 gene editing of soybean *AITRs* showed moderately enhanced salt tolerance [33], but ABA sensitivity was slightly enhanced in the *gmaitr* mutants [33], opposite to that of the Arabidopsis *aitr* mutants [12], indicating that AITRs in different plants may have different functions, at least in some aspects.

We provide evidence in this study that NtAITRs shared similar features with AITRs characterized so far in other plants including Arabidopsis, tomato, cotton and soybean [12,32,33], i.e., *NtAITRs* are ABA response genes, as their expression levels were increased in response to ABA treatment (Figure 3), and NtAITRs function as transcription repressors, as the NtAITRs examined repressed the expression of the reporter gene in transfected protoplasts (Figure 4). It should be noted that due to the low expression level of *NtAITR2* and high nucleotide identity of the *NtAITRs* gene pairs (Figure S3), we failed to generated constructs for *NtAITR2* and *NtAITR5*, therefore transcription repression activities of NtAITR2 and NtAITR5 were not examined in this study. However, considering that NtAITRs shared high amino acid identity and similarity, and all have a partially conserved LxLxL motif (Figure 1), it is reasonable to assume that NtAITR2 and NtAITR5 can also function as transcription repressors, yet experimental evidence are required to examine if this is indeed the case.

Our evidence in this study also show that even though NtAITRs shared similar features with AITRs in other plants, a difference in regulating ABA sensitivity was observed, as examined with the *ntaitr12356* quintuple mutants generated by CRISPR/Cas9 gene editing (Figure 6). Greatly increased ABA sensitivity was observed in the *ntaitr12356* quintuple mutants as compared with the wild type tobacco plants (Figure 8), opposite to that observed in the Arabidopsis *aitr* mutants, which showed greatly decreased ABA sensitivity [12,13]. On the other hand, NtAITRs may have similar functions as GmAITRs in regulating ABA sensitivity, as slightly enhanced ABA sensitivity was observed in the *gmaitr* mutants [33]. There are six genes in Arabidopsis but eight in cotton encoding AITRs [12,33], and there are also eight in tobacco (Figure 1). Considering that the maximum decrease in ABA sensitivity in Arabidopsis *aitr* mutants was observed in *aitr256* triple mutants [13], and the *ntaitr* mutants used this study are *ntaitr12356* quintuple mutants (Figure 6), whereas increased ABA sensitivity observed the *gmaitr* mutants may be only caused by the loss of function of two *GmAITR* genes [33], it is very likely that high-order mutants of *GmAITRs* may also result in greatly increased ABA sensitivity in soybean, i.e., NtAITRs and GmAITRs may have similar functions in regulating ABA sensitivity.

Our results show that so far, all the AITRs examined in different plants negatively regulate abiotic stress tolerance, Arabidopsis *aitr236* triple and high-order mutants showed greatly enhanced tolerance to drought and salt [12,13], *gmaitr* mutants showed moderately enhanced tolerance to salt [33], and *ntaitr12356* quintuple mutants showed greatly enhanced tolerance to drought (Figure 7). These results provided evidence that *ATIRs* in crops indeed could be good targets for CRISPR/Cas9 gene editing to improve abiotic stress tolerance.

As an industrial crop with great economic importance, tobacco's yield and quality are affected by abiotic stresses [36,37,39–41]. Breeding of tobacco plants with enhanced abiotic stress tolerance will increase their yield and quality. As ABA affects plant responses to abiotic stresses via regulating ABA response genes [6–13], characterization of unknown-function ABA response genes may enable the identification of novel regulators of abiotic stress tolerance. Our evidence in this study indicates that NtAITRs are this kind of regulators. As the *ntaitr12356* quintuple mutants showed enhanced sensitivity to ABA (Figure 8), it is very likely that the functions of NtAITRs in regulating abiotic stress tolerance in tobacco are due to its feedback regulating roles of ABA responses. Yet, detailed mechanisms of NtAITRs in regulating abiotic stress tolerance in tobacco need to be further investigated.

Most importantly, our evidence in this study clearly indicates that it is not only possible but also practical to used *NtAITRs* as targets for CRISPR/Cas9 editing to improve abiotic stress tolerance in tobacco. First, coding sequences of *NtAITRs* shared high nucleotide identity (Figure S3), therefore it will be easier to identify target sequences that can target at least a pair of *NtAITR* genes. As an example, we identified four target sequences that can be used to target five *NtAITR* genes (Figure 5). Second, it is not difficult to generate transgene-free gene-edited mutants for *NtAITRs*, which may avoid the difficulty of using transgenic plants in production. Even though not all the target sites were edited, we obtained mutants with five *NtAITR* genes being edited (Figure 6), and we successfully isolated transgene-free mutants by using flowering phenotypes caused by the *FT* expression cassette in the *CRISPR/Cas9* construct (Figure 5). Third, greatly enhanced drought tolerance was observed in mutants with only some of the *NtAITR* genes being edited. The eight NtAITR proteins shared high amino acid sequence identity as well as conserved domains (Figures 1 and S2), suggesting that they may have redundant functions. However, greatly enhanced drought tolerance was already observed in the *ntaitr12356* quintuple mutants (Figure 7), indicating that it may be not necessary to edit all the *NtAITRs* to obtain tobacco plants with enhance abiotic stress tolerance. Yet, it will be of great interest to examine if further enhanced abiotic tolerance can be achieved if all the *NtAITRs* are being edited and if the loss of function of all *NtAITRs* may have any fitness cost.

In summary, our results in this study show that the expression of *NtAITRs* is induced by ABA and that NtAITRs function as transcription repressors and they positively regulate ABA sensitivity but negatively regulate drought tolerance in tobacco. As a result, *NtAITRs* can be used as targets for CRISPR/Cas9 gene-editing-based molecular breeding to improve abiotic stress tolerance in tobacco.

4. Materials and Methods

4.1. Identification of AITRs in Tobacco

To identify NtAITRs in tobacco, the entire amino acid sequences of Arabidopsis AITRs were used for Protein BLASTing on NCBI (https://blast.ncbi.nlm.nih.gov/Blast.cgi, accessed on 30 October 2018) with *N. tabacum* as selected organism. The full-length amino acid sequences of the retrieved NtAITRs were then used in reiterative BLASTing until no more NtAITRs were identifiable. The coding sequences of the *NtAITRs* were obtained by using the locus numbers to conduct gene searching on the KEGG GENES Database (https://www.genome.jp/kegg/genes.html, accessed on 30 October 2018), and corresponding genome sequences were obtained by BLASTing the *N. tabacum* genome with the coding sequences on the Sol Genomics Network (https://solgenomics.net/organism/Nicotiana_tabacum/genome, accessed on 30 October 2018). Full-length amino acid sequences of obtained NtAITRs were subjected to phylogenetic analysis (http://hylogeny.fr/phylo_cgi/index.cgi, accessed on 1 September 2022) with AITRs from Arabidopsis and soybean [12,33], and motif analysis was performed using MEME with default settings [44].

4.2. Plant Materials and Growth Conditions

The K326 tobacco was used as wild type for plant transformation and as control for ABA and drought treatment and drought tolerance assays. The *ntaitr12356* quintuple mutants were generated by CRISPR/Cas9 gene editing. The non-edited plant wild type-I was segregated from the offspring of the T0 plant from which the *ntaitr12356-c3* quintuple mutant was isolated and was also used as a control for ABA and drought treatment. The Col wild type Arabidopsis was used for protoplast isolation. The tobacco seedlings used for RNA isolation and ABA treatments were grown on $\frac{1}{2}$ MS plates, and tobacco plants used for plant transformation, drought treatment and drought tolerance assays were grown in soil pots at 22–24 °C in a growth room with a 16/8 h (light/dark) photoperiod, as described previously [45].

4.3. ABA and Drought Treatment, RNA Isolation and RT-PCR

To examine the expression of *NtAITRs* and ABA signaling component genes in response to ABA, 10-day-old tobacco seedlings were treated with 50 μM ABA for 4 h in darkness on a shaker at 40 rpm and frozen in liquid N_2 immediately for RNA isolation. To examine the expression of *NtAITRs* in response to drought treatment, 20-day-old soil-grown K326 wild type tobacco seedlings were watered with 20% PFG6000 and seedlings were collected 6 h after and frozen in liquid N_2 immediately for RNA isolation.

Total RNA was isolated by using a Plant RNA kit (OMEGA BIO-TEK, Norcross, GA, USA) following the manufacturer's instructions, and 1–2 μg total RNA was subjected to first-strand cDNA synthesis by using an EasyScript First-strand DNA Synthesis Super Mix (TransGen Biotech, Beijing, China). RT-PCR analysis was performed by using the synthesized cDNA as templates. The expression of the *NtGAPDH* gene was used as an internal reference control [46]. The primers used were listed in Table S1.

4.4. Constructs

The reporter construct of *LexA-Gal4:GUS* and the effector constructs *GD* and *LD-VP* were as described previously [47,48].

To generate *35S:GD-NtAITRs* constructs for protoplast transfection assays, the full-length open reading frame (ORF) of the *NtAITRs* was amplified by RT-PCR using RNA isolated from tobacco seedlings and cloned in frame with an N-terminal GD tag into the *pUC19* vector with the *35S* promoter [47,48].

In order to generate CRISPR/Cas9 constructs for *NtAITRs* gene editing, a total of four different target sequences within the exons of *NtAITRs* that may be simultaneously target five genes were selected on CRISPRscan (http://www.crisprscan.org/?page=sequence, accessed on 30 October 2018), and target specificity was evaluated on cas-offinder (http://www.rgenome.net/cas-offinder, accessed on 30 October 2018). The four target sequences are the following: 5′-GCACCTAAACCGGTCGCCGG-3′ for *NtAITR1* and *NtAITR3*, 5′-GCTCCTAAACCAGCCGCTGG-3′ for *NtAITR5* and *NtAITR6*, 5′-GCCGTGTGATGTGTGGAAGA-3′ for *NtAITR1*, *NtAITR2* and *NtAITR3*, and 5′-GCCGTGTGATGTGTGGAGAA-3′ for *NtAITR5* and *NtAITR6*. The four targets were inserted in to the *pHEE* vector with an *FT* expression cassette as reported previously [42]. The primers used were listed in Table S1.

4.5. Plasmid Isolation, Protoplast Isolation and Transient Transfection

Plasmid isolation, protoplast isolation and transfection, and GUS activity assays were performed by following the procedure described previously [12,48–50]. Briefly, plasmid DNA of the reporter and effector constructs was isolated by using an EndoFree Plasmid Maxi Kit (OMEGA BIO-TEK). Protoplasts were isolated from leaves collected from ~3-week-old Col wild-type Arabidopsis and co-transfected with the *LexA-Gal4:GUS* reporter, the *LD-VP* and the *GD-NtAITRs* effector plasmid DNA. The transfected protoplasts were incubated for 20–22 h under darkness at room temperature. GUS activities were measured by using a SynergyTM HT fluorescence microplate reader (BioTEK, Winooski, VT, USA).

4.6. Generation of Transgene-Free ntaitr Mutants

Five- to six-week-old K326 tobacco plants were used for Agrobacterium *GV3101*-mediated stable transformation by infiltration, as described previously [45]. T0 transgenic seedlings generated were transferred into soil pots and grown in a growth room. The plants with early flowering phenotypes were selected, and gene editing status was detected by PCR amplification and sequencing of the DNA sequence fragments of *NtAITR1*, *NtAITR2*, *NtAITR3*, *NtAITR5* and *NtAITR6*. Transgene-free homozygous mutant plants were isolated in T1 or T2 progeny based on flowering phenotype and sequencing, as previously described [42,51].

4.7. Seed Germination and Seedling Greening Assays

Seed germination and cotyledon greening assays were used to examine ABA sensitivity of the *ntaitr12356* mutants obtained, and the K326 wild type and the wild type-I tobacco were used as controls. Healthy and uniform-sized seeds of the K326 wild type, the wild type-I and the *ntaitr12356* mutants were surface sterilized and germinated on 1/2 MS plates in the presence or absence of 2 μM ABA. Germinated seeds were counted daily after the transfer. Pictures were taken 14 days after the transfer, and seedlings with green cotyledons were counted.

4.8. Drought Tolerance Assays

Drought tolerance assays of soil-grown tobacco were performed by using the procedures described previously [52–54] with some modification. Seeds of the K326 wild type, the wild type I and the *ntaitr12356* mutants were sown in soil trays and grown for 10 days with sufficient watering. The plants were then subjected to drought treatment by withholding watering for 20 days, and watering was resumed. Pictures were taken before drought treatment, after drought treatment and 5 days after watering was resumed.

Supplementary Materials: The following are available online at https://www.mdpi.com/article/10.3390/ijms232315268/s1.

Author Contributions: Conceptualization, S.W.; investigation, G.L., Y.M., X.W. (Xiaoping Wang), N.C., D.M., S.C., W.W., X.W. (Xutong Wang), X.H. and L.Y.; data curation, G.L., Y.M. and S.W.; writing—original draft preparation, S.W. and G.L.; writing—review and editing, G.L. and S.W.; project administration, S.W.; funding acquisition, G.L. and S.W. All authors have read and agreed to the published version of the manuscript.

Funding: This work was supported by Natural Science Foundation of Shandong Province, grant number ZR2020MC089, National Natural Science Foundation of China, grant number 32071938, and a startup funding from Linyi University, grant number LYDX2019BS039.

Institutional Review Board Statement: Not applicable.

Informed Consent Statement: Not applicable.

Data Availability Statement: All data obtained were presented in this article and the supporting information.

Acknowledgments: We thank all our lab members for their helpful discussion and suggestions.

Conflicts of Interest: The authors declare no conflict of interest.

References

1. Wang, W.; Vinocur, B.; Altman, A. Plant responses to drought, salinity and extreme temperatures: Towards genetic engineering for stress tolerance. *Planta* **2003**, *218*, 1–14. [CrossRef] [PubMed]
2. Fujita, M.; Fujita, Y.; Noutoshi, Y.; Takahashi, F.; Narusaka, Y.; Yamaguchi-Shinozaki, K.; Shinozaki, K. Crosstalk between abiotic and biotic stress responses: A current view from the points of convergence in the stress signaling networks. *Curr. Opin. Plant Biol.* **2006**, *9*, 436–442. [CrossRef] [PubMed]
3. Boyer, J.S. Plant productivity and environment. *Science* **1982**, *218*, 443–448. [CrossRef] [PubMed]
4. Ghosh, D.; Xu, J. Abiotic stress responses in plant roots: A proteomics perspective. *Front. Plant Sci.* **2014**, *5*, 6. [CrossRef] [PubMed]
5. Yamaguchi-Shinozaki, K.; Shinozaki, K. Transcriptional regulatory networks in cellular responses and tolerance to dehydration and cold stresses. *Annu. Rev. Plant Biol.* **2006**, *57*, 781–803. [CrossRef]
6. Rodriguez, P.L.; Leube, M.P.; Grill, E. Molecular cloning in Arabidopsis thaliana of a new protein phosphatase 2C (PP2C) with homology to ABI1 and ABI2. *Plant Mol. Biol.* **1998**, *38*, 879–883. [CrossRef]
7. Gosti, F.; Beaudoin, N.; Serizet, C.; Webb, A.A.; Vartanian, N.; Giraudat, J. ABI1 protein phosphatase 2C is a negative regulator of abscisic acid signaling. *Plant Cell* **1999**, *11*, 1897–1910. [CrossRef]
8. Fujii, H.; Verslues, P.E.; Zhu, J.K. Identification of two protein kinases required for abscisic acid regulation of seed germination, root growth, and gene expression in Arabidopsis. *Plant Cell* **2007**, *19*, 485–494. [CrossRef]
9. Umezawa, T.; Nakashima, K.; Miyakawa, T.; Kuromori, T.; Tanokura, M.; Shinozaki, K.; Yamaguchi-Shinozaki, K. Molecular basis of the core regulatory network in ABA responses: Sensing, signaling and transport. *Plant Cell Physiol.* **2010**, *51*, 1821–1839. [CrossRef]
10. Guo, J.; Yang, X.; Weston, D.J.; Chen, J.G. Abscisic acid receptors: Past, present and future. *J. Integr. Plant Biol.* **2011**, *53*, 469–479. [CrossRef]
11. Dong, T.; Park, Y.; Hwang, I. Abscisic acid: Biosynthesis, inactivation, homoeostasis and signalling. *Essays Biochem.* **2015**, *58*, 29–48.
12. Tian, H.; Chen, S.; Yang, W.; Wang, T.; Zheng, K.; Wang, Y.; Cheng, Y.; Zhang, N.; Liu, S.; Li, D.; et al. A novel family of transcription factors conserved in angiosperms is required for ABA signalling. *Plant Cell Environ.* **2017**, *40*, 2958–2971. [CrossRef] [PubMed]
13. Chen, S.; Zhang, N.; Zhou, G.; Hussain, S.; Ahmed, S.; Tian, H.; Wang, S. Knockout of the entire family of AITR genes in Arabidopsis leads to enhanced drought and salinity tolerance without fitness costs. *BMC Plant Boil.* **2021**, *21*, 137. [CrossRef] [PubMed]
14. Fujita, Y.; Nakashima, K.; Yoshida, T.; Katagiri, T.; Kidokoro, S.; Kanamori, N.; Umezawa, T.; Fujita, M.; Maruyama, K.; Ishiyama, K.; et al. Three SnRK2 protein kinases are the main positive regulators of abscisic acid signaling in response to water stress in Arabidopsis. *Plant Cell Physiol.* **2009**, *50*, 2123–2132. [CrossRef] [PubMed]
15. Park, S.Y.; Peterson, F.C.; Mosquna, A.; Yao, J.; Volkman, B.F.; Cutler, S.R. Agrochemical control of plant water use using engineered abscisic acid receptors. *Nature* **2015**, *520*, 545–548. [CrossRef]

16. Yoshida, T.; Fujita, Y.; Maruyama, K.; Mogami, J.; Todaka, D.; Shinozaki, K.; Yamaguchi-Shinozaki, K. Four Arabidopsis AREB/ABF transcription factors function predominantly in gene expression downstream of SnRK2 kinases in abscisic acid signaling in response to osmotic stress. *Plant Cell Environ.* **2015**, *38*, 35–49. [CrossRef]
17. Zhao, Y.; Chan, Z.; Gao, J.; Xing, L.; Cao, M.; Yu, C.; Hu, Y.; You, J.; Shi, H.; Zhu, Y.; et al. ABA receptor PYL9 promotes drought resistance and leaf senescence. *Proc. Natl. Acad. Sci. USA* **2016**, *113*, 1949–1954. [CrossRef]
18. Bueso, E.; Rodriguez, L.; Lorenzo-Orts, L.; Gonzalez-Guzman, M.; Sayas, E.; Muñoz-Bertomeu, J.; Ibañez, C.; Serrano, R.; Rodriguez, P.L. The single-subunit RING-type E3 ubiquitin ligase RSL1 targets PYL4 and PYR1 ABA receptors in plasma membrane to modulate abscisic acid signaling. *Plant J.* **2014**, *80*, 1057–1071. [CrossRef]
19. Seo, K.I.; Lee, J.H.; Nezames, C.D.; Zhong, S.; Song, E.; Byun, M.O.; Deng, X.W. ABD1 is an Arabidopsis DCAF substrate receptor for CUL4-DDB1-based E3 ligases that acts as a negative regulator of abscisic acid signaling. *Plant Cell* **2014**, *26*, 695–711. [CrossRef]
20. Yu, F.; Lou, L.; Tian, M.; Li, Q.; Ding, Y.; Cao, X.; Wu, Y.; Belda-Palazon, B.; Rodriguez, P.L.; Yang, S.; et al. ESCRT-I component VPS23A affects ABA signaling by recognizing ABA receptors for endosomal degradation. *Mol. Plant.* **2016**, *9*, 1570–1582. [CrossRef]
21. Belda-Palazon, B.; Rodriguez, L.; Fernandez, M.A.; Castillo, M.C.; Anderson, E.A.; Gao, C.; Gonzalez-Guzman, M.; Peirats-Llobet, M.; Zhao, Q.; De Winne, N.; et al. FYVE1/FREE1 interacts with the PYL4 ABA receptor and mediates its delivery to the vacuolar degradation pathway. *Plant Cell* **2016**, *28*, 2291–2311. [CrossRef] [PubMed]
22. Stone, S.L.; Williams, L.A.; Farmer, L.M.; Vierstra, R.D.; Callis, J. KEEP ON GOING, a RING E3 ligase essential for Arabidopsis growth and development, is involved in abscisic acid signaling. *Plant Cell* **2006**, *18*, 3415–3428. [CrossRef] [PubMed]
23. Chen, Y.T.; Liu, H.X.; Stone, S.; Callis, J. ABA and the ubiquitin E3 ligase KEEP ON GOING affect proteolysis of the Arabidopsis thaliana transcription factors ABF1 and ABF3. *Plant J.* **2013**, *75*, 965–976. [CrossRef]
24. Liu, H.; Stone, S.L. Abscisic acid increases Arabidopsis ABI5 transcription factor levels by promoting KEG E3 ligase self-ubiquitination and proteasomal degradation. *Plant Cell* **2010**, *22*, 2630–2641. [CrossRef] [PubMed]
25. Liu, H.; Stone, S.L. Cytoplasmic degradation of the Arabidopsis transcription factor abscisic acid insensitive 5 is mediated by the RING-type E3 ligase KEEP ON GOING. *J. Biol. Chem.* **2013**, *288*, 20267–20279. [CrossRef] [PubMed]
26. Kong, L.; Cheng, J.; Zhu, Y.; Ding, Y.; Meng, J.; Chen, Z.; Xie, Q.; Guo, Y.; Li, J.; Yang, S.; et al. Degradation of the ABA co-receptor ABI1 by PUB12/13 U-box E3 ligases. *Nat. Commun.* **2015**, *6*, 8630. [CrossRef]
27. Jung, C.; Seo, J.S.; Han, S.W.; Koo, Y.J.; Kim, C.H.; Song, S.I.; Nahm, B.H.; Choi, Y.D.; Cheong, J.J. Overexpression of AtMYB44 enhances stomatal closure to confer abiotic stress tolerance in transgenic Arabidopsis. *Plant Physiol.* **2008**, *146*, 623–635. [CrossRef]
28. Liu, S.; Hu, Q.; Luo, S.; Li, Q.; Yang, X.; Wang, X.; Wang, S. Expression of wild-type PtrIAA14.1, a poplar Aux/IAA gene causes morphological changes in Arabidopsis. *Front. Plant Sci.* **2015**, *6*, 388. [CrossRef]
29. Huang, Y.; Feng, C.Z.; Ye, Q.; Wu, W.H.; Chen, Y.F. Arabidopsis WRKY6 transcription factor acts as a positive regulator of abscisic acid signaling during seed germination and early seedling development. *PLoS Genetics* **2016**, *12*, e1005833. [CrossRef] [PubMed]
30. Wang, X.; Wang, W.; Wang, Y.; Zhou, G.; Liu, S.; Li, D.; Adnan; Hussain, S.; Ahmed, S.; Zhang, C.; et al. AIW1 and AIW2, two ABA-induced WD40 repeat-containing transcription repressors function redundantly to regulate ABA and salt responses in Arabidopsis. *J. Plant Interact.* **2020**, *15*, 196–206. [CrossRef]
31. Cheng, Y.; Ma, Y.; Zhang, N.; Lin, R.; Yuan, Y.; Tian, H.; Hussain, S.; Chen, S.; Yang, W.; Li, Y.; et al. The R2R3 MYB transcription factor MYB71 regulates abscisic acid response in Arabidopsis. *Plants* **2022**, *11*, 1369. [CrossRef] [PubMed]
32. Wang, T.; Dong, Q.; Wang, W.; Chen, S.; Cheng, Y.; Tian, H.; Li, X.; Hussain, S.; Wang, L.; Gong, L.; et al. Evolution of AITR family genes in cotton and their functions in abiotic stress tolerance. *Plant Biol.* **2021**, *23* (Suppl. 1), 58–68. [CrossRef] [PubMed]
33. Wang, T.; Xun, H.; Wang, W.; Ding, X.; Tian, H.; Hussain, S.; Dong, Q.; Li, Y.; Cheng, Y.; Wang, C.; et al. Mutation of *GmAITR* genes by CRISPR/Cas9 genome editing results in enhanced salinity stress tolerance in soybean. *Front. Plant Sci.* **2021**, *12*, 779598. [CrossRef] [PubMed]
34. Andrianov, V.; Borisjuk, N.; Pogrebnyak, N.; Brinker, A.; Dixon, J.; Spitsin, S.; Flynn, J.; Matyszczuk, P.; Andryszak, K.; Laurelli, M.; et al. Tobacco as a production platform for biofuel: Overexpression of Arabidopsis DGAT and LEC2 genes increases accumulation and shifts the composition of lipids in green biomass. *Plant Biotechnol. J.* **2010**, *8*, 277–287. [CrossRef] [PubMed]
35. Barla, F.G.; Kumar, S. Tobacco biomass as a source of advanced biofuels. *Biofuels* **2016**, *10*, 335–346. [CrossRef]
36. de las Mercedes Dana, M.; Pintor-Toro, J.A.; Cubero, B. Transgenic tobacco plants overexpressing chitinases of fungal origin show enhanced resistance to biotic and abiotic stress agents. *Plant Physiol.* **2006**, *142*, 722–730. [CrossRef]
37. Wei, W.; Zhang, Y.; Han, L.; Guan, Z.; Chai, T. A novel WRKY transcriptional factor from *Thlaspi caerulescens* negatively regulates the osmotic stress tolerance of transgenic tobacco. *Plant Cell Rep.* **2008**, *27*, 795–803. [CrossRef]
38. Yang, C.; Wang, R.; Gou, L.; Si, Y.; Guan, Q. Overexpression of *Populus trichocarpa* mitogen-activated protein kinase kinase4 enhances salt tolerance in tobacco. *Int. J. Mol. Sci.* **2017**, *18*, 2090. [CrossRef]
39. Gao, Y.; Yang, J.; Duan, W.; Ma, X.; Qu, L.; Xu, Z.; Yang, Y.; Xu, J. NtRAV4 negatively regulates drought tolerance in *Nicotiana tabacum* by enhancing antioxidant capacity and defence system. *Plant Cell Rep.* **2022**, *41*, 1775–1788. [CrossRef]
40. Sun, H.; Sun, X.; Wang, H.; Ma, X. Advances in salt tolerance molecular mechanism in tobacco plants. *Hereditas* **2020**, *157*, 5. [CrossRef]
41. Xie, H.; Bai, G.; Lu, P.; Li, H.; Fei, M.; Xiao, B.; Chen, X.; Tong, Z.; Wang, Z.; Yang, D. Exogenous citric acid enhances drought tolerance in tobacco (*Nicotiana tabacum*). *Plant Biol.* **2022**, *24*, 333–343. [CrossRef] [PubMed]

42. Cheng, Y.; Zhang, N.; Hussain, S.; Ahmed, S.; Wang, S. Integration of a FT expression cassette into CRISPR/Cas9 construct enables fast generation and easy identification of transgene-free mutants in Arabidopsis. *PLoS ONE* **2019**, *14*, e0218583. [CrossRef] [PubMed]
43. Bai, G.; Xie, H.; Yao, H.; Li, F.; Chen, X.; Zhang, Y.; Xiao, B.; Yang, J.; Li, Y.; Yang, D.H. Genome-wide identification and characterization of ABA receptor PYL/RCAR gene family reveals evolution and roles in drought stress in Nicotiana tabacum. *BMC Genomics* **2019**, *20*, 575. [CrossRef] [PubMed]
44. Bailey, T.L.; Boden, M.; Buske, F.A.; Frith, M.; Grant, C.E.; Clementi, L.; Ren, J.; Li, W.W.; Noble, W.S. MEME SUITE: Tools for motif discovery and searching. *Nucleic Acids Res.* **2009**, *37*, W202–W208. [CrossRef] [PubMed]
45. Sparkes, I.A.; Runions, J.; Kearns, A.; Hawes, C. Rapid, transient expression of fluorescent fusion proteins in tobacco plants and generation of stably transformed plants. *Nat. Protoc.* **2006**, *1*, 2019–2025. [CrossRef] [PubMed]
46. Wang, Z.; Wang, S.; Wu, M.; Li, Z.; Liu, P.; Li, F.; Chen, Q.; Yang, A.; Yang, J. Evolutionary and functional analyses of the 2-oxoglutarate-dependent dioxygenase genes involved in the flavonoid biosynthesis pathway in tobacco. *Planta* **2019**, *249*, 543–561. [CrossRef]
47. Tiwari, S.B.; Hagen, G.; Guilfoyle, T.J. Aux/IAA proteins contain a potent transcriptional repression domain. *Plant Cell* **2004**, *16*, 533–543. [CrossRef]
48. Wang, S.; Tiwari, S.B.; Hagen, G.; Guilfoyle, T.J. Auxin response factor 7 restores the expression of auxin-responsive genes in mutant Arabidopsis leaf mesophyll protoplasts. *Plant Cell* **2005**, *17*, 1979–1993. [CrossRef]
49. Wang, X.; Wang, X.; Hu, Q.; Dai, X.; Tian, H.; Zheng, K.; Wang, X.; Mao, T.; Chen, J.G.; Wang, S. Characterization of an activation-tagged mutant uncovers a role of GLABRA2 in anthocyanin biosynthesis in Arabidopsis. *Plant J.* **2015**, *83*, 300–311. [CrossRef]
50. Dai, X.; Zhou, L.; Zhang, W.; Cai, L.; Guo, H.; Tian, H.; Schiefelbein, J.; Wang, S. A single amino acid substitution in the R3 domain of GLABRA1 leads to inhibition of trichome formation in Arabidopsis without affecting its interaction with GLABRA3. *Plant Cell Environ.* **2016**, *39*, 897–907. [CrossRef] [PubMed]
51. Liu, Y.; Zeng, J.; Yuan, C.; Guo, Y.; Yu, H.; Li, Y.; Huang, C. Cas9-PF, an early flowering and visual selection marker system, enhances the frequency of editing event occurrence and expedites the isolation of genome-edited and transgene-free plants. *Plant Biotechnol. J.* **2019**, *17*, 1191–1193. [CrossRef] [PubMed]
52. Chen, Y.; Han, Y.; Zhang, M.; Zhou, S.; Kong, X.; Wang, W.; Min, Y.Z. Overexpression of the wheat expansin gene *TaEXPA2* improved seed production and drought tolerance in transgenic tobacco plants. *PLoS ONE* **2016**, *11*, e0153494. [CrossRef] [PubMed]
53. Xia, Z.; Xu, Z.; Wei, Y.; Wang, M. Overexpression of the maize sulfite oxidase increases sulfate and GSH Levels and enhances drought tolerance in transgenic tobacco. *Front. Plant Sci.* **2018**, *9*, 298. [CrossRef] [PubMed]
54. Li, X.; Wang, Q.; Guo, C.; Sun, J.; Li, Z.; Wang, Y.; Yang, A.; Pu, W.; Guo, Y.; Gao, J.; et al. NtNAC053, a novel NAC transcription factor, confers drought and salt tolerances in tobacco. *Front. Plant Sci.* **2022**, *13*, 817106. [CrossRef]

International Journal of
Molecular Sciences

MDPI

Review

Biological Roles of Lipids in Rice

Kun Zhou [1,†], Zhengliang Luo [1,†], Weidong Huang [1], Zemin Liu [1], Xuexue Miao [1], Shuhua Tao [1], Jiemin Wang [1], Jian Zhang [2,*], Shiyi Wang [2] and Xiaoshan Zeng [1,*]

[1] Hunan Rice Research Institute, Hunan Academy of Agricultural Sciences, Changsha 410125, China; zhzhkp@163.com (K.Z.); luozl@hunaas.cn (Z.L.); hwd4@163.com (W.H.); dhlzmin@163.com (Z.L.); miaoxuexue410@163.com (X.M.); taoshuhua2020@163.com (S.T.); wangjiemin2009@163.com (J.W.)

[2] State Key Lab of Rice Biology and Breeding, China National Rice Research Institute, Hangzhou 311400, China; 82101225170@caas.cn

* Correspondence: zhangjian@caas.cn (J.Z.); zengxiaoshan2@163.com (X.Z.)

† These authors contributed equally to this work.

Abstract: Lipids are organic nonpolar molecules with essential biological and economic importance. While the genetic pathways and regulatory networks of lipid biosynthesis and metabolism have been extensively studied and thoroughly reviewed in oil crops such as soybeans, less attention has been paid to the biological roles of lipids in rice, a staple food for the global population and a model species for plant molecular biology research, leaving a considerable knowledge gap in the biological roles of lipids. In this review, we endeavor to furnish a current overview of the advancements in understanding the genetic foundations and physiological functions of lipids, including triacylglycerol, fatty acids, and very-long-chain fatty acids. We aim to summarize the key genes in lipid biosynthesis, metabolism, and transcriptional regulation underpinning rice's developmental and growth processes, biotic stress responses, abiotic stress responses, fertility, seed longevity, and recent efforts in rice oil genetic improvement.

Keywords: rice (*Oryza sativa* L.); lipids; genetic improvement; grain quality; pollen fertility; seed germination; seed longevity; biotic stress; abiotic stress; yield

1. Introduction

Lipids are organic nonpolar molecules containing hydrogen, carbon, and oxygen atoms, which provide the framework for the structure and function of living cells. As a major type of lipid, triglycerides are stored as fat in adipose cells, serving as the high-density energy-storage depot for organisms and providing thermal insulation. Some lipids, like steroid hormones, act as chemical messengers between cells, tissues, and organs [1]. In terms of cellular function, lipids are integral components of cell membranes, contributing to their structure, flexibility, and permeability. They are also involved in cell signaling and the regulation of membrane-bound enzymes [2]. Plant-derived or vegetable oils are of significant importance to human society due to their versatile and wide-ranging applications. These oils are crucial dietary fats for human consumption and are extensively utilized in culinary practices and food production. They also serve as essential raw materials for the production of biofuels, offering sustainable alternatives to traditional fossil fuels. In the cosmetic and personal care industry, plant-derived oils are prized for their emollient and moisturizing properties, making them valuable ingredients in skincare products, soaps, and hair care formulations. Furthermore, these oils play a vital role in industrial processes, contributing to manufacturing paints, lubricants, and various commercial products [3]. Their diverse applications highlight the pivotal role of plant-derived oils in both consumer products and industrial processes.

Given the biological and economic importance of plant-derived lipids, the genetic pathways and regulatory networks of lipid biosynthesis and metabolism have been extensively studied and thoroughly reviewed, particularly in oil crops such as soybeans, rapeseed, and sunflowers [4–6]. Rice is a staple food for more than half of the world's population, particularly in Asia. The lipids in rice grains comprise triglycerides (TAGs), phospholipids, fatty acids, and other bioactive lipid components, with a total amount ranging from 2 to 3%, much lower than oil crops [7]. The lipids are concentrated in rice embryos and brans, essentially determining milling, appearance, and cooking quality [7]. In addition, lipids are the essential building blocks of cellular membranes and act as signal molecules to modulate various biological processes [2]. Despite the full recognition of the significant roles of lipids in rice, less attention has been paid to lipids when compared to proteins and nucleic acids in the rice research community, leaving a considerable knowledge gap in the biological roles of lipids [8]. Fortunately, recent molecular genetic research has uncovered a number of rice genes involved in lipid biosynthesis, metabolism, and signaling pathways underpinning various biological processes. This review endeavors to furnish a current overview of the advancements in understanding the genetic foundation and physiological functions of lipids in the context of rice's developmental and growth processes, biotic stress responses, abiotic stress responses, and fertility, as well as seed longevity (Figure 1; Table 1).

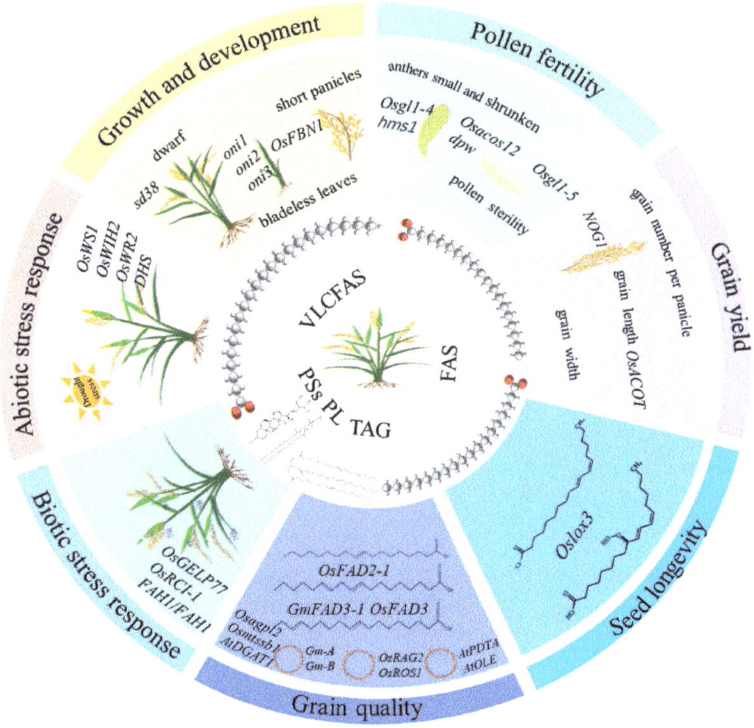

Figure 1. Significant roles of lipids in rice. FAS: fatty acids; PSs: phytosterols; PL: Phospholipid; TAG: triacylglycerol; VLCFAS: very-long-chain fatty acids.

Table 1. A summary of the lipid-related genes and their functions in rice.

Gene	LOC ID	Annotation	Phenotype Description	References
WSL1	*LOC_Os06g39750*	β–ketoacyl CoA synthase	The mutants had reduced growth, leaf fusion, sparse wax crystals, and low fertility.	[9]
ONI1	*LOC_Os03g08360*	β–ketoacyl CoA synthase	The mutants produced small shoots and aberrant outer epidermal cells and ceased to grow after germination.	[10]
ONI2	*LOC_Os10g28060*	β–ketoacyl CoA synthase	The mutants had tiny shoots, fused leaves and ceased growing after germination.	[11]
ONI3	*LOC_Os09g19930*	Long-chain fatty acid ω-alcohol dehydrogenase	The mutants had severe growth defects with fused neighboring organs and finally became lethal.	[12,13]
OsFBN1	*LOC_Os09g04790*	Probable plastid-lipid-associated protein	Over-expression of *OsFBN1* exhibited more tillers, short panicles, poor grain-filling percentage, and JA levels under heat stress.	[14]
SD38	*LOC_Os10g33370*	β–ketoacyl CoA synthase	The mutants' cell length, 1000-grain weight, and seed setting rate were significantly reduced.	[15]
OsGL1-5	*LOC_Os10g33250*	Glossy1-homologous gene	The mutants showed pollen sterility and significant defects in the biosynthesis of VLCFAs.	[16]
OsACOS12	*LOC_Os04g24530*	Acyl-CoA synthetase	The mutants had normal vegetative growth but white anthers that were shorter and smaller and produced no mature pollen grains	[17,18]
DPW	*LOC_Os03g07140*	Fatty Acyl Carrier Protein Reductase	The mutants display defected anther and degenerated pollen grains with an irregular exine	[19]
PEM1	*LOC_Os02g09920*	Methyl-CpG-binding domain family member	*pem1* anthers became small and shrunken, with 30% lower viable pollen grains	[20]
HMS1	*LOC_Os03g12030*	β-ketoacyl-CoA synthase	The mutants that exhibited male sterility at low humidity but were fully fertile at high humidity	[21]
OsGL1-4	*LOC_Os02g40784*	Glossy1-homologous gene	The mutants significantly reduced the content of VLC alkanes (C25 and C27), finally leading to sterile pollen grains	[22]
NOG1	*LOC_Os01g54860*	Enoyl-CoA hydratase/isomerase	*NOG1* can significantly increase the grain yield of commercial high-yield varieties.	[23]
OsACOT	*LOC_Os04g35590*	Acyl-CoA thioesterase	Over-expression of the miR1432-resistant form of *OsACOT* resulted in an increased yield of approximately 50%.	[24]
AtWRI1	*AT3G54320*	AP2/EREBP transcription factor	Over-expression of *AtWRI1* in rice increased fatty acid content by 30–40% in vegetative organs but decreased fatty acid content in the endosperm.	[25]

Table 1. *Cont.*

Gene	LOC ID	Annotation	Phenotype Description	References
RAG2	*LOC_Os07g11380*	α-amylase/trypsin inhibitor	Over-expression of *RAG2* gave rise to a 10–30% increase in lipid content and higher storage protein levels.	[26]
OsAGPL2	*LOC_Os01g44220*	ADP-glucose pyrophosphorylase large subunit	The oil content of mutant grains reached over 10%.	[27]
Gm-A *Gm-B*	*U09118* *U09119*	24 kDa oleosin	Transgenic rice showed greater numbers of smaller oil bodies, and lipid content increased by 36.93% and 46.06%, respectively	[28]
OsROS1	*LOC_Os01g11900*	DNA demethylase	The number of cell layers in the anthers increased significantly, while the oil content of the rice increased by 48%, and the vitamin E content doubled	[29]
AtPDAT *AtDGAT1* *AtWRI1* *AtOLE*	*AT5G13640* *AT2G19450* *AT3G54320* *AT4G25140*	Phospholipid:diacylglycerol acyltransferase Diacylglycerol acyltransferase AP2/EREBP transcription factor Oleosin1	Increased TAG, oleic acid, palmitic acid, and total oil contents by 26%, 28%, 27%, and 70%	[30]
OsAGPL2 *OsMTSSB1* *AtDGAT1*	***LOC_Os01g44220*** ***LOC_Os05g43440*** ***AT2G19450***	ADP-glucose pyrophosphorylase large subunit Mitochondrion-targeted single-stranded DNA-binding protein Diacylglycerol acyltransferase	The lipid content increased from approximately 2% to around 12%,	[31]
OsFAD2-1	*LOC_Os02g48560*	ω-6 fatty acid desaturase	Mutant showed an increase in oleic acid and a reduction in linoleic and palmitic acids in the grains	[32,33]
GmFAD3-1 *OsFAD3*	*GLYMA_14G194300v4* *LOC_Os12g01370*	microsomal omega-3-fatty acid desaturase ω-3 fatty acid desaturase gene	GmFAD3-1 and OsFAD3 over-expression lines consistently increased 23.8- and 27.9-fold across multiple generations.	[34]
OsLOX3	*LOC_Os03g49260*	lipoxygenase gene	Suppressing *LOX3* expression in rice endosperm successfully extended rice seed longevity	[35]
OsWS1	*LOC_Os04g40590*	Membrane-bound O-acyl transferase gene	*OsWS1* over-expression lines displayed 35% higher VLCFA contents, denser wax papillae around the stoma, and a higher survival rate upon water-deficit treatment.	[36]
OsWIH2	*LOC_Os03g31679*	cysteine-rich and transmembrane domain-containing protein WIH2	Over-expression of *OsWIH2* positively regulate rice drought resistance by alleviating water loss, reducing reactive oxygen species (ROS) accumulation, and altering wax content	[37]
DHS	*LOC_Os02g45780*	RING-type E3 ubiquitin ligase	Over-expression of DHS significantly reduced the cuticular wax contents and conferred drought-hypersensitive.	[38]

Table 1. *Cont.*

Gene	LOC ID	Annotation	Phenotype Description	References
OsWR2	*LOC_Os06g40150*	Ethylene response factor	Over-expression of *OsWR2* increased the total cuticular wax in leaves and panicles, decreased water loss, and enhanced drought resistance.	[39]
OsFAH1 ***OsFAH2***	*LOC_Os12g43363* *LOC_Os03g56820*	fatty acid 2-hydroxylase	*fah1fah2* harbors fewer microdomains and higher susceptibility to rice blast fungus infection	[40]
OsGELP77	*LOC4340180*	GDSL esterase/lipase gene	Elevated expression of *OsGELP77* or pyramiding of a natural elite haplotype significantly increased resistance to various pathogens	[41]
OsLOX11/OsRCI-1	*LOC_Os12g37260*	lipoxygenase gene	Over-expression of *OsRCI-1* elevated the levels of JA, jasmonate-isoleucine, and trypsin protease inhibitors, which decreased colonization, fecundity, and mass of the BHP insects	[42]

2. Roles of Lipids in Rice Growth and Development

In very-long-chain fatty acids (VLCFAs) biosynthesis, β-ketoacyl CoA synthase (E.C.2.3.1.119, KCS) is the key enzyme. Knock-out of WSL1 encoding a typical KCS protein resulted in pleiotropic phenotypes, including reduced growth, leaf fusion, sparse wax crystals, and low fertility. The observed phenotype might be attributed to the significant reduction of VLCFA precursors of C20–C24 and total cuticular wax load on wsl1 leaf blades and sheaths [9].

From a series of rice onion mutants showing small cabbage-like shoots or bladeless leaves, three lipid-related genes involved in the shoot apical meristem were fine-mapped [43]. As a fatty acid elongase (β-ketoacyl CoA synthase), ONI1 biochemically synthesizes VLCFAs in the outermost epidermal cell layer [10]. Similarly, ONI2 catalyzes the first step of elongation reactions of a carbon chain of VLCFAs. Disruption of ONI2 led to tiny shoots in which leaves were fused and ceased growing after germination [11]. ONI3 was shown to be the ortholog of *Arabidopsis* HOTHEAD, containing glucose-methanol-choline (GMC) oxidoreductase and NAD(P)-binding Rossmann-like domains. It is functionally involved in the biosynthesis of long-chain fatty acids. Compared to *oni1* and *oni2*, *oni3/mini1* showed more severe growth defects with fused neighboring organs and finally became lethal after germination [12,13].

Fibrillins (FBNs) are a conserved plastid-lipid-associated protein (PAP) family modulating lipid metabolism. In rice, Fibrillin 1 (OsFBN1), a chloroplast-localized protein, was found to specifically bind C18 and C20 fatty acids in vitro. Over-expression lines of OsFBN1 exhibited more tillers, short panicles, poor grain-filling percentage, and JA levels than wild-type and RNAi- silencing lines under heat stress. More interestingly, more plastoglobules, which are defined as a hub of lipid metabolism in the chloroplast [14], were observed in the over-expression lines, suggesting the essential role of rice *OsFBN1* in plastoglobule formation and plant growth [44].

Zhang et al. (2022) identified a rice semi-dwarf mutant, *semi-dwarf 38* (*sd38*), with significantly reduced cell length. They revealed that *SD38* encodes a fatty acid elongase responsible for the synthesis of C24:0 VLCFAs (VLCFAs). Exogenous application of VLCFA (C24:0) or ethephon could partially reverse the dwarf phenotype of *sd38* [15].

3. Lipids Control Rice Pollen Fertility

The outer layer of the anther, called the anther cuticle, and the pollen wall's outer layer, known as pollen exine, are essential lipid layers for the development of the male reproductive organ [45]. The anther cuticle comprises cutin and cuticle wax, while the exine is composed of sporopollenin, a highly resistant biopolymer derived from fatty acids [46]. It is well established that defective lipid synthesis in the cuticle or exine leads to nonviable pollen and sterile plants. To date, numerous genes related to lipid metabolism that cause male sterility have been identified, and their functions have been elucidated in rice and other plant species [45].

Wax-deficient anther1 (*Wda1*), designated as *OsGL1-5*, was cloned from a T-DNA insertional mutant showing pollen sterility and significant defects in the biosynthesis of VLCFAs in both layers. It was observed that epicuticular wax crystals were absent in the outer layer of the anther, and pollen exine formation was compromised in the mutant anthers [16]. Besides *Wda1*, rice acyl-CoA synthetase5 (OsACOS5) is also implicated in pollen exine development and fertility [17,18].

In plant cuticles, fatty alcohols synthesized by fatty acyl-CoA reductase are important fatty constituents [46]. DPW (Defective Pollen Wall) has been identified as a novel fatty acid reductase mediating the production of 1-hexadecanol. DPW and its *Arabidopsis* ortholog MS2 are functionally conserved, as their corresponding mutants display defected anther and degenerated pollen grains with an irregular exine [19,47]. In *dpw* anthers, cutin monomers were dramatically reduced with an altered composition of cuticular wax, soluble fatty acids, and alcohols. The work showcased the link between primary fatty alcohol synthesis, anther cuticle, and pollen sporopollenin biosynthesis in monocots and dicots [19].

GDSL esterases and lipases, featured by the conserved motif Gly-Asp-Ser-Leu, are a subfamily of hydrolytic/lipolytic enzymes [48]. Zhao et al. (2020) map-based cloned an endoplasmic reticulum-localized GDSL lipase gene *RMS2* from an irradiation-induced mutant population. *rms2* exhibited complete male sterility, while the vegetative growth was normal. Biochemically, *RMS2* possesses lipid hydrolase activity, while the dysfunction of *RMS2* led to significant changes in the content of 16 lipid components and numerous other metabolites. Interestingly, *RMS2* might be a key node in the rice male fertility regulatory network, since master male fertility regulators Undeveloped Tapetum1 (UDT1) and Persistent Tapetal Cell1 (PTC1) could activate its transcription [49].

A recent study revealed that lipid metabolism might be subject to epigenetic regulation [20]. *PEM1* (*pollen expressed MBD-like 1*) encoding a methyl-CpG-binding domain protein is involved in the pollen exine development. pem1 anthers became small and shrunken with 30% lower viable pollen grains when compared to the WT. Unlike many of the other male sterile mutants showing retarded tapetum degradation, *pem1* mutants had regular PCD progress of the tapetum. However, they exhibited abnormal Ubisch body formation, delayed exine occurrence, defective exines, and increased anther cuticles [20]. Additionally, a novel alpha integrin-like protein DPW3 (Defective Pollen Wall3) has been associated with pollen wall formation and fertility, although the detailed molecular mechanism requires further exploration [50].

Humidity-sensitive genic male sterility (HGMS) lines have great potential in hybrid rice breeding. However, HGMS-related genes have yet to be identified. Chen et al. (2020) isolated a rice HGMS mutant that exhibited male sterility at low humidity but was fully fertile at high humidity [21]. The causal gene *HUMIDITY-SENSITIVE GENIC MALE STERILITY 1* (*HMS1*) encodes a putative β-ketoacyl-CoA synthase. Through interacting with its co-factor HMS1-INTERACTING PROTEIN (HMS1I), HMS1 promotes the conversion of C24:0 and C26:0 fatty acids into C26:0 and C28:0 fatty acids on the pollen wall, which helps to protect pollen from dehydration under low-humidity conditions. In a yeast system, HMS1 and HMS1I were found to catalyze the biosynthesis of VLCFAs longer than C24. Notably, the authors developed an HGMS line using the *HMS1* gene with convertible pollen fertility under different humidity conditions, suggesting that *HMS1* can be potentially used for hybrid breeding in indica and japonica rice.

Unlike other reported *OsGLs* involved in drought resistance, *OsGL1-4/CER1* was identified as a critical regulator in male reproductive development. Ni et al. (2018) revealed that *OsGL1-4/CER1* functions in VLC alkanes biosynthesis, underpinning anther development and plastids differentiation. *OsGL1-4/CER1* transcription was robustly detected in the developed tapetum and bicellular pollen cells. Meanwhile, disruption of *OsGL1-4/CER1* significantly reduced the content of VLC alkanes (C25 and C27), finally leading to sterile pollen grains containing fewer amyloplasts [22]. A few years later, two independent groups both found that the pollen sterility of cer1 is attributed to the excessively fast dehydration at anthesis and defective adhesion and hydration under normal conditions, which could be recovered by artificial high humidity [51,52]. Ni et al. (2021) further specified that the lipid composition of tryphine is the primary reason for such humidity-sensitive genic male sterility, since the fertility could be recovered by co-pollination with mixed *OsCER1*Cas mutant and maize pollens [52].

4. Lipids Involves in the Regulation of Grain Yield

Enoyl-CoA hydratase (ECH) catalyzes the second step in fatty acid metabolism's physiologically important beta-oxidation pathway [53]. A Chinese research group found that *NUMBER OF GRAINS 1* (*NOG1*) encoding an enoyl-CoA hydratase/isomerase regulates grain number per panicle without penalty on the panicle number and grain weight. A 12-bp indel in the promoter affected the transcription of *NOG1*, which finally contributed to the variations in total fatty acids and linolenic acid (LA, C18:3) among different varieties. In addition to the fatty acids, *NOG1* is involved in the biosynthesis of JA, since C18:3 is the synthetic precursor of JA. Introgression of *NOG1* in modern cultivars achieved up to a 25.8% increase in grain yield, showing the massive potential of lipid-related genes in crop genetic improvement [23].

Zhao et al. (2019) identified a seed-predominantly expressed microRNA, miR1432. The suppression of miR1432 facilitated grain filling rate and could increase grain yield up to 17.14% in field trials. In their search for the downstream targets of miR1432, the authors discovered that rice Acyl-CoA thioesterase (*OsACOT*) is a genuine target gene negatively regulated by miR1432-directed cleavage. *OsACOT* serves as a crucial enzyme in the fatty acid desaturation and elongation pathway, particularly in the conversion of the fatty acid 16:0 into 18:2. Intriguingly, over-expression of miR1432-resistant form of *OsACOT* mimicked miR1432 suppression lines and resulted in an increased yield of approximately 50% [24].

5. Grain Quality Is Affected by Lipid Contents and Components

Although rice grains possess a relatively low lipid content, the lipids in rice bran and grains exhibit a balanced composition of saturated (e.g., C16:0) and unsaturated (e.g., C18:1 and C18:2) fatty acids, which presents rice as a valuable and healthful oil source for human consumption [54]. Over the past few decades, substantial efforts have been dedicated to investigating the genetic underpinnings of rice lipid metabolism, with a particular emphasis on fatty acids. Employing a classic QTL mapping strategy, three QTLs, qRFC-1, qRFC-2, and qRFC-5, that accounted for 44% of the fat content variation were mapped in a doubled haploid population [55]. Later, two other groups did similar work, identifying nearly 20 QTLs contributing to grain lipid content. Nevertheless, no candidate genes underlying the QTLs were uncovered [56,57]. Ying et al. (2012) mapped 29 fatty acid content-associated QTLs, including eight QTLs that were repeatedly detected across multiple years and 11 QTLs harboring orthologs of key *Arabidopsis* lipid metabolism enzymes. Interestingly, a strong QTL for oleic (18:1) and linoleic (18:2) acids was associated with a rice ortholog of a gene encoding acyl-CoA: diacylglycerol acyltransferase (DGAT) and another for palmitic acid (16:0) mapped similarly to the acyl-ACP thioesterase (*FatB*) gene ortholog [58]. In 2021, a genome-wide association study was conducted in a diverse panel of 533 cultivated rice accessions and identified 46 loci controlling oil composition and concentration. Genes involved in lipid metabolism, namely *PAL6*, *LIN6*, *MYR2*, and

ARA6, are significant contributors to the natural variance in oil composition within rice subpopulations. Notably, *qPAL6* exhibits a substantial effect, explaining 49.86% and 57.44% of the phenotypic variance in the C16:0 composition. Similarly, qLIN6 accounts for 28.56% and 55.31% of the phenotypic variance in C18:2 conversion from C18:1. Furthermore, *qMYR2* encodes a myristoyl-ACP thioesterase responsible for the hydrolysis of specific acyl-ACPs, leading to the production of saturated fatty acids. Disruption of *qMYR2* results in lowered C14:0 and C16:0 but increased C18:2. Additionally, *qARA6*, annotated as a 3-ketoacyl-CoA synthase, modulates the biosynthesis of very-long-chain fatty acids (C20–C22). Functional impairment of *qARA6* significantly reduces the C20:0, C20:1, C20:2, and C22:0 components [59].

The economic significance of rice oil has led to the implementation of various strategies aimed at enhancing the lipid content and healthy fatty acid components through genetic means. One key strategy involves boosting lipid synthesis efficiency by utilizing lipid rate-limiting enzymes or regulator genes from rice or other plant species. WRINKLED1 (WRI1) is a master regulator of seed oil biosynthesis in *Arabidopsis*. Over-expression of *AtWRI1* in rice increased fatty acid content by 30–40% in vegetative organs but decreased fatty acid content in the endosperm, whereas the endosperm-specific expression of *CoWRI1* from coconut showed minor effects on rice oil content [25,60]. Over-expression of *RAG2*, a member of 14-to-16-kDa α-amylase/trypsin inhibitors in rice, gave rise to a 10–30% increase in lipid content and higher storage protein levels [26].

However, given that the overall carbon flux to starch is dominant over lipid biosynthesis, it is not surprising that minor effects could be observed when only lipid biosynthesis is enhanced [31]. As suggested by a report on *Chlamydomonas reinhardtii*, in which lipid content increased by 3.5-fold when an essential starch synthesis gene was knocked out [61], blocking the starch biosynthesis might be an efficient strategy to divert the carbon flux to lipid biosynthesis. Indeed, Wei et al. (2017) knocked out an ADP-glucose pyrophosphorylase (*AGP*) large subunit gene to have the oil content in rice grains over 10% [27].

The protein oleosin is notably abundant in the oil bodies of plant seeds, and it serves a crucial function in the accumulation of lipids. By introducing two soybean oleosin genes that encode 24 kDa proteins in rice, under the control of an embryo-specific rice promoter *REG-2*, the transgenic rice showed more oil bodies in smaller size and a significant increase in seed lipid content. Specifically, the lipid content increased by 36.93% and 46.06% compared to the control, while the fatty acid profiles of triacylglycerols remained unchanged [28]. In addition, modifying *OsROS1* and *OsMitssb1* significantly increased the thickness of rice aleurone layer cells, which could finally lead to an over-48% increase in grain oil content, representing an alternative strategy in rice lipid metabolic engineering [29,62].

In pursuit of achieving a balanced carbon flux between carbohydrates and lipids in rice, Izadi-Darbandi et al. (2020) undertook the metabolic engineering of rice grains by integrating *Arabidopsis* genes such as *AtWRI1*, *AtOle*, diacylglycerol acyltransferase (*AtDGAT*), and phospholipid:diacylglycerol acyltransferase (*AtPDAT*) to facilitate the final step of TAG assembly. This intervention substantially increased TAG, oleic acid, palmitic acid, and total oil contents by 26%, 28%, 27%, and 70%, respectively [30]. In 2023, Liu et al. published a significant case of multigene engineering aimed at augmenting lipid content in rice grains. The goal was achieved by the concurrent knock-out of two rice genes, *AGPL2* and *Mtssb1*, and the endosperm-specific expression of *DGAT1* from *Arabidopsis*. These modifications effectively redirected carbon flux from starch to lipid, resulting in a substantial increase in oil biosynthesis, elevating the lipid content from approximately 2% to around 12%. Meanwhile, the fatty acid compositions were significantly altered with more beneficial components such as GLA. A lipid content of 12% is very close to oil crops like soybean, which represented the highest lipid level ever reported in rice or other starchy-type grains [31]. This work highlighted a practical approach for genetically improving oil contents in rice and other crops with starchy grains.

Rice oil comprises approximately 18% palmitic acid, 36% oleic acid, and 37% linoleic acid. Linoleic acid is particularly susceptible to non-enzymatic oxidation due to its oxidative instability [63]. The high oleic acid content in rice has consistently been a desirable characteristic due to its enhanced stability and beneficial nutritional profile. It is well known that the microsomal omega-6 fatty acid desaturase (FAD2) catalyzes the conversion of oleic acid to linoleic acid in seed oil. In *Arabidopsis* and other cops, knock-out or knock-down of *FAD2* successfully produced high oleic and low linoleic oil seeds [64,65]. In rice, the suppression of *FAD2-1*, one of the transcripts of *FAD2* predominantly expressed in seeds, through RNA interference (RNAi), has been observed to lead to an increase in oleic acid and a reduction in linoleic and palmitic acids in the grains. This finding suggests that *OsFAD2-1* represents a viable target for enhancing the genetic profile of fatty acids in rice [32,33]. In an attempt to enhance the α-linolenic acid (ALA) content and to reduce the ratio of linoleic acid (LA) to ALA, Liu et al. (2012) systematically introduced six ω-3 (Δ-15) fatty acid desaturase (FAD) genes from rice and soybean into the rice. The ALA contents in the seeds of *GmFAD3-1* and *OsFAD3* over-expression lines exhibited a consistent increase of 23.8- and 27.9-fold across multiple generations. Comparative analysis revealed that the endosperm-specific promoters GluC and REG outperformed the constitutive maize Ubi-1 promoter in enhancing ALA levels in rice embryos and bran [34,66].

6. Lipids Regulate Seed Longevity in Rice

The longevity and viability of seeds are closely associated with the presence of lipid peroxidation products, specifically malondialdehyde and acetaldehyde. These products can lead to cell damage and intoxication by reacting with macromolecules. Linolenic acid (LNA) and LA are the most significant polyunsaturated fatty acids. The peroxidation of LNA and LA is typically linked to the breakdown of cell structure, the formation of cytotoxic products, and the release of volatile decomposition products, all of which contribute to the deterioration of rice seed longevity. Lipoxygenases, a conserved family of fatty acid dioxygenases, facilitate the addition of oxygen to polyunsaturated fatty acids such as LNA and LA. Among the three embryo-derived LOX isoenzymes, LOX3 is at a dominant level and has been linked to fatty acid peroxidation during seed storage and stored grain quality [67]. Xu et al. (2015) found that LOX3 is capable of producing 9-hydroperoxyoctadecadienoic acid (9-HPOD) using LA as substrate. Meanwhile, suppressing *LOX3* expression in rice endosperm successfully extended rice seed longevity under either artificial aging or natural aging conditions, whereas the major agronomic traits remain unchanged [35]. In addition, the knock-down of *LOX3* could also effectively protect the functional component β-carotene from deterioration in the carotenoid-enriched golden rice [68].

7. Lipids Related to Abiotic Stress Response in Rice

Cuticular wax on the leaf aerial surfaces is the outermost diffusion barrier against the uncontrolled loss or uptake of water and gases [69]. To most plants, cuticular waxes are complex mixtures of primarily VLCFAs, hydrocarbons, alcohols, flavonoids, and so on [46], though the proportions of the major constituents may vary from each other. Because the defective accumulation or altered composition of wax could be visually detected, screening of cuticular wax mutants like eceriferum (cer) in *Arabidopsis* and Glossy1 in maize has been carried out for decades [46,70]. In rice, Islam et al. (2009) systematically identified 11 homologs of maize GL1, designated as *OsGL1-1* to *OsGL1-11*. According to phylogenetic analysis, most of the OsGL1s could be induced by abiotic stresses and might be involved in cuticular or epicuticular wax biosynthesis [71]. At least, *osgl1-1*, *osgl1-2*, *osgl1-3*, and *osgl1-6* have been found to participate in drought stress response in rice. Compared to the WT, the *osgl1-2* exhibited less wax crystallization, a thinner cuticular layer, and reduced total cuticular wax amount, leading to higher sensitivity to drought stress at the reproductive stage. In contrast, the over-expression lines showed the opposite phenotypes [71]. *OsGL1-1* was annotated as a sterol desaturase or short-chain dehydrogenase/reductase. *OsGL1-3* protein has a conserved fatty acid hydroxylase domain (FAH domain) and a WAX2 C-terminal domain, while *OsGL1-6* gene is a member of the fatty aldehyde decarbonylase gene family. The *osgl1-1*, *osgl1-3* and *osgl1-6* mutants showed very similar phenotypes to *osgl1-2*, with decreased cuticular wax deposition, thinner cuticular membranes, and hypersensitivity to drought stress [72–74].

It was found that *osa-miR1848* regulates *OsWS1*, a membrane-bound O-acyl transferase gene member, to determine wax biosynthesis [36]. *OsWS1* and *osa-miR1848* exhibited a negative time-and-spatial co-expression pattern, especially under water-deficit treatment, suggesting antagonistic roles of the two factors. Compared to the control, *OsWS1* over-expression lines displayed 35% higher VLCFA contents, denser wax papillae around the stoma, and a higher survival rate upon water-deficit treatment. Conversely, *OsWS1*-RNAi and *osa-miR1848* over-expression plants exhibited opposing changes.

OsWIH2 is another reported fatty acid synthesis enzyme that positively regulates rice drought resistance by alleviating water loss and reactive oxygen species (ROS) accumulation and altering wax content [37]. *OsWIH2* works as a complex with HOTHEAD (HTH), which is an α-alcohol dehydrogenase catalyzing the biosynthesis of long-chain α-,ω-dicarboxylic fatty acids (LCFAs) in the cutin and wax biosynthesis pathway [75]. Meanwhile, drought-inducible transcription factor OsbHLH130 could activate the transcription of *OsWIH2*, thus forming a bHLH130-OsWIH2-HTH regulatory module in cuticular wax biosynthesis and drought response.

Besides the wax or cuticle synthesis enzymes, several genes involved in the regulation of wax biosynthesis and drought tolerance have been identified. These include the transcription factor *CFL1*, *OsWR2*, and *DWA1* [39,76–78]. Wang et al. (2018) reported that DROUGHT HYPERSENSITIVE (DHS) negatively regulates wax biosynthesis by promoting the turnover of an HD-ZIP IV family member ROC4, thereby influencing drought tolerance [38]. Over-expression of DHS significantly reduced the cuticular wax contents and conferred drought hypersensitivity. As a RING-type E3 ligase, DHS functions in mediating the ubiquitination of ROC4 in vivo, subsequently leading to its degradation via the ubiquitin/26S proteasome-mediated pathway. Notably, the DHS-ROC4 module has been identified as directly targeting Os-BDG, a putative regulator of wax biosynthesis. Thus, this study elucidates a precisely regulated molecular mechanism that links wax biosynthesis with drought stress response in rice, offering valuable insights for developing drought-tolerant rice cultivars.

Xiang et al. (2022) identified *Salt Tolerance and Heading Date 1* (*STH1*) from the African rice species Oryza glaberrima. *STH1* encodes an α/β hydrolase family member primarily localized in the major sites of peroxisomal β-oxidation, including the cytoplasm and mitochondria. It catalyzes the hydrolytic degradation of unsaturated fatty acids to maintain fatty acid metabolic homeostasis and membrane fluidity, thus controlling rice salt, heat, and osmotic tolerance. Correspondingly, rice seedlings showed a more significant growth acceleration in terms of shoot length when grown on a mannitol-containing medium supplemented with the fatty acid 11-eicosenoic acid, implying that metabolic flux redirection by blocking fatty acids from degradation can enhance plant tolerance to various abiotic stresses. *STH1HP46* is found to be a rare allele in modern Asian cultivars, which shows great promise in boosting grain yield under salt stress [79].

Very few favorable thermo-tolerance genes have been identified, mostly working through reactive oxygen species scavenging, cytotoxic protein elimination, and unfolded protein renaturation [80,81]. Kan et al. (2022) revealed a novel "TT2-SCT1-OsWR2" thermal tolerance mechanism linking G protein, Ca2+ sensing, and wax metabolism [82]. Known as a Gγ subunit protein, *TT2* was cloned as a major QTL contributing to heat tolerance. Heat-treated plants carrying a weak allele of *TT2* had enhanced thermo-tolerance, exhibiting less-severe reductions in wax content compared with plants with the normal *TT2* allele. The authors have validated that the heat signal increases cytosolic Ca2+, which subsequently undergoes conversion into an active form facilitated by CaM. Following this, CaM binds to SCT1 to repress its activity, thereby decreasing the transcription of *OsWR2*, the pivotal regulator of wax biosynthesis [39]. Consequently, this perturbation results in reduced wax content and induces thermo-sensitivity. Field trial experiment results showed that *TT2* governs rice thermo-tolerance in both vegetative and reproductive stages. Under heat stress, NIL-TT2HPS32 plants significantly outperformed NIL-TT2HJX plants with 54.7% higher grain yield, showing promising potential in breeding thermo-tolerant rice varieties.

8. Rice Biotic Stress Response Is Mediated by Lipids

Microdomains, also called membrane rafts, are small, heterogeneous, liquid-ordered domains composed mainly of sphingolipids and sterols [83]. Since plant defense-related proteins primarily congregate in plasma membrane microdomains, innate immunity is associated with microdomains' functions. By knocking down two fatty acid 2-hydroxylases catalyzing the 2-hydroxylation of sphingolipid fatty acids, Nagano et al. (2016) obtained *fah1 fah2* genetic lines harboring fewer microdomains and higher susceptibility to rice blast fungus infection. *Os-RbohB* and *Os-RbohH* are microdomains-located NADPH oxidases involved in the pathogen defense process. They revealed that microdomains are required for the dynamics of the GTPase Rac1 and respiratory burst oxidase homologs (Rbohs) in response to chitin elicitors, regulating chitin-induced immunity through ROS signaling mediated by the Rac1-RbohB pathway [40].

Many plant phosphatidylinositol phosphates, particularly PtdIns(4,5)P2, have been demonstrated to be disease-susceptibility factors [84]. From a rice lesion mimic mutant with broad-spectrum disease resistance, Sha et al. (2023) cloned a cytidine diphosphate diacylglycerol synthase gene *RBL1* involved in phospholipid biosynthesis. *rbl1* had drastically reduced the amount of phosphatidylinositol and its derivative PtdIns(4,5)P2, which is enriched in the biotrophic interfacial complex and extra-invasive hyphal membrane structure associate with effector secretion and fungal infection. Intriguingly, the authors generated a valuable allele named *RBL1Δ12*, which confers broad-spectrum disease resistance to 10 *M. oryzae* field strains, five *Xoo* strains, and two rice false smut *Ustilaginoidea virens* strains in model rice variety backgrounds. At the same time, no yield penalty was observed in multiple field trials [85]. The work demonstrated the feasibility of balancing rice growth and immunity via genetically editing lipid-related genes.

In addition to its functions in seed longevity, Lipoxygenase 3 (LOX3) may negatively influence rice blast disease resistance. *lox3* mutants grow normally with high levels of methyl-linolenate and reactive oxygen species (ROS). After *M. oryzae* infection, *lox3* plants exhibited serious blast symptoms and reduced defense response, accompanied by increased ROS-mediated cell death. Moreover, exogenous JA repressed hyphal expansion and ROS-mediated cell death, finally alleviating blast symptoms in the mutant [86].

A recent publication described the functional characterization of *OsGELP77*, a major QTL contributing to broad-spectrum resistance in rice. *OsGELP77* is annotated as an ER-localized, GDSL-type lipase with typical lipase activity. Based on its pathogen- and JA-inducible expression pattern, *OsGELP77* was found to positively regulate resistance to various pathogens, possibly through modulating glycerolipid metabolism, glycerophospholipid metabolism, and sphingolipid metabolism as well as JA homeostasis. In addition, *OsWRKY45* may work upstream of *OsGELP77* to trigger rice immunity. Elevated expression of *OsGELP77* or pyramiding of a natural elite haplotype *OsGELP77*Hap3 significantly increased resistance to various pathogens to achieve higher yield [41].

Oxylipins are a big family of lipid compounds well documented for their function in insect resistance in rice. *LOX11* (Lipoxygenase-11)/*OsRCI-1* is a chloroplast-localized protein gene in rice showing induced expression by brown planthopper (BPH) infestation. Over-expression of *OsRCI-1* elevated the levels of JA, jasmonate-isoleucine, and trypsin protease inhibitors, which imposed decreased colonization, fecundity, and mass of the BPH insects when compared with the WT. Moreover, the decreased attractiveness to BPH and enhanced attractiveness to the parasitoid of *oeRCI* plants correlated with higher levels of BPH-induced 2-heptanone, 2-heptanol, and 2-heptanone. These results indicate that *OsRCI-1* is involved in herbivore-induced JA bursts and plays a role in plant defense [42].

9. Perspectives

This comprehensive review delineates recent progress in elucidating the genetic underpinnings and physiological functions of lipids in rice, a model species for monocots and cereals. Despite the aforementioned advancements, our knowledge of the biological roles of lipids in rice remains relatively rudimentary and fragmented compared to that in *Arabidopsis* and other oil crops. The bulk of published rice cases hinge on robust correlations between phenotype and variations in lipid content or components. However, the molecular mechanisms underlying the relationship between lipid variation and phenotypic diversity remains largely unexplored. One example that has been thoroughly documented is the role of *STH1* in catalyzing the hydrolytic degradation of unsaturated fatty acids, which is essential for maintaining membrane fluidity [79]. Meanwhile, a large number of rice lipid-related genes have been characterized via a forward genetics approach utilizing randomly collected mutants and map-based cloning, which resulted in a fragmented comprehension of the rice lipid regulation network. Thus, there is a need to allocate additional resources to systematically explore the gene families associated with lipid biosynthesis, metabolism, and signaling. Finally, as has been shown in a few cases in rice [87], the advancement of new mass spectrometric techniques in lipidomic research has the potential to significantly enhance the identification of novel lipid components within the pathways and networks of cellular lipids in rice biological systems.

Author Contributions: Conceptualization, K.Z., Z.L. (Zhengliang Luo), J.Z., S.W. and X.Z. investigation, W.H., Z.L. (Zemin Liu) and X.M.; writing—original draft preparation, K.Z., J.Z., S.W. and X.Z.; writing—review and editing, K.Z., S.W. and X.Z.; supervision, X.Z., S.T. and J.W.; funding acquisition, K.Z. All authors have read and agreed to the published version of the manuscript.

Funding: This work was funded by the agricultural science and technology innovation fund project of Hunan Province (China) (2023CX09), Central Public-interest Scientific Institution Basal Research Fund (No. Y2023PT17), National Key Research and Development Program of China (2023YFD1201201) and Changsha Natural Science Foundation (kq2208128).

Data Availability Statement: Not applicable.

Conflicts of Interest: The authors declare that the research was conducted in the absence of any commercial or financial relationships that could be construed as potential conflicts of interest.

References

1. Gurr, M.I.; Harwoord, J.L.; Frayn, K.N.; Murphy, D.J.; Michell, R.H. *Lipids: Biochemistry, Biotechnology and Health*, 6th ed.; Wiley-Blackwell: Hoboken, NJ, USA, 2016.
2. Suh, M.C.; Kim, H.U.; Nakamura, Y. Plant lipids: Trends and beyond. *J. Exp. Bot.* **2022**, *73*, 2715–2720.
3. Tao, B. Industrial Applications for Plant Oils and Lipids. In *Bioprocessing for Value-Added Products from Renewable Resources*; Yang, S.-T., Ed.; Elsevier: Amsterdam, The Netherlands, 2007.
4. Harwood, J.L.; Guschina, I.A. Regulation of lipid synthesis in oil crops. *FEBS Lett.* **2013**, *587*, 2079–2081.
5. Qi, W.; Lu, H.; Zhang, Y.; Cheng, J.; Huang, B.; Lu, X.; Sheteiwy, M.S.; Kuang, S.; Shao, H. Oil crop genetic modification for producing added value lipids. *Crit. Rev. Biotechnol.* **2020**, *40*, 777–786.
6. Sagun, J.V.; Yadav, U.P.; Alonso, A.P. Progress in understanding and improving oil content and quality in seeds. *Front. Plant Sci.* **2023**, *14*, 1116894.
7. Zhou, Z.K.; Robards, K.; Helliwell, S.; Blanchard, C. Composition and functional properties of rice. *Int. J. Food Sci. Technol.* **2002**, *37*, 849–868.
8. Li, P.; Chen, Y.H.; Lu, J.; Zhang, C.Q.; Liu, Q.Q.; Li, Q.F. Genes and Their Molecular Functions Determining Seed Structure, Components, and Quality of Rice. *Rice* **2022**, *15*, 18.
9. Yu, D.M.; Ranathunge, K.; Huang, H.; Pei, Z.Y.; Franke, R.; Schreiber, L.; He, C. Wax Crystal-Sparse Leaf1 encodes a beta-ketoacyl CoA synthase involved in biosynthesis of cuticular waxes on rice leaf. *Planta* **2008**, *228*, 675–685. [PubMed]
10. Ito, Y.; Kimura, F.; Hirakata, K.; Tsuda, K.; Takasugi, T.; Eiguchi, M.; Nakagawa, K.; Kurata, N. Fatty acid elongase is required for shoot development in rice. *Plant J.* **2011**, *66*, 680–688.
11. Tsuda, K.; Akiba, T.; Kimura, F.; Ishibashi, M.; Moriya, C.; Nakagawa, K.; Kurata, N.; Ito, Y. ONION2 Fatty Acid Elongase is Required for Shoot Development in Rice. *Plant Cell Physiol.* **2013**, *54*, 209–217.
12. Fang, Y.X.; Hu, J.; Xu, J.; Yu, H.P.; Shi, Z.Y.; Xiong, G.S.; Zhu, L.; Zeng, D.; Zhang, G.; Gao, Z.; et al. Identification and characterization of Mini1, a gene regulating rice shoot development. *J. Integr. Plant Biol.* **2015**, *57*, 151–161. [PubMed]

13. Akiba, T.; Hibara, K.I.; Kimura, F.; Tsuda, K.; Shibata, K.; Ishibashi, M.; Moriya, C.; Nakagawa, K.; Kurata, N.; Itoh, J.I.; et al. Organ fusion and defective shoot development in oni3 mutants of rice. *Plant Cell Physiol.* **2014**, *55*, 42–51.
14. Shanmugabalaji, V.; Zita, W.; Collombat, J.; Kessler, F. Plastoglobules: A hub of lipid metabolism in the chloroplast. *Adv. Bot. Res.* **2022**, *101*, 91–119.
15. Zhang, X.B.; Wang, Y.; Wang, X.W.; Zhu, Z.; Zhang, X.F.; Jia, L.Q.; Li, Y.Y.; Tian, W.J.; Chen, H.Y.; Zhu, X.Y.; et al. A very-long-chain fatty acid synthesis gene, *SD38*, influences plant height by activating ethylene biosynthesis in rice. *Plant J.* **2022**, *112*, 1084–1097.
16. Jung, K.H.; Han, M.J.; Lee, D.Y.; Lee, Y.S.; Schreiber, L.; Franke, R.; Faust, A.; Yephremov, A.; Saedler, H.; Kim, Y.W.; et al. *Wax-deficient anther1* is involved in cuticle and wax production in rice anther walls and is required for pollen development. *Plant Cell* **2006**, *18*, 3015–3032. [PubMed]
17. Yang, X.J.; Liang, W.Q.; Chen, M.J.; Zhang, D.B.; Zhao, X.X.; Shi, J.X. Rice fatty acyl-CoA synthetase OsACOS12 is required for tapetum programmed cell death and male fertility. *Planta* **2017**, *246*, 105–122.
18. Li, Y.L.; Li, D.D.; Guo, Z.L.; Shi, Q.S.; Xiong, S.X.; Zhang, C.; Zhu, J.; Yang, Z.N. *OsACOS12*, an orthologue of Arabidopsis acyl-CoA synthetase5, plays an important role in pollen exine formation and anther development in rice. *BMC Plant Biol.* **2016**, *16*, 256.
19. Shi, J.; Tan, H.X.; Yu, X.H.; Liu, Y.Y.; Liang, W.Q.; Ranathunge, K.; Franke, R.B.; Schreiber, L.; Wang, Y.; Kai, G.; et al. *Defective Pollen Wall* Is Required for Anther and Microspore Development in Rice and Encodes a Fatty Acyl Carrier Protein Reductase. *Plant Cell* **2011**, *23*, 2225–2246.
20. Song, Y.Y.; Tang, Y.Y.; Liu, L.T.; Xu, Y.Y.; Wang, T. The methyl-CpG-binding domain family member PEM1 is essential for Ubisch body formation and pollen exine development in rice. *Plant J.* **2022**, *111*, 1283–1295. [PubMed]
21. Chen, H.Q.; Zhang, Z.G.; Ni, E.D.; Lin, J.W.; Peng, G.Q.; Huang, J.L.; Zhu, L.Y.; Deng, L.; Yang, F.F.; Luo, Q.; et al. HMS1 interacts with HMS1I to regulate very-long-chain fatty acid biosynthesis and the humidity-sensitive genic male sterility in rice (*Oryza sativa*). *New Phytol.* **2020**, *225*, 2077–2093.
22. Ni, E.D.; Zhou, L.Y.; Li, J.; Jiang, D.G.; Wang, Z.H.; Zheng, S.Y.; Qi, H.; Zhou, Y.; Wang, C.M.; Xiao, S.; et al. *OsCER1* Plays a Pivotal Role in Very-Long-Chain Alkane Biosynthesis and Affects Plastid Development and Programmed Cell Death of Tapetum in Rice (*Oryza sativa* L.). *Front. Plant Sci.* **2018**, *9*, 1217.
23. Huo, X.; Wu, S.; Zhu, Z.F.; Liu, F.X.; Fu, Y.C.; Cai, H.W.; Sun, X.Y.; Gu, P.; Xie, D.X.; Tan, L.B.; et al. NOG1 increases grain production in rice. *Nat. Commun.* **2017**, *8*, 1497. [PubMed]
24. Zhao, Y.-F.; Peng, T.; Sun, H.-Z.; Teotia, S.; Wen, H.-L.; Du, Y.-X.; Zhang, J.; Li, J.Z.; Tnag, G.L.; Xue, H.W.; et al. miR1432-*OsACOT* (Acyl-CoA thioesterase) module determines grain yield via enhancing grain filling rate in rice. *Plant Biotechnol. J.* **2019**, *17*, 712–723. [PubMed]
25. Yang, J.X.; Tian, R.C.; Gao, Z.Q.; Yang, H.B. Characterization of *AtWRI1* in fatty acids and starch synthesis in rice. *Biosci. Biotechnol. Biochem.* **2019**, *83*, 1807–1814.
26. Zhou, W.; Wang, X.; Zhou, D.; Ouyang, Y.; Yao, J. Overexpression of the 16-kDa α-amylase/trypsin inhibitor RAG2 improves grain yield and quality of rice. *Plant Biotechnol. J.* **2017**, *15*, 568–580.
27. Wei, X.; Jiao, G.; Lin, H.; Sheng, Z.; Shao, G.; Xie, L.; Tang, S.; Xu, Q.; Hu, P. *GRAIN INCOMPLETE FILLING 2* regulates grain filling and starch synthesis during rice caryopsis development. *J. Integr. Plant Biol.* **2017**, *59*, 134–153. [PubMed]
28. Liu, W.X.; Liu, H.L.; Qu, L.Q. Embryo-specific expression of soybean oleosin altered oil body morphogenesis and increased lipid content in transgenic rice seeds. *Theor. Appl. Genet.* **2013**, *126*, 2289–2297.
29. Liu, J.; Wu, X.; Yao, X.; Yu, R.; Larkin, P.J.; Liu, C.-M. Mutations in the DNA demethylase *OsROS1* result in a thickened aleurone and improved nutritional value in rice grains. *Proc. Natl. Acad. Sci. USA* **2018**, *115*, 11327–11332.
30. Izadi-Darbandi, A.; Younessi-Hamzekhanlu, M.; Sticklen, M. Metabolically engineered rice biomass and grain using genes associated with lipid pathway show high level of oil content. *Mol. Biol. Rep.* **2020**, *47*, 7917–7927.
31. Liu, X.; Li, Z.; Ying, J.; Shu, Y.; Liu, W.; Li, G.; Chen, L.; Luo, J.; Wang, S.; Wang, Y.; et al. Multi-gene engineering boosts oil content in rice grains. *Plant Commun.* **2024**, *5*, 100736.
32. Zaplin, E.S.; Liu, Q.; Li, Z.; Butardo, V.M.; Blanchard, C.L.; Rahman, S. Production of high oleic rice grains by suppressing the expression of the *OsFAD2-1* gene. *Funct. Plant Biol.* **2013**, *40*, 996–1004.
33. Tiwari, G.J.; Liu, Q.; Shreshtha, P.; Li, Z.Y.; Rahman, S. RNAi-mediated down-regulation of the expression of *OsFAD2-1*: Effect on lipid accumulation and expression of lipid biosynthetic genes in the rice grain. *BMC Plant Biol.* **2016**, *16*, 189.
34. Yin, Z.J.; Liu, H.L.; Dong, X.B.; Tian, L.H.; Xiao, L.; Xu, Y.N.; Qu, L.Q. Increasing α-linolenic acid content in rice bran by embryo-specific expression of ω3/Δ15-desaturase gene. *Mol. Breed.* **2014**, *33*, 987–996.
35. Xu, H.B.; Wei, Y.D.; Zhu, Y.S.; Lian, L.; Xie, H.G.; Cai, Q.H.; Chen, Q.S.; Lin, Z.P.; Wang, Z.H.; Xie, H.; et al. Antisense suppression of *LOX3* gene expression in rice endosperm enhances seed longevity. *Plant Biotechnol. J.* **2015**, *13*, 526–539.
36. Xia, K.F.; Ou, X.J.; Gao, C.Z.; Tang, H.D.; Jia, Y.X.; Deng, R.F.; Xu, X.L.; Zhang, M.Y. OsWS1 involved in cuticular wax biosynthesis is regulated by osa-miR1848. *Plant Cell Environ.* **2015**, *38*, 2662–2673. [PubMed]
37. Gu, X.Y.; Gao, S.X.; Li, J.; Song, P.Y.; Zhang, Q.; Guo, J.F.; Wang, X.Y.; Han, X.Y.; Wang, X.J.; Zhu, Y.; et al. The bHLH transcription factor regulated gene *OsWIH2* is a positive regulator of drought tolerance in rice. *Plant Physiol. Biochem.* **2021**, *169*, 269–279.
38. Wang, Z.Y.; Tian, X.J.; Zhao, Q.Z.; Liu, Z.Q.; Li, X.F.; Ren, Y.K.; Tang, J.Q.; Fang, J.; Xu, Q.J.; Bu, Q.Y. The E3 Ligase DROUGHT HYPERSENSITIVE Negatively Regulates Cuticular Wax Biosynthesis by Promoting the Degradation of Transcription Factor ROC4 in Rice. *Plant Cell* **2018**, *30*, 228–244.

39. Zhou, X.Y.; Jenks, M.A.; Liu, J.; Liu, A.L.; Zhang, X.W.; Xiang, J.H.; Zou, J.; Peng, Y.; Chen, X.B. Overexpression of Transcription Factor OsWR2 Regulates Wax and Cutin Biosynthesis in Rice and Enhances its Tolerance to Water Deficit. *Plant Mol. Biol. Rep.* **2014**, *32*, 719–731.
40. Nagano, M.; Ishikawa, T.; Fujiwara, M.; Fukao, Y.; Kawano, Y.; Kawai-Yamada, M.; Shimamoto, K. Plasma Membrane Microdomains Are Essential for Rac1-RbohB/H-Mediated Immunity in Rice. *Plant Cell* **2016**, *28*, 1966–1983.
41. Zhang, M.J.; Chen, D.; Tian, J.J.; Cao, J.B.; Xie, K.B.; He, Y.Q.; Yuan, M. *OsGELP77*, a QTL for broad-spectrum disease resistance and yield in rice, encodes a GDSL-type lipase. *Plant Biotechnol. J.* **2024**, *22*, 1352–1371. [PubMed]
42. Liao, Z.H.; Wang, L.; Li, C.Z.; Cao, M.J.; Wang, J.N.; Yao, Z.L.; Zhou, S.Y.; Zhou, G.X.; Zhang, D.Y.; Lou, Y.G. The lipoxygenase gene is involved in the biosynthesis of herbivore-induced JAs and regulates plant defense and growth in rice. *Plant Cell Environ.* **2022**, *45*, 2827–2840. [PubMed]
43. Tsuda, K.; Ito, Y.; Yamaki, S.; Miyao, A.; Hirochika, H.; Kurata, N. Isolation and mapping of three rice mutants that showed ectopic expression of KNOX genes in leaves. *Plant Sci.* **2009**, *177*, 131–135.
44. Li, J.J.; Yang, J.; Zhu, B.H.; Xie, G.S. Overexpressing *OsFBN1* enhances plastoglobule formation, reduces grain-filling percent and jasmonate levels under heat stress in rice. *Plant Sci.* **2019**, *285*, 230–238.
45. Wan, X.Y.; Wu, S.W.; Li, Z.W.; An, X.L.; Tian, Y.H. Lipid Metabolism: Critical Roles in Male Fertility and Other Aspects of Reproductive Development in Plants. *Mol. Plant* **2020**, *13*, 955–983. [PubMed]
46. Kunst, L.; Samuels, A.L. Biosynthesis and secretion of plant cuticular wax. *Prog. Lipid Res.* **2003**, *42*, 51–80. [PubMed]
47. Aarts, M.G.M.; Hodge, R.; Kalantidis, K.; Florack, D.; Wilson, Z.A.; Mulligan, B.J.; Scott, R.; Pereira, A. The *Arabidopsis MALE STERILITY 2* protein shares similarity with reductases in elongation/condensation complexes. *Plant J.* **1997**, *12*, 615–623.
48. Akoh, C.C.; Lee, G.C.; Liaw, Y.C.; Huang, T.H.; Shaw, J.F. GDSL family of serine esterases/lipases. *Prog. Lipid Res.* **2004**, *43*, 534–552.
49. Zhao, J.; Long, T.; Wang, Y.F.; Tong, X.H.; Tang, J.; Li, J.L.; Wang, H.M.; Tang, L.Q.; Li, Z.Y.; Shu, Y.Z.; et al. *RMS2* Encoding a GDSL Lipase Mediates Lipid Homeostasis in Anthers to Determine Rice Male Fertility. *Plant Physiol.* **2020**, *182*, 2047–2064.
50. Mondol, P.C.; Xu, D.W.; Duan, L.; Shi, J.X.; Wang, C.H.; Chen, X.F.; Chen, M.J.; Hu, J.P.; Liang, W.Q.; Zhang, D.B. Defective Pollen Wall 3 (DPW3), a novel alpha integrin-like protein, is required for pollen wall formation in rice. *New Phytol.* **2020**, *225*, 807–822.
51. Yu, B.; Liu, L.T.; Wang, T. Deficiency of very long chain alkanes biosynthesis causes humidity-sensitive male sterility via affecting pollen adhesion and hydration in rice. *Plant Cell Environ.* **2019**, *42*, 3340–3354.
52. Ni, E.D.; Deng, L.; Chen, H.Q.; Lin, J.W.; Ruan, J.M.; Liu, Z.L.; Zhuang, C.X.; Zhou, H. *OsCER1* regulates humidity-sensitive genic male sterility through very-long-chain (VLC) alkane metabolism of tryphine in rice. *Funct. Plant Biol.* **2021**, *48*, 461–468.
53. Agnihotri, G.; Liu, H.W. Enoyl-CoA hydratase: Reaction, mechanism, and inhibition. *Bioorganic Med. Chem.* **2003**, *11*, 9–20.
54. Krishna, A.G.G.; Hemakumar, K.H.; Khatoon, S. Study on the composition of rice bran oil and its higher free fatty acids value. *J. Am. Oil Chem. Soc.* **2006**, *83*, 117–120.
55. Hu, Z.L.; Li, P.; Zhou, M.Q.; Zhang, Z.H.; Wang, L.X.; Zhu, L.H.; Zhu, Y.G. Mapping of quantitative trait loci (QTLs) for rice protein and fat content using doubled haploid lines. *Euphytica* **2004**, *135*, 47–54.
56. Liu, W.J.; Zeng, J.; Jiang, G.H.; He, Y.Q. QTLs identification of crude fat content in brown rice and its genetic basis analysis using DH and two backcross populations. *Euphytica* **2009**, *169*, 197–205.
57. Qin, Y.; Kim, S.M.; Zhao, X.; Lee, H.S.; Jia, B.Y.; Kim, K.M.; Eun, M.Y.; Sohn, J.K. QTL detection and MAS selection efficiency for lipid content in brown rice (*Oryza sativa* L.). *Genes Genom.* **2010**, *32*, 506–512.
58. Ying, J.Z.; Shan, J.X.; Gao, J.P.; Zhu, M.Z.; Shi, M.; Lin, H.X. Identification of Quantitative Trait Loci for Lipid Metabolism in Rice Seeds. *Mol. Plant* **2012**, *5*, 865–875. [PubMed]
59. Zhou, H.; Xia, D.; Li, P.B.; Ao, Y.T.; Xu, X.D.; Wan, S.S.; Li, Y.H.; Wu, B.; Shi, H.; Wang, K.Y.; et al. Genetic architecture and key genes controlling the diversity of oil composition in rice grains. *Mol. Plant* **2021**, *14*, 456–469.
60. Sun, R.; Ye, R.; Gao, L.; Zhang, L.; Wang, R.; Mao, T.; Zheng, Y.S.; Li, D.D.; Lin, Y.J. Characterization and Ectopic Expression of *CoWRI1*, an *AP2/EREBP* Domain-Containing Transcription Factor from Coconut (*Cocos nucifera L.*) Endosperm, Changes the Seeds Oil Content in Transgenic Arabidopsis thaliana and Rice (*Oryza sativa L.*). *Front. Plant Sci.* **2017**, *8*, 63.
61. Li, Y.T.; Han, D.X.; Hu, G.R.; Sommerfeld, M.; Hu, Q.A. Inhibition of starch synthesis results in overproduction of lipids in *Chlamydomonas reinhardtii*. *Biotechnol. Bioeng.* **2010**, *107*, 258–268.
62. Li, D.Q.; Wu, X.B.; Wang, H.F.; Feng, X.; Yan, S.J.; Wu, S.Y.; Liu, J.X.; Yao, X.F.; Bai, A.N.; Zhao, H.; et al. Defective mitochondrial function by mutation in *THICK ALEURONE 1* encoding a mitochondrion-targeted single-stranded DNA-binding protein leads to increased aleurone cell layers and improved nutrition in rice. *Mol. Plant* **2022**, *15*, 1638–1639.
63. Zhou, Z.; Robards, K.; Helliwell, S.; Blanchard, C. Ageing of stored rice: Changes in chemical and physical attributes. *J. Cereal. Sci.* **2002**, *35*, 65–78.
64. Stoutjesdijk, P.A.; Singh, S.P.; Liu, Q.; Hurlstone, C.J.; Waterhouse, P.A.; Green, A.G. hpRNA-mediated targeting of the Arabidopsis *FAD2* gene gives highly efficient and stable silencing. *Plant Physiol.* **2002**, *129*, 1723–1731. [PubMed]
65. Liu, Q.; Singh, S.P.; Green, A.G. High-stearic and high-oleic cottonseed oils produced by hairpin RNA-mediated post-transcriptional gene silencing. *Plant Physiol.* **2002**, *129*, 1732–1743. [PubMed]
66. Liu, H.L.; Yin, Z.J.; Xiao, L.; Xu, Y.N.; Qu, L.Q. Identification and evaluation of ω-3 fatty acid desaturase genes for hyperfortifying α-linolenic acid in transgenic rice seed. *J. Exp. Bot.* **2012**, *63*, 3279–3287.

67. Suzuki, Y.; Ise, K.; Li, C.Y.; Honda, I.; Iwai, Y.; Matsukura, U. Volatile components in stored rice [*Oryza sativa (L.)*] of varieties with and without lipoxygenase-3 in seeds. *J. Agric. Food Chem.* **1999**, *47*, 1119–1124.
68. Gayen, D.; Ali, N.; Sarkar, S.N.; Datta, S.K.; Datta, K. Down-regulation of lipoxygenase gene reduces degradation of carotenoids of golden rice during storage. *Planta* **2015**, *242*, 353–363.
69. Kerstiens, G. Water transport in plant cuticles: An update. *J. Exp. Bot.* **2006**, *57*, 2493–2499.
70. Jenks, M.A.; Tuttle, H.A.; Eigenbrode, S.D.; Feldmann, K.A. Leaf Epicuticular Waxes of the Eceriferum Mutants in Arabidopsis. *Plant Physiol.* **1995**, *108*, 369–377. [PubMed]
71. Islam, M.A.; Du, H.; Ning, J.; Ye, H.Y.; Xiong, L.Z. Characterization of Glossy1-homologous genes in rice involved in leaf wax accumulation and drought resistance. *Plant Mol. Biol.* **2009**, *70*, 443–456.
72. Qin, B.X.; Tang, D.; Huang, J.; Li, M.; Wu, X.R.; Lu, L.L.; Wang, K.J.; Yu, H.X.; Chen, J.M.; Gu, M.H.; et al. Rice *OsGL1-1* is involved in leaf cuticular wax and cuticle membrane. *Mol. Plant* **2011**, *4*, 985–995.
73. Zhou, X.Y.; Li, L.Z.; Xiang, J.H.; Gao, G.F.; Xu, F.X.; Liu, A.L.; Zhang, X.W.; Peng, Y.; Chen, X.B.; Wan, X.Y. *OsGL1-3* is Involved in Cuticular Wax Biosynthesis and Tolerance to Water Deficit in Rice. *PLoS ONE* **2015**, *10*, e116676.
74. Zhou, L.Y.; Ni, E.D.; Yang, J.W.; Zhou, H.; Liang, H.; Li, J.; Jiang, D.G.; Wang, Z.H.; Liu, Z.L.; Zhuang, C.X. Rice *OsGL1-6* is involved in leaf cuticular wax accumulation and drought resistance. *PLoS ONE* **2013**, *8*, e65139.
75. Lolle, S.J.; Hsu, W.; Pruitt, R.E. Genetic analysis of organ fusion in. *Genetics* **1998**, *149*, 607–619. [PubMed]
76. Wu, R.H.; Li, S.B.; He, S.; Wassmann, F.; Yu, C.H.; Qin, G.J.; Schreiber, L.; Qu, L.J.; Gu, H.Y. CFL1, a WW Domain Protein, Regulates Cuticle Development by Modulating the Function of HDG1, a Class IV Homeodomain Transcription Factor, in Rice and *Arabidopsi*. *Plant Cell* **2011**, *23*, 3392–3411. [PubMed]
77. Mawlong, I.; Ali, K.; Kurup, D.; Yadav, S.; Tyagi, A. Isolation and characterization of an AP2/ERF-type drought stress inducible transcription factor encoding gene from rice. *J. Plant Biochem. Biotechnol.* **2014**, *23*, 42–51.
78. Zhu, X.Y.; Xiong, L.Z. Putative megaenzyme DWA1 plays essential roles in drought resistance by regulating stress-induced wax deposition in rice. *Proc. Natl. Acad. Sci. USA* **2013**, *110*, 17790–17795. [PubMed]
79. Xiang, Y.H.; Yu, J.J.; Liao, B.; Shan, J.X.; Ye, W.W.; Dong, N.Q.; Guo, T.; Kan, Y.; Zhang, H.; Yang, Y.B.; et al. An a/f3 hydrolase family member negatively regulates salt tolerance but promotes flowering through three distinct functions in rice. *Mol. Plant* **2022**, *15*, 1908–1930. [PubMed]
80. Zhu, J.K. Abiotic Stress Signaling and Responses in Plants. *Cell* **2016**, *167*, 313–324. [PubMed]
81. Zhang, J.Y.; Li, X.M.; Lin, H.X.; Chong, K. Crop Improvement Through Temperature Resilience. *Annu. Rev. Plant Biol.* **2019**, *70*, 753–780. [PubMed]
82. Kan, Y.; Mu, X.R.; Zhang, H.; Gao, J.; Shan, J.X.; Ye, W.W.; Lin, H.X. TT2 controls rice thermotolerance through SCT1-dependent alteration of wax biosynthesis. *Nat. Plants* **2022**, *8*, 53–67. [PubMed]
83. Mongrand, S.; Stanislas, T.; Bayer, E.M.F.; Lherminier, J.; Simon-Plas, F. Membrane rafts in plant cells. *Trends Plant Sci.* **2010**, *15*, 656–663. [PubMed]
84. Qin, L.; Zhou, Z.Q.; Li, Q.; Zhai, C.; Liu, L.J.; Quilichini, T.D.; Gao, P.; Kessler, S.A.; Jaillais, Y.; Datla, R.; et al. Specific Recruitment of Phosphoinositide Species to the Plant-Pathogen Interfacial Membrane Underlies Arabidopsis Susceptibility to Fungal Infection. *Plant Cell* **2020**, *32*, 1665–1688. [PubMed]
85. Sha, G.; Sun, P.; Kong, X.J.; Han, X.Y.; Sun, Q.P.; Fouillen, L.; Zhao, J.; Li, Y.; Yang, L.; Wang, Y.; et al. Genome editing of a rice CDP-DAG synthase confers multipathogen resistance. *Nature* **2023**, *618*, 1017–1023. [PubMed]
86. Su, S.Y.; Tang, P.; Zuo, R.B.; Chen, H.F.; Zhao, T.Q.; Yang, S.M.; Yang, J. Exogenous Jasmonic Acid Alleviates Blast Resistance Reduction Caused by LOX3 Knockout in Rice. *Biomolecules* **2023**, *13*, 1197. [CrossRef]
87. Liu, X.; Li, Z.; Ouyang, B.; Wang, F.; Lan, D.; Wang, Y. Lipidomics analysis of rice bran during storage unveils mechanisms behind dynamic changes in functional lipid molecular species. *Food Chem.* **2024**, *447*, 138946. [PubMed]

MDPI AG
Grosspeteranlage 5
4052 Basel
Switzerland
Tel.: +41 61 683 77 34

International Journal of Molecular Sciences Editorial Office
E-mail: ijms@mdpi.com
www.mdpi.com/journal/ijms

www.ingramcontent.com/pod-product-compliance
Lightning Source LLC
Chambersburg PA
CBHW061511210926
51850CB00008B/1886
* 9 7 8 3 7 2 5 8 6 4 7 8 2 *